Asia Conference on Mechanical and Aerospace Engineering (ACMAE 2017)

MATEC Web of Conferences Volume 151 (2018)

Yokohama, Japan
29-31 December 2017

Editors:

O. Adiguzel
I. McAndrew
Y. Yokoi
V. Koryanov

ISBN: 978-1-5108-5873-2

Printed from e-media with permission by:

Curran Associates, Inc.
57 Morehouse Lane
Red Hook, NY 12571

proceedings.com

Some format issues inherent in the e-media version may also appear in this print version.

Printed by Curran Associates, Inc. (2018)

For additional information, please contact EDP Sciences – Web of Conferences
at the address below.

EDP Sciences – Web of Conferences
17, Avenue du Hoggar
Parc d'Activité de Courtabœuf
BP 112
F-91944 Les Ulis Cedex A
France

Phone: +33 (0) 1 69 18 75 75
Fax: +33 (0) 1 69 28 84 91

contact-edps@webofconferences.org

Additional copies of this publication are available from:

Curran Associates, Inc.
57 Morehouse Lane
Red Hook, NY 12571 USA
Phone: 845-758-0400
Fax: 845-758-2633
Email: curran@proceedings.com
Web: www.proceedings.com

TABLE OF CONTENTS

INVESTIGATIONS OF BENDING SINGULARITY ORDERS IN A V-NOTCHED COMPOSITE LAMINATE PLATE BASED ON THE RESSINER-MINDLIN THEORY 1
Chung-De Chen

INVESTIGATION OF MICRO HARDNESS, COOLING RATE AND MICROSTRUCTURE OF ATIG WELDED SAMPLES OF AL-SIC COMPOSITE 7
Sivachidambaram Pichumani, Raghuraman Srinivasan, Venkatraman Ramamoorthi

NONLINEAR DYNAMIC RESPONSES OF A HONEYCOMB SANDWICH PLATE SUBJECT TO TRANSVERSE EXCITATIONS 12
Dongmei Wang, Wei Zhang , Minghui Yao, Yinli Liu

MATERIAL PROPERTY AND STRESS FIELD NEAR NOTCH TIP IN STEEL PLATE UNDER HIGH ELECTRIC CURRENT 17
Thomas J.-C. Liu, Ji-Fu Tseng, Chun-Der Cheng

COMPUTATIONAL MODELLING OF AN ORGANIC RANKINE CYCLE (ORC) WASTE HEAT RECOVERY SYSTEM FOR AN AIRCRAFT ENGINE 21
S. Saadon

COMPARISON OF EFFICIENCY OF CONVENTIONAL SHAPED CIRCULAR AND ELLIPTICAL SHAPED COMBUSTOR 27
Dharmahinder S. Chand, Daamanjyot Barara, Gautam Ganesh, Suraj Anand

EXPERIMENTAL AND NUMERICAL DESIGN AND OPTIMIZATION OF A COUNTER-FLOW HEAT EXCHANGER 33
Salman Bahrami, Mohammad H. Rahimian

FUNDAMENTAL STUDY ON OPERATIONAL PARAMETERS OF DIAPHRAGMLESS SHOCK TUBE 37
Masanori Nishiyama, Masato Taguchi, Masashi Kashitani

EXPERIMENTAL RESEARCH ABOUT SHOCK WAVE IN A 1+1/2 COUNTER-ROTATING TURBINE 41
Chao Li, Huo-Xing Liu, Zhi-Hong Zhou

A TIME-DOMAIN SUBSTRUCTURE SYNTHESIS FOR FINITE ROTATIONS OF FLEXIBLE MECHANICAL SYSTEMS 47
Shulin Chen, Li Liu, Zhaoyue Chen

ANALYSIS OF THE PROBLEMS EXISTING IN MECHANICAL STRUCTURE DESIGN OF THE LEGGED MOBILE ROBOT 52
Mingjuan Xie, Chunyan Zhang, Xinglu Yin

A COMPUTATIONAL STUDY OF THE AERODYNAMIC PERFORMANCE OF A DRAGONFLY FOREWING IN GLIDING FLIGHT 57
Huixiang Liu, Guoyi He, Chengyu Ma, Qi Wang, Yun Luo

RESEARCH OF BOUNDARY LAYER TRANSITION CHARACTERISTICS INDUCED BY THREE-DIMENSIONAL DISCRETE ROUGHNESS 61
Feng Li, Chao Gao, Zijie Zhao, Xudong Ren

SYNCHRONOUS SURFACE PRESSURE AND VELOCITY MEASUREMENTS OF STANDARD MODEL IN HYPERSONIC FLOW 66
Zhijun Sun, Yunsong Gu, Keming Cheng, Di Peng, Yingzheng Liu

MEASUREMENT OF MUTUAL INTERFERENCE SOUND OF COLUMNAR OBJECTS IN AIR FLOW 71
Shohji Hamada, Yoshifumi Yokoi

THE GRADATIONAL ROUTE PLANNING FOR AIRCRAFT STEALTH PENETRATION BASED ON GENETIC ALGORITHM AND SPARSE A-STAR ALGORITHM 75
Maoquan Li, Yunfei Zhang, Shihao Li

RESEARCH AND APPLICATION OF VIRTUAL SIMULATION TECHNOLOGY IN THE AEROSPACE BEARING DESIGN AND MANUFACTURE 80
Jiangshan Liu, Ming Chen

STRUCTURE DESIGN OF AN INNOVATIVE ADAPTIVE VARIABLE CAMBER WING 86
An-Min Zhao, Zou Hui, De-Shan Liu, Bing-Fei Liu

HIGH ACCURACY POSITIONING USING JET THRUSTERS FOR QUADCOPTER 90
Chenhuan Pi, Jinkwan Lin, Cheng Stone

AERODYNAMIC ANALYSIS OF LOW SPEED WING DESIGN USING TAGUCHI L9 ORTHOGONAL ARRAY 96
Kenneth Witcher, Ian McAndrew, Elena Vishnevskaya
FACTORIAL EXPERIMENT ON ROUTE STRUCTURES USING SAAM ON A SPECIFIED AREA IN ASEAN 101
S. M. Phyoe, J. W. Goh, R. Guo, Z. W. Zhong
FLUID SIMILARITY CRITERIONS IN AIRCRAFT LANDING GEAR 105
Xiang-yong Liu, Wan-li Li, Zhi-qiang Xu
COMPARING THE PERFORMANCE OF REFERENCE TRAJECTORY MANAGEMENT AND CONTROLLER RECONFIGURATION IN ATTITUDE FAULT TOLERANT CONTROL 110
Rouzbeh Moradi, Alireza Alikhani, Mohsen F. Jegarkandi
DESIGN OF A SYMMETRICAL QUAD-ROTOR BIPLANE TAIL-SITTER AIRCRAFT WITHOUT CONTROL SURFACES AND EXPERIMENTAL VERIFICATION 115
Hongyu Wang, Xun Zhao, Hui Bai, Cunyue Lu, Baomin Zhang, Cong Li
CONCEPT DEVELOPMENT OF CONTROL SYSTEM FOR PERSPECTIVE UNMANNED AERIAL VEHICLES 121
Vsevolod V. Koryanov, Tatiana V. Kokuytseva, Alexey G. Toporkov, Stepan N. Iljukhin, I. O. Akimov, Noa Mohamado, Victoria Da-Poian
UAV DELIVERY MONITORING SYSTEM 126
Khin T. San, Sun J. Mun, Yeong H. Choe, Yoon S. Chang
SIMULATION AND EFFECTIVENESS ANALYSIS ON ONE VERSUS ONE BEYOND VISUAL RANGE AIR COMBAT 132
Haoyu Liu, Yunfei Zhang, Shihao Li
AN AIRBORNE TARGET LOCALIZATION ALGORITHM BASED ON UKF AND OPTIMAL MANEUVER STRATEGY 137
Chunxiao Zhang, Xiaofeng Shi
CONSTRUCTION AND APPLICATION RESEARCH OF KNOWLEDGE GRAPH IN AVIATION RISK FIELD 142
Qian Zhao, Qing Li, Jingqian Wen
ADAPTIVE COMMAND FILTERED INTEGRATED GUIDANCE AND CONTROL FOR HYPERSONIC VEHICLE WITH MAGNITUDE, RATE AND BANDWIDTH CONSTRAINTS 148
Liang Wang, Weihua Zhang, Ke Peng, Donghui Wang
RESEARCH ON HUMAN-ERROR FACTORS OF CIVIL AIRCRAFT PILOTS BASED ON GREY RELATIONAL ANALYSIS 157
Yundong Guo , Youchao Sun, Si Chen
A METHOD OF CODING FOR AEROSPACE PRODUCT QUALITY DNA 161
Jiayu Yao
Author Index

PREFACE

It is our great pleasure to welcome you to the 2017 Asia Conference on Mechanical and Aerospace Engineering (ACMAE 2017) which will be held in Yokohama, Japan during December 29-31, 2017. ACMAE 2017 is derived from the conference ICMAE (International Conference on Mechanical and Aerospace Engineering), which focuses on Asia's research, achievements, markets, industry.

The major goal of the conference is to bring academic scientists, engineers, industry researchers together to exchange and share their experiences and research results, and discuss the practical challenges encountered and the solutions adopted. Distinguished experts and professors will be invited as keynote speakers to deliver the latest information in their respective expertise areas. It will be a golden opportunity for the students, researchers and engineers to interact with the experts and specialists to get their advice or consultation on technical matters, sales and marketing strategies.

This proceedings present a selection from papers submitted to the conference from universities, research institutes and industries. All of the papers were subjected to peer-review by conference committee members and international reviewers. The papers selected for this volume depended on their quality and their relevancy to the conference. The volume tends to present to the readers the recent advances in the field of mechanical and aerospace engineering and various related areas, which include material performance analysis and testing, aeronautical machinery equipment and powerplant, fluid mechanics and computation, aircraft structure design and control, aerospace engineering and applications

We would like to thank all the authors who have contributed to this volume and also to the organizing committee, reviewers, speakers, chairpersons, sponsors and all the conference participants for their support to ACMAE 2017.

Conference Co-chair

Prof. Ian McAndrew,
Embry Riddle Aeronautical University, UK

edp sciences

MATEC Web of Conferences

Statement of Peer review

In submitting conference proceedings to MATEC Web of Conferences, I certify to the Publisher that I adhere to the **Policy on Publishing Integrity** of the journal in order to safeguard good scientific practice in publishing.

1. All articles have been subjected to peer review administered by the proceedings editors.
2. Reviews have been conducted by expert referees, who have been requested to provide unbiased and constructive comments aimed, whenever possible, at improving the work.
3. Proceedings editors have taken all reasonable steps to ensure the quality of the materials they publish and their decision to accept or reject a paper for publication has been based only on the merits of the work and the relevance to the journal.

Title, date and place of the conference

ACMAE 2017

OSAKA DECEMBER 29-31, 2017

Proceedings editor(s):

PROF IAN MCANDREW

PROF YOSHIFUMI YOKOI

Date and editor's signature

5th Feb, 2018

PROF IAN MCANDREW

edp sciences

MATEC Web of Conferences

Statement of Peer review

In submitting conference proceedings to MATEC Web of Conferences, I certify to the Publisher that I adhere to the **Policy on Publishing Integrity** of the journal in order to safeguard good scientific practice in publishing.

1. All articles have been subjected to peer review administered by the proceedings editors.
2. Reviews have been conducted by expert referees, who have been requested to provide unbiased and constructive comments aimed, whenever possible, at improving the work.
3. Proceedings editors have taken all reasonable steps to ensure the quality of the materials they publish and their decision to accept or reject a paper for publication has been based only on the merits of the work and the relevance to the journal.

Title, date and place of the conference

2017 Asia Conference on Mechanical and Aerospace Engineering (ACMAE 2017)
29-31 December, 2017
Yokohama, Japan

Proceedings editor(s):

Dr. Vsevolod V. Koryanov

Date and editor's signature 04 February 2018

17, avenue du Hoggar - PA de Courtabœuf – BP 112 - 91944 Les Ulis Cedex A (France)
Tél. : 33 (0)1 69 18 75 75 - Fax : 33(0)1 69 07 45 17 – www.edpsciences.org

edp sciences

MATEC Web of Conferences

Statement of Peer review

In submitting conference proceedings to MATEC Web of Conferences, I certify to the Publisher that I adhere to the **Policy on Publishing Integrity** of the journal in order to safeguard good scientific practice in publishing.

1. All articles have been subjected to peer review administered by the proceedings editors.
2. Reviews have been conducted by expert referees, who have been requested to provide unbiased and constructive comments aimed, whenever possible, at improving the work.
3. Proceedings editors have taken all reasonable steps to ensure the quality of the materials they publish and their decision to accept or reject a paper for publication has been based only on the merits of the work and the relevance to the journal.

Title, date and place of the conference

2017 Asia Conference on Mechanical and Aerospace Engineering (ACMAE), December 29-31, 2017, Yokohama, Japan

Proceedings editor(s):

Prof Dr. Osman Adiguzel

Date and editor's signature February 6, 2018

Investigations of Bending Singularity Orders in a V-notched Composite Laminate Plate Based on the Ressiner-Mindlin Theory

Chung-De Chen

Department of Mechanical Engineering, National Cheng Kung University, Tainan 70101, Taiwan

Abstract. In this paper, the bending singularity at the apex of a V-notched composite laminate plate is investigated. The anisotropy of the laminate is modeled by the Stroh formalism. Based on the eigenfunction expansion method, the bending singularity orders can be determined by solving an eigenvalue problem numerically. The singularity orders depend on the plate angle, material orientation, material anisotropy and the laminate stacking sequence. The comparison cases show that the material orientation should avoid in order to reduce the bending singularity. The layers near the free surfaces have more significant effects on the singularity order. The findings presented in this paper are helpful in the design of the composite laminate with V-notch.

1 Introduction

The V-notch structures are common seen in many engineering applications. High stress concentrations occur near the notch. For sharp notch tip, the stress singularities occur mathematically. The stress singularity at sharp notch tip was first investigated by Williams [1]. He considered a notched plate subjected to inplane loads. He introduced the eigenfunction expansion method to find the asymptotic stress field near the notch tip. The concept of the stress singularity orders was also introduced by Williams. Theocaris introduced the complex formulation to solve the singularity order in plane elasticity [2]. After these pioneers' efforts, the problems related to stress singularities were extended to anisotropic materials [3], muli-material wedges [4] and functional materails [5].

For bending singularities in notched plate subjected out-of-plane bending, Williams was also the first researcher to investigate such problem [6]. His results were based on the classical plate theory. However, Hartranft and Sih proved that the classical plate theory lead to an incorrect result of singular stresses [7]. They found that for cracked problem, the first order shear deformation theory (Ressiner-Midlin plate theory) would be a better alternative. Huang investigated the bending singularity based on the Ressiner-Mindlin plate theory [8]. Recently, Chen investigated the bending and shear singularity of notched plate [9]. In Chen's results, the stress intensities and the singularity orders were solved simultaneously based on the Mellin transform.

For formulations of anisotropic materials, two systems, The Leknitskii formalisms and Stroh formalisms were proposed [10]. These two formulations were also utilized to solve the stress singularity problems in notched structures [11, 12]. Recently, Chen investigated bending singularities of an anisotropic notched plate [13]. He used the Mellin transform to solve the stress intensities and the results agreed with the finite element calculations.

In the papers regarding to bending stress singularities [8, 9, 13], the plate is restricted to be homogeneous. To the best of the author's knowledge, no attempt can be found in the open literature for bending singularities in notched laminated plate. In this paper, we extend the homogeneous anisotropic plate to composite laminate plates. The anisotropy of the laminate is modeled by the Stroh formalism. Based on the eigenfunction expansion method, the bending singularity is investigated.

This paper is organized as follows. In the Introduction section, a literature review regarding to the stress singularities is presented. The motivation of this study is also given in this chapter. The second chapter gives a brief review of the Ressiner-Mindlin composite laminate. The governing equations in terms of displacement and rotations are derived in detail. In the third chapter, a laminate plate with sharp notch is considered. The Stroh formalism is used to model the anisotropy of the material. The eigenfunction expansion method is introduced to determine the bending singularity. In the chapter of Results and Discussions, comparison cases are considered and the numerical results are presented. Discussions on the strength of bending singularity orders for various conditions are presented in this chapter. Finally, concluding remarks are given in the last chapter.

2 Formulations of ressiner-mindlin composite laminate

Consider a Ressiner-Mindlin laminate shown in Fig. 1. The displacements in rectangular coordinate

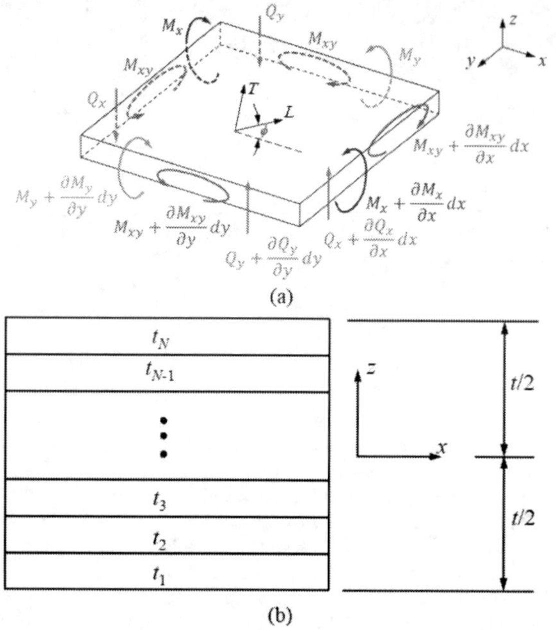

Figure 1. A Ressiner-Mindlin anisotropic laminate: (a) laminate subjected to moments and shear forces; (b) laminate stacking order.

$$u_x(x,\ y,\ z)=u_{x0}(x,\ y)-z\psi_x(x,\ y) \quad (1a)$$

$$u_y(x,\ y,\ ,z)=u_{y0}(x,\ y)-z\psi_y(x,\ y) \quad (1b)$$

$$w=w(x,\ y) \quad (1c)$$

According to the strain-displacement relation, the strains in rectangular coordinate

$$\varepsilon_x=-z\frac{\partial\psi_x}{\partial x}\,,\ \varepsilon_y=-z\frac{\partial\psi_y}{\partial y}\,,\ \varepsilon_z=0\,,$$

$$\gamma_{xy}=-z\left(\frac{\partial\psi_x}{\partial y}+\frac{\partial\psi_y}{\partial x}\right),\ \gamma_{xz}=\frac{\partial w}{\partial x}-\psi_x\,,\ \gamma_{yz}=\frac{\partial w}{\partial y}-\psi_y \quad (2)$$

The constitutive relation is

$$\begin{Bmatrix}\sigma_x\\ \sigma_y\\ \tau_{yz}\\ \tau_{xz}\\ \tau_{xy}\end{Bmatrix}=\begin{Bmatrix}-z\left(Q_{11}\frac{\partial\psi_x}{\partial x}+Q_{12}\frac{\partial\psi_y}{\partial y}+Q_{16}\left(\frac{\partial\psi_x}{\partial y}+\frac{\partial\psi_y}{\partial x}\right)\right)\\ -z\left(Q_{12}\frac{\partial\psi_x}{\partial x}+Q_{12}\frac{\partial\psi_y}{\partial y}+Q_{26}\left(\frac{\partial\psi_x}{\partial y}+\frac{\partial\psi_y}{\partial x}\right)\right)\\ Q_{44}\left(\frac{\partial w}{\partial y}-\psi_y\right)+Q_{45}\left(\frac{\partial w}{\partial x}-\psi_x\right)\\ Q_{45}\left(\frac{\partial w}{\partial y}-\psi_y\right)+Q_{55}\left(\frac{\partial w}{\partial x}-\psi_x\right)\\ -z\left(Q_{16}\frac{\partial\psi_x}{\partial x}+Q_{26}\frac{\partial\psi_y}{\partial y}+Q_{66}\left(\frac{\partial\psi_x}{\partial y}+\frac{\partial\psi_y}{\partial x}\right)\right)\end{Bmatrix} \quad (3)$$

where Q_{ij} denotes the reduced elastic constants, which is given by

$$Q_{ij}=\begin{cases}C_{ij}-\dfrac{C_{i3}C_{3j}}{C_{33}},\ ij=11,\ 22,\ 12,\ 16,\ 26,\ 66\\ C_{ij},\ ij=44,\ 55,\ 45\end{cases} \quad (4)$$

Here C_{ij} denotes the elastic constant. For N-layer laminate, the moments and shear forces are defined as

$$M_x=-\sum_{k=1}^{N}\int_{h_{k-1}-t/2}^{h_k-t/2}z\sigma_x^{(k)}dz\,,\ M_y=-\sum_{k=1}^{N}\int_{h_{k-1}-t/2}^{h_k-t/2}z\sigma_y^{(k)}dz\,,$$

$$M_{xy}=-\sum_{k=1}^{N}\int_{h_{k-1}-t/2}^{h_k-t/2}z\tau_{xy}^{(k)}dz \quad (5a)$$

$$Q_x=\sum_{k=1}^{N}\int_{h_{k-1}-t/2}^{h_k-t/2}\tau_{xz}^{(k)}dz\,,\ Q_y=\sum_{k=1}^{N}\int_{h_{k-1}-t/2}^{h_k-t/2}\tau_{yz}^{(k)}dz \quad (5b)$$

where h is the thickness of the plate and

$$h_k=\begin{cases}0,\ k=0\\ \sum_{j=1}^{k}t_j,\ k\neq 0\end{cases} \quad (6)$$

The stacking order is shown in Figure 1(b). Substitution of Eq. (5) into (3) gives

$$\begin{Bmatrix}M_x\\ M_y\\ M_{xy}\end{Bmatrix}=\begin{bmatrix}D_{11} & D_{12} & D_{16}\\ D_{12} & D_{22} & D_{26}\\ D_{16} & D_{26} & D_{66}\end{bmatrix}\begin{Bmatrix}\partial\psi_x/\partial x\\ \partial\psi_y/\partial y\\ \partial\psi_x/\partial y+\partial\psi_y/\partial x\end{Bmatrix} \quad (7a)$$

$$\begin{Bmatrix}Q_y\\ Q_x\end{Bmatrix}=\begin{bmatrix}k_4^2F_{44} & k_4k_5F_{45}\\ k_4k_5F_{45} & k_5^2F_{55}\end{bmatrix}\begin{Bmatrix}\dfrac{\partial w}{\partial y}-\psi_y\\ \dfrac{\partial w}{\partial x}-\psi_x\end{Bmatrix} \quad (7b)$$

where k_4 and k_5 denote the shear corrected factors, and

$$D_{11}=\frac{1}{3}\sum_{k=1}^{N}\left[\left(h_k-\frac{t}{2}\right)^3-\left(h_{k-1}-\frac{t}{2}\right)^3\right]Q_{11}^{(k)}\,,$$

$$D_{12}=\frac{1}{3}\sum_{k=1}^{N}\left[\left(h_k-\frac{t}{2}\right)^3-\left(h_{k-1}-\frac{t}{2}\right)^3\right]Q_{12}^{(k)}\,,$$

$$D_{22}=\frac{1}{3}\sum_{k=1}^{N}\left[\left(h_k-\frac{t}{2}\right)^3-\left(h_{k-1}-\frac{t}{2}\right)^3\right]Q_{11}^{(k)}\,,$$

$$D_{16}=\frac{1}{3}\sum_{k=1}^{N}\left[\left(h_k-\frac{t}{2}\right)^3-\left(h_{k-1}-\frac{t}{2}\right)^3\right]Q_{16}^{(k)}\,,$$

$$D_{26}=\frac{1}{3}\sum_{k=1}^{N}\left[\left(h_k-\frac{t}{2}\right)^3-\left(h_{k-1}-\frac{t}{2}\right)^3\right]Q_{26}^{(k)}\,, \quad (8a)$$

$$F_{44}=\sum_{k=1}^{N}Q_{44}^{(k)}t_k\,,\ F_{55}=\sum_{k=1}^{N}Q_{55}^{(k)}t_k\,,\ F_{45}=\sum_{k=1}^{N}Q_{45}^{(k)}t_k \quad (8b)$$

The equilibrium equations in terms of moments and shear forces are

$$\frac{\partial M_x}{\partial x}+\frac{\partial M_{xy}}{\partial y}+Q_x=0\,,\ \frac{\partial M_{xy}}{\partial x}+\frac{\partial M_y}{\partial y}+Q_y=0\,,$$

$$\frac{\partial Q_x}{\partial x}+\frac{\partial Q_y}{\partial y}=0 \tag{9}$$

Substitution of Eq. (7) into Eq. (9) yields

$$D_{11}\frac{\partial^2\psi_x}{\partial x^2}+D_{12}\frac{\partial^2\psi_y}{\partial x\partial y}+D_{16}\frac{\partial^2\psi_x}{\partial x\partial y}+D_{16}\frac{\partial^2\psi_y}{\partial x^2}$$
$$+D_{16}\frac{\partial^2\psi_x}{\partial x\partial y}+D_{26}\frac{\partial^2\psi_y}{\partial y^2}+D_{66}\frac{\partial^2\psi_x}{\partial y^2}+D_{66}\frac{\partial^2\psi_y}{\partial x\partial y}$$
$$+k_4k_5F_{45}\left(\frac{\partial w}{\partial y}-\psi_y\right)+k_{55}^2F_{55}\left(\frac{\partial w}{\partial x}-\psi_x\right)=0 \tag{10a}$$

$$D_{16}\frac{\partial^2\psi_x}{\partial x^2}+D_{26}\frac{\partial^2\psi_y}{\partial x\partial y}+D_{66}\frac{\partial^2\psi_x}{\partial x\partial y}+D_{66}\frac{\partial^2\psi_y}{\partial x^2}$$
$$+D_{12}\frac{\partial^2\psi_x}{\partial x\partial y}+D_{22}\frac{\partial^2\psi_y}{\partial y^2}+D_{26}\frac{\partial^2\psi_x}{\partial y^2}+D_{26}\frac{\partial^2\psi_y}{\partial x\partial y} \tag{10b}$$
$$+k_{44}^2F_{44}\left(\frac{\partial w}{\partial y}-\psi_y\right)+k_4k_5F_{45}\left(\frac{\partial w}{\partial x}-\psi_x\right)=0$$

$$k_4k_5F_{45}\left(\frac{\partial^2 w}{\partial y\partial x}-\frac{\partial\psi_y}{\partial x}\right)+k_5^2F_{55}\left(\frac{\partial^2 w}{\partial x^2}-\frac{\partial\psi_x}{\partial x}\right)$$
$$+k_4^2F_{44}\left(\frac{\partial^2 w}{\partial y^2}-\frac{\partial\psi_y}{\partial y}\right)+k_4k_5F_{45}\left(\frac{\partial^2 w}{\partial x\partial y}-\frac{\partial\psi_x}{\partial y}\right)=0 \tag{10c}$$

3. Determination of the bending singularities

Fig. 2 shows a anisotropic plate with a sharp V-notch. The plate angle is α (notch opening angle is $2\pi-\alpha$). For bending moment singularity, the transverse shear force τ_{xz} and τ_{yz} are finite at the corner. It is noted that the bending moments are of the order of $\partial\psi_x/\partial x$, and the transverse shear forces are of the order of $(\partial w/\partial x-\psi_x)$. Therefore, for $r\to 0$, the term $\partial\psi_x/\partial x$ approaches infinity, the term $\partial^2\psi_x/\partial x^2$ approaches infinity, and the term $(\partial w/\partial x-\psi_x)$ is finite. It suggests that, for $r\to 0$, the last two terms regarded to $(\partial w/\partial x-\psi_x)$ and $(\partial w/\partial y-\psi_y)$ in Eqs. (10a) and (10b) can be neglected. Eqs. (10a) and (10b) can be written as

$$D_{11}\frac{\partial^2\psi_x}{\partial x^2}+D_{12}\frac{\partial^2\psi_y}{\partial x\partial y}+D_{16}\frac{\partial^2\psi_x}{\partial x\partial y}+D_{16}\frac{\partial^2\psi_y}{\partial x^2}+D_{16}\frac{\partial^2\psi_x}{\partial x\partial y}$$
$$+D_{26}\frac{\partial^2\psi_y}{\partial y^2}+D_{66}\frac{\partial^2\psi_x}{\partial y^2}+D_{66}\frac{\partial^2\psi_y}{\partial x\partial y}=0 \tag{11}$$

$$D_{16}\frac{\partial^2\psi_x}{\partial x^2}+D_{26}\frac{\partial^2\psi_y}{\partial x\partial y}+D_{66}\frac{\partial^2\psi_x}{\partial x\partial y}+D_{66}\frac{\partial^2\psi_y}{\partial x^2}$$
$$+D_{12}\frac{\partial^2\psi_x}{\partial x\partial y}+D_{22}\frac{\partial^2\psi_y}{\partial y^2}+D_{26}\frac{\partial^2\psi_x}{\partial y^2}+D_{26}\frac{\partial^2\psi_y}{\partial x\partial y}=0 \tag{12}$$

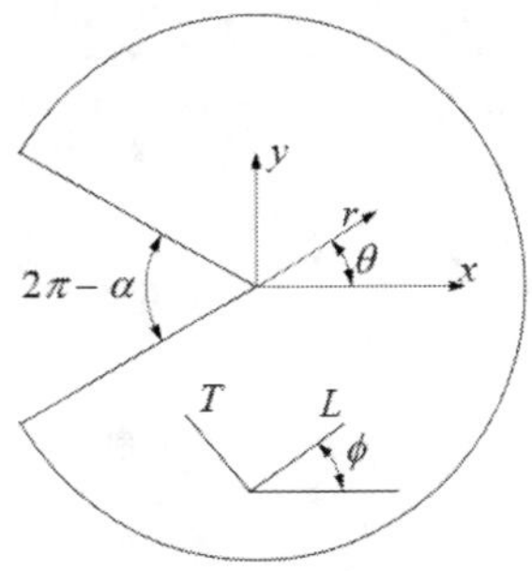

Figure 2. An anisotropic plate with V-notch.

Introducing the complex function f_1 and f_2, the rotations and moments can be represented by

$$\psi_x=2\,\mathrm{Re}[f_1'+f_2'],\ \psi_y=2\,\mathrm{Re}[\lambda_1f_1'+\lambda_2f_2'] \tag{13}$$

$$M_x=2\,\mathrm{Re}\left[\sum_{k=1}^{2}\left(D_{11}+D_{12}\lambda_kp_k+D_{16}(p_k+\lambda_k)\right)f_k''\right] \tag{14a}$$

$$M_y=2\,\mathrm{Re}\left[\sum_{k=1}^{2}\left(D_{12}+D_{22}\lambda_kp_k+D_{26}(p_k+\lambda_k)\right)f_k''\right] \tag{14b}$$

$$M_{xy}=2\,\mathrm{Re}\left[\sum_{k=1}^{2}\left(D_{16}+D_{26}\lambda_kp_k+D_{66}(p_k+\lambda_k)\right)f_k''\right] \tag{14c}$$

Where

$$\lambda_k=-\frac{D_{11}+2D_{16}p_k+D_{66}p_k^2}{D_{16}+(D_{12}+D_{66})p_k+D_{26}p_k^2}=-\frac{D_{16}+(D_{12}+D_{66})p_k+D_{26}p_k^2}{D_{66}+2D_{26}p_k+D_{22}p_k^2} \tag{15}$$

and p_k is the roots of the characteristic equation

$$\det\left[\mathbf{Q}+p\left(\mathbf{R}+\mathbf{R}^T\right)+p^2\mathbf{T}\right]=0 \tag{16}$$

Where

$$\mathbf{Q}=\begin{bmatrix}D_{11} & D_{16}\\ D_{16} & D_{66}\end{bmatrix},\ \mathbf{R}=\begin{bmatrix}D_{16} & D_{12}\\ D_{66} & D_{26}\end{bmatrix},\ \mathbf{T}=\begin{bmatrix}D_{66} & D_{26}\\ D_{26} & D_{22}\end{bmatrix} \tag{17}$$

Assume the complex potentials have the form

$$f_k=a_k\zeta_k^{-s+2}+b_k\zeta_k^{-\bar{s}+2}=a_kr^{-s+2}\xi_k^{-s+2}+b_kr^{-\bar{s}+2}\xi_k^{-\bar{s}+2} \tag{18}$$

where s is the eigenvalue and

$$\zeta_k=x+p_ky \tag{19}$$
$$\xi_k=m+p_kn \tag{20}$$

in which

$$m = \cos\theta \quad (21a)$$

$$n = \sin\theta \quad (21b)$$

Substitution of Eq. (18) into Eq. (14) gives

$$M_x = 2\mathrm{Re}\left[(-s+2)(-s+1)r^{-s}\sum_{k=1}^{2}\left(m_x^{(k)}\xi_k^{-s}a_k + \overline{m}_x^{(k)}\overline{\xi}_k^{-s}\overline{b}_k\right)\right] \quad (22a)$$

$$M_y = 2\mathrm{Re}\left[(-s+2)(-s+1)r^{-s}\sum_{k=1}^{2}\left(m_y^{(k)}\xi_k^{-s}a_k + \overline{m}_y^{(k)}\overline{\xi}_k^{-s}\overline{b}_k\right)\right] \quad (22b)$$

$$M_{xy} = 2\mathrm{Re}\left[(-s+2)(-s+1)r^{-s}\sum_{k=1}^{2}\left(m_{xy}^{(k)}\xi_k^{-s}a_k + \overline{m}_{xy}^{(k)}\overline{\xi}_k^{-s}\overline{b}_k\right)\right] \quad (22c)$$

Consider the boundary conditions

$$M_\theta(\theta = \alpha/2) = 0,\ M_\theta(\theta = -\alpha/2) = 0,$$
$$M_{r\theta}(\theta = \alpha/2) = 0,\ M_{r\theta}(\theta = -\alpha/2) = 0 \quad (23)$$

Where

$$M_\theta = n^2 M_x + m^2 M_y - 2mnM_{xy},$$
$$M_{r\theta} = -mnM_x + mnM_y + (m^2 - n^2)M_{xy} \quad (24)$$

Substitution of Eq. (22) into (23) gives a system of equations in terms of matrix form, i.e.,

$$\mathbf{Ma} = \mathbf{0} \quad (25)$$

where the 4x4 matrix **M** and 4x1 vector **a** are

$$\mathbf{a} = \begin{bmatrix} a_1 & \overline{b}_1 & a_2 & \overline{b}_2 \end{bmatrix}^T \quad (26)$$

$$\mathbf{M} = \begin{bmatrix} M_{11} & M_{12} & M_{13} & M_{14} \\ M_{21} & M_{22} & M_{23} & M_{24} \\ M_{31} & M_{32} & M_{33} & M_{34} \\ M_{41} & M_{42} & M_{43} & M_{44} \end{bmatrix} \quad (27)$$

$$M_{11} = m_\theta^{(1)}\xi_1^{-s}\Big|_{\theta=\alpha/2},\ M_{12} = \overline{m}_\theta^{(1)}\overline{\xi}_1^{-s}\Big|_{\theta=\alpha/2},$$
$$M_{13} = m_\theta^{(2)}\xi_2^{-s}\Big|_{\theta=\alpha/2},\ M_{14} = \overline{m}_\theta^{(2)}\overline{\xi}_2^{-s}\Big|_{\theta=\alpha/2},$$
$$M_{21} = m_\theta^{(1)}\xi_1^{-s}\Big|_{\theta=-\alpha/2},\ M_{22} = \overline{m}_\theta^{(1)}\overline{\xi}_1^{-s}\Big|_{\theta=-\alpha/2},$$
$$M_{23} = m_\theta^{(2)}\xi_2^{-s}\Big|_{\theta=-\alpha/2},\ M_{24} = \overline{m}_\theta^{(2)}\overline{\xi}_2^{-s}\Big|_{\theta=-\alpha/2}$$
$$M_{31} = m_{r\theta}^{(1)}\xi_1^{-s}\Big|_{\theta=\alpha/2},\ M_{32} = \overline{m}_{r\theta}^{(1)}\overline{\xi}_1^{-s}\Big|_{\theta=\alpha/2},$$
$$M_{33} = m_{r\theta}^{(2)}\xi_2^{-s}\Big|_{\theta=\alpha/2},\ M_{34} = \overline{m}_{r\theta}^{(2)}\overline{\xi}_2^{-s}\Big|_{\theta=\alpha/2},$$
$$M_{41} = m_{r\theta}^{(1)}\xi_1^{-s}\Big|_{\theta=-\alpha/2},\ M_{42} = \overline{m}_{r\theta}^{(1)}\overline{\xi}_1^{-s}\Big|_{\theta=-\alpha/2},$$
$$M_{43} = m_{r\theta}^{(2)}\xi_2^{-s}\Big|_{\theta=-\alpha/2},\ M_{44} = \overline{m}_{r\theta}^{(2)}\overline{\xi}_2^{-s}\Big|_{\theta=-\alpha/2} \quad (28)$$

For nontrivial solution, we have

$$\det[\mathbf{M}] = 0 \quad (29)$$

The eigenvalue s and the associated bending singularity order $\omega = -s$ can be determined by solving Eq. (29). Mathematically, there are infinity numbers of eigenvalues. Due to the requirement of singularity and the finite value of rotation at the notch tip, only the eigenvalues in the range $0 < s < 1$ are of interest.

4. Results and discussions

In this paper, two materials are considered, the material properties of which are
Material A:

$E_L = 10(1 + \varepsilon)$ GPa, $E_T = E_Z = 10$ GPa,
$\nu_{LT} = \nu_{LZ} = \nu_{TZ} = 0.25$, $G_{LT} = G_{LZ} = G_{TZ} = 4$ Gpa

Material B

$E_L = 132.8$ GPa, $E_T = E_Z = 10.76$ GPa,
$G_{LT} = G_{LZ} = 5.65$ GPa, $G_{TZ} = 3.61$ GPa,
$\nu_{LT} = 0.24$, $\nu_{LZ} = \nu_{LT}$, $\nu_{TZ} = 0.49$

In the above material properties, the subscript L, T and Z denote the fiber direction, transverse direction and out-of-plane direction, respectively, E_L, E_T and E_Z are the Young's moduli along L, T and Z directions, respectively, and ν_{LT}, ν_{LZ} and ν_{TZ} are the Poisson's ratios. The laminate is composed of 8layers of lamina. The layer stacking sequence is denoted as $[\phi_1/\phi_2/\phi_3/\phi_4]_s$, where subscript s denotes the symmetric laminate, ϕ_1 to ϕ_4 denote the angle between x-axis and L-direction for the first 4 layers. In material A, an anisotropic parameter ε is introduced. The degree of the anisotropy increses as ε is large. When ε is zero, the material degenerates to isotropic.Fig. 3 shows the variations of the bending singularity orders for material A laminate. The stacking sequence is $[0°/0°/0°/0°]_s$. The results reveal that when $\alpha < 180°$, no singularity is found. Under some plate angle, two eigenvalues are found in the ragne $0 < s < 1$. Taking the case with very small ε ($\varepsilon = 10^{-9}$) as an example, only one eigenvalue exists when $\alpha < 258°$. When $\alpha > 258°$, two eigenvalues are found. The results also show that for stronger anisotropy, the singularity order becomes weaker. When plate angle $\alpha = 360°$, it becomes a cracked plate. The singularity order becomes -0.5, which coincides with the classical square-root singularity of a crack structure.

Fig. 4 shows the variations of the bending singularity orders for material A laminate with a stacking sequence of $[90°/90°/90°/90°]_s$. In this case, the bending singularity is stronger than that shown in Fig. 3 for anisotropic plate. It is noted that for very small ε ($\varepsilon = 10^{-9}$), the results are identical to those shown in Fig. 3. For example, both the two figures show the same value of $\omega_1 = -0.4552$. It is also noted that this value agree with the isotropic case appeared in the literature [9].

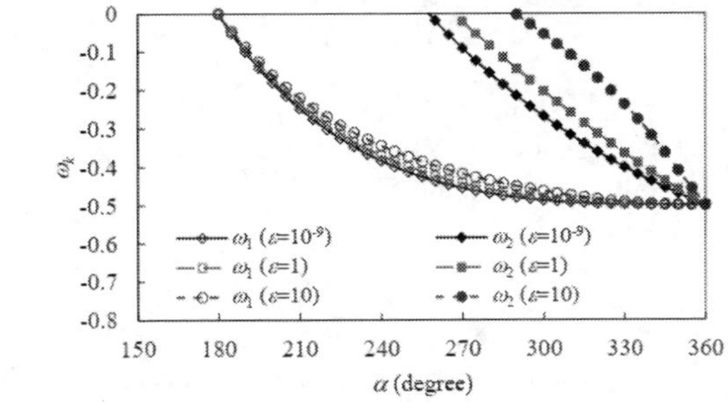

Figure 3. Variations of bending singularity orders for various plate angle α. The stacking sequence of the laminate is $[\,0^\circ/0^\circ/0^\circ/0^\circ\,]_s$ with mat erial A

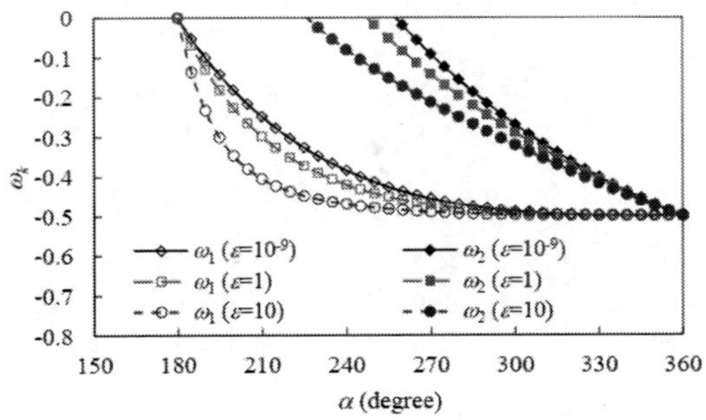

Figure 4. Variations of bending singularity orders for various plate angle α. The stacking sequence of the laminate is $[\,90^\circ/90^\circ/90^\circ/90^\circ\,]_s$ with material A.

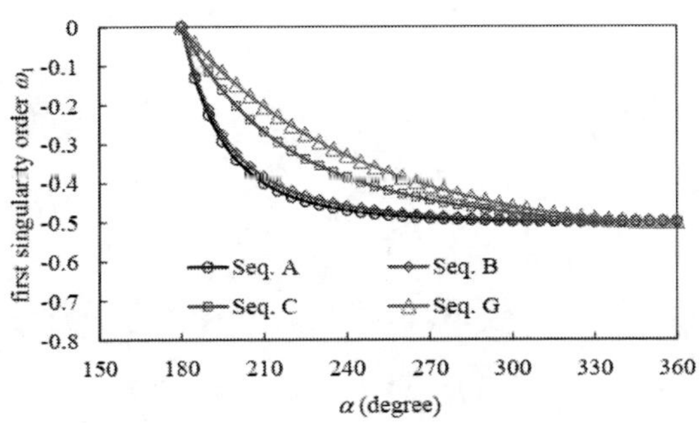

Figure 5. The first bending singularity order for Seq. A, B, C and G with material B and for various plate angle α.

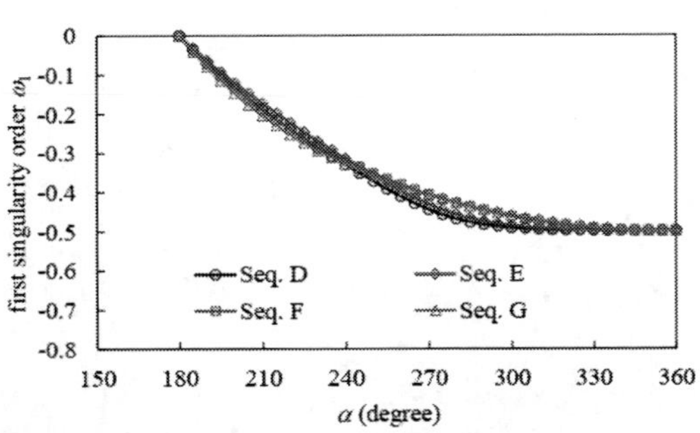

Figure 6. The first bending singularity order for Seq. D, E, F and G with material B and for various plate angle α.

Table 1. Stacking sequences

Seq.	Stacking squence
A	$[\,90^\circ/90^\circ/90^\circ/90^\circ\,]_s$
B	$[\,90^\circ/90^\circ/0^\circ/0^\circ\,]_s$
C	$[\,0^\circ/0^\circ/90^\circ/90^\circ\,]_s$
D	$[\,45^\circ/45^\circ/45^\circ/45^\circ\,]_s$
E	$[\,45^\circ/45^\circ/0^\circ/0^\circ\,]_s$
F	$[\,0^\circ/0^\circ/45^\circ/45^\circ\,]_s$
G	$[\,0^\circ/0^\circ/0^\circ/0^\circ\,]_s$

Seven stacking sequences, Seq. A to Seq. G, are considered in this study, as listed in Table 1. Fig. 5 shows the effects of stacking sequence on the bending singularity orders for material B laminate. In this figure, we consider Seq. A, B, C and G. For Seq. A, all the eight layers have the same fiber orientation of $\alpha = 90^\circ$. For Seq. B, the material orientations of layers 3 to 6 are replaced by 0°. For all considered cases, no singularity is found when $\alpha < 180^\circ$. It is seen that the first singularity orders for Seq. B are weaker than those of Seq. A. However, the differences are not significant. For Seq. C, the material orientations of first 2 and last 2 layers are replaced by 0°. From the results shown in Fig. 5, it is seen that the singularity orders of Seq. C are significantly weaker than the previous two sequences. From the results shown in Figs. 3 and 4, we see that the singularity orders tend to be stronger when material orientation angle is 90°. For Seq. G, all the eight layers have the same fiber orientation ($\alpha = 0^\circ$). This stacking sequence has the weakest singularity in Fig. 5. For plate bending, the layers near the middle plane have little effect on the plate stiffness. It can be explained that the singularity orders of Seq. B has little difference from those of Seq. A.

Fig. 6 shows the comparisons of the bending singularity for various plate angles. The stacking sequences considered in this figure are Seq. D, E, F and G. Comparing to Seq. A, B, and C, the non-zero fiber orientations are replaced by 45° in Seq. D, E and F. The numerical results show that no significant difference is observed for the four stacking sequences. The observations from Figs. 5 and 6 suggest that the fiber orientation should avoid 90° in order to reduce the bending singularity.

5. Conclusions

In this paper, the bending singularities near the apex of the V-notch in an anisotropic Ressiner-Mindlin composite laminate are investigated. The constitutive relation is derived based on the lamination theory. The Stroh-like formalism is introduced to model the anisotropy of the material. Based on the eigenfunction expansion method, the bending singularity orders can be determined by solving an eigenvalue problem numerically. The numerical investigations reveal that the bending singularity orders depend on the plate angle, material orientation, material anisotropy and the laminate stacking sequence. The material orientation should avoid 90° in order to reduce the bending singularity, especially for the lamina near the free surface. The findings presented in this paper are helpful in the design of the composite laminate with V-notch.

Acknowledgement

The author is grateful to the Ministry of Science and Technology of Taiwan for financial support through grant MOST 106-2221-E-006-098-.

References

1. M.L. Williams: J. Appl. Mech. **19** 526–528 (1952)
2. P.S. Theocaris: Int. J. Eng. Sci. **12** 107–120 (1974)
3. F. Delale: Int. J. Solids Struct. **20** 31–40 (1984)
4. H.P. Chen: Int. J. Solids Struct. **35** 1057–1073 (1998)
5. C.H. Chue and C.D. Chen: Int. J. Solids Struct. **39** 3131–3158 (2002)
6. M.L. Williams, In: Proceeding of First US National Congress of Applied Mechanics (1952), p. 325–329.
7. R.J. Hartranft and G.C. Sih: J. Math. Phys. **47** 371–90 (1968)
8. C.S. Huang: Int. J. Mech. Sci. **45** 1–20 (2003)
9. C.D. Chen: Appl. Math. Model. **39** 7657–7679 (2015)
10. T.C.T. Ting: Anisotropic Elasticity: Theory and Applications (Oxford University Press, New York, 1996).
11. T.C.T. Ting: Int. J. Solids Struct. **17** 1057–1068 (1981)
12. C.D. Chen: Int. J. Solids Struct. **43** 957–981 (2006)
13. C.D. Chen: Appl. Math. Model. 55 183–204 (2018)

Investigation of Micro Hardness, Cooling Rate and Microstructure of ATIG Welded samples of Al-SiC composite

Sivachidambaram Pichumani[1], Raghuraman Srinivasan[2], Venkatraman Ramamoorthi[3].

School of Mechanical Engineering, SASTRA Univeristy, Thanjavur, India – 613401.

Abstract. Activated TIG welding has been performed on Al – 8% SiC composite 5mm plate with various fluxes such as Al_2O_3, MnO_2, CaO, MgO, SiO_2 & TiO_2, to study & analyze the Microstructure, Micro hardness and cooling rate. Correlation study between micro hardness, microstructure and cooling rate for Constant Current TIG welding and Activated TIG welding on Al-SiC composite are also carried out to analyze the relation between the effect of cooling rate on microstructure & the effect of microstructure on micro hardness. The experimental results of ATIG welding on Al-SiC composite shows fine grain weld microstructure on ATIG – SiO_2 & ATIG – TiO_2, which results in higher micro hardness. Micro hardness values are taken in different locations of weld surface at 1mm, 2mm & 3mm below the weld surface and the same is also observed along the weld zone to heat affected zone upto 12mm for the center of the weldment. Minimum micro hardness values found in ATIG – MnO_2, ATIG – CaO & ATIG – MgO are due to intermediate micro structure between coarse and fine in heat affected zone. ATIG – Al_2O_3 weld zone & heat affected zone and heat affected zone of ATIG – MnO_2, ATIG – CaO & ATIG – MgO shows coarse microstructure leading to reduction in micro hardness value. Cooling rate for the different CCTIG & ATIG welding are recorded and correlation between the micro structures are studied. Coarse micro structure in weld zone and heat affected zone have least cooling rate whereas fine micro structure in weld zone resulted at higher cooling rate. Heat affected zone strongly depends on temperature gradient between the weld center and weldment's heat affected zone.

1. Introduction

Gas Tungsten Arc Welding (GTAW) also known as Tungsten Inert Gas (TIG) welding showed high quality weld than other arc welding process. Since non ferrous metals requires quality weld with reduced defects and improved mechanical properties of the weld, TIG welding is considered to be more suitable and appropriate welding process. However, its shallow weld penetration causes lesser production rate. Increase in weld passes for thick sections lead to higher input in the weld which reduces the weld strength [1] due to the formation of coarse grain microstructure [2].

To reduce the heat input into weld the Activated TIG welding (ATIG) a new variant was found out. ATIG welding process was developed in early 1960's by the Paton Welding Institute, Ukraine to improve the weld penetration [3, 4]. By improving the weld penetration, number weld passes get reduced by reducing the weld pass the heat input given the weld material is also get reduced [5]. The activated flux TIG welding may have two types of mechanisms that are mostly accepted, first one is based on the reverse Marangoni convection effect[6-8], and the other one is based on arc construction effect [9, 10].

Active flux coated on the weld increases the penetration by following the theory of reverse Marangonic effect proposed by Helipe & Roper[11-13] and arc constriction effect as stated by Lucas & Howse[14,15]. Due to the constricted arc & reverse Marangonic effect the weld penetration is observed [16].

2. Experiments

Autogenous welding was performed on Al - 8% SiC composite material with a plate thickness of 5mm using ADOR CHAMPTIG 300AD welding machine [17]. Other welding parameters considered during welding are provided in Table 1. Welded samples are shown in Figure 1-2.

Table 1: Welding parameter during CCTIG and ATIG welding

Parameter	Condition
Current Type	AC Current
Constant Current TIG	110 A
Electrode Diameter	3.2 mm
Electrode Material	2% Th - Tungsten Electrode
Arc Length	2 mm
Arc Voltage	18V
Welding Speed	2 mm/s
Argon Flow Rate	18 l/min
Active Layer Coating	0.5-1 mg/cm^2
Active fluxes	Al_2O_3, CaO, MgO, SiO_2, TiO_2 & MnO_2
Heat Input	990w

During welding, cooling rate at 10mm away weld zone was calculated from time - temperature profile acquired using data acquisition system (DAQ) through LabView software coupled with National Instrument NI cDAQ 9174 kit which employs a temperature acquiring module -NI 9211 with K type thermocouple.

Using standard metallographic procedure the welded samples were prepared for microstructure observation from microscopic image analyzer. Vickers micro hardness of different region such as weld zone, heat affect zone and base material were measured with load of 500g with 10 seconds as indentation time performed using Shimadzu micro hardness tester.

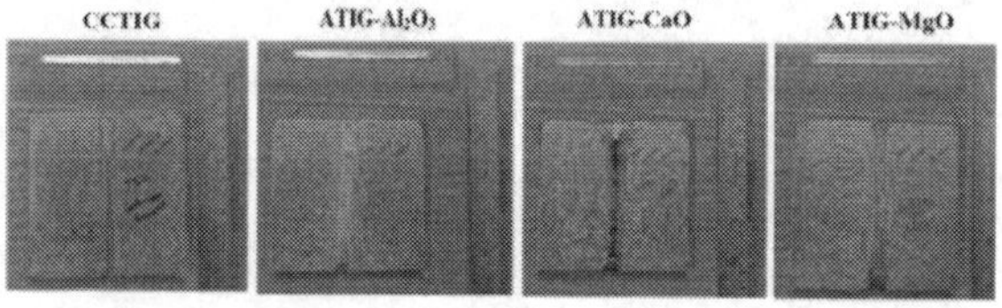

Figure 1: Welded samples of CCTIG, ATIG-Al_2O_3, ATIG-CaO and ATIG-MgO

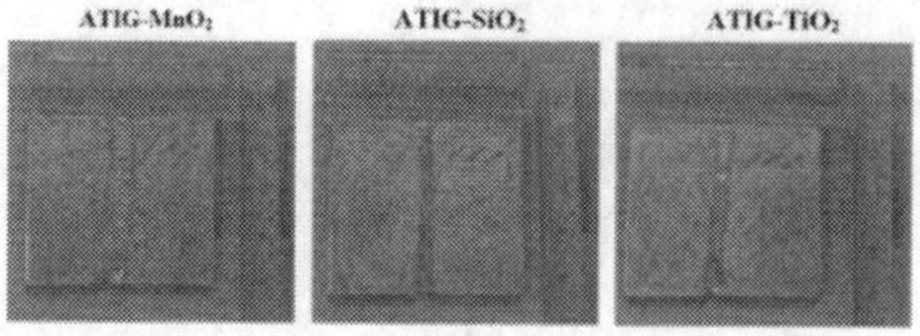

Figure 2: Welded samples of ATIG-TiO_2, ATIG-SiO_2 and ATIG-MnO_2

3. Results & Discussion

3.1 Micro hardness – CCTIG & ATIG welded samples

Table 2 shows the average micro hardness value of CCTIG & ATIG weldment on weld centre, 3mm away from weld centre & 7mm away from weld centre. Standard deviation is also tabulated. Figure 3, shows the micro hardness 1mm below the weld surface. Figure 4 shows the micro hardness value 2mm below the weld surface. Figure 5 shows the micro hardness values of 3mm below weld surface. Here micro hardness values are observed in different locations from weld centre to 12mm away from the weld centre. Micro hardness value did not have much change after 10mm away from the weld centre which is more or less equal to the base material micro hardness.

Table 2: CCTIG & ATIG welded samples micro hardness

Welding conditions	Weld centre (HV)		3mm away from weld centre (HV)		7mm away from weld centre (HV)	
	Average	Standard deviation	Averag e	Standard deviation	Avera ge	Standard deviation
CCTIG	**60.21**	0.61	**57.61**	1.19	**53.92**	2.23
ATIG-Al_2O_3	**60.73**	1.86	**57.00**	1.82	**52.58**	2.10
ATIG-CaO	**57.84**	1.72	**55.89**	1.44	**54.58**	1.97
ATIG-MgO	**62.86**	2.41	**58.50**	2.15	**53.93**	1.63
ATIG-SiO2	**63.12**	2.41	**60.50**	3.29	**57.00**	4.11
ATIG-TiO_2	**62.22**	1.06	**58.38**	1.58	**53.72**	2.69
ATIG-MnO_2	**61.42**	2.51	**58.12**	1.20	**54.33**	4.24

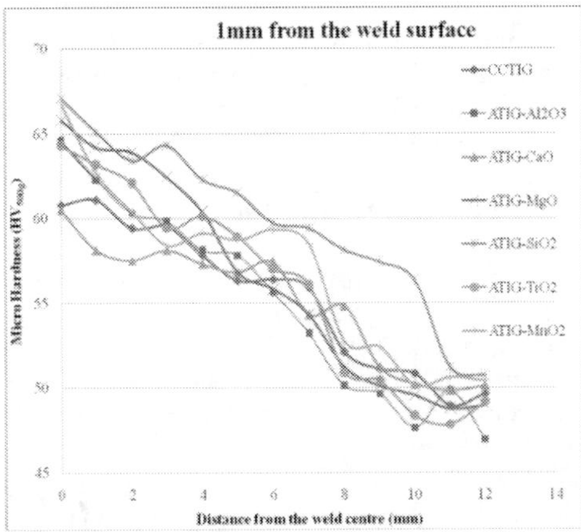

Figure 3: Micro hardness of CCTIG & ATIG welded samples 1mm below the weld surface

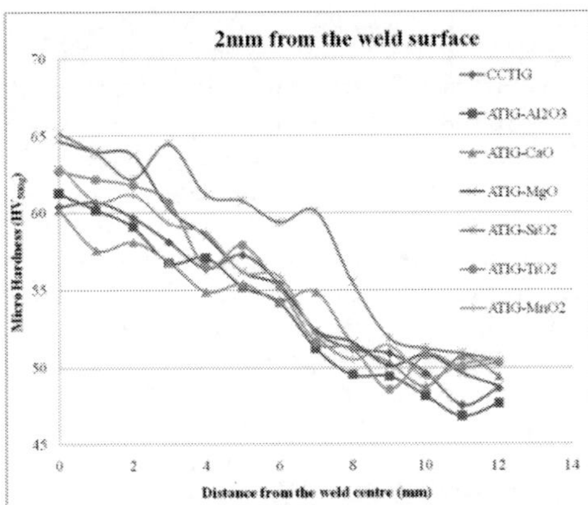

Figure 4: Micro hardness of CCTIG & ATIG welded samples 2mm below the weld surface

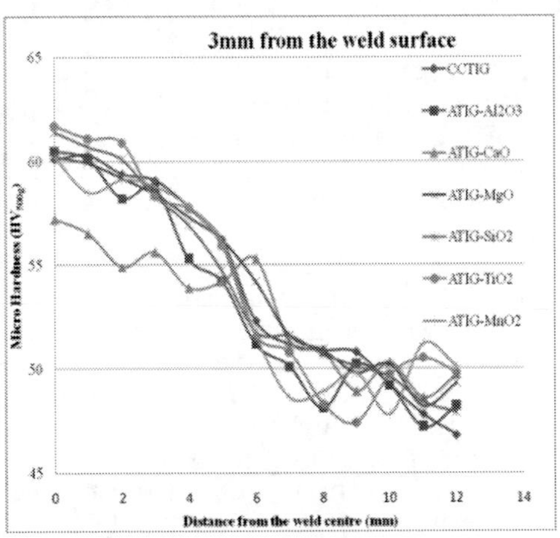

Figure 5: Micro hardness of CCTIG & ATIG welded samples 3mm below the weld surface

3.2 Microstructure – Base material, CCTIG & ATIG welded samples

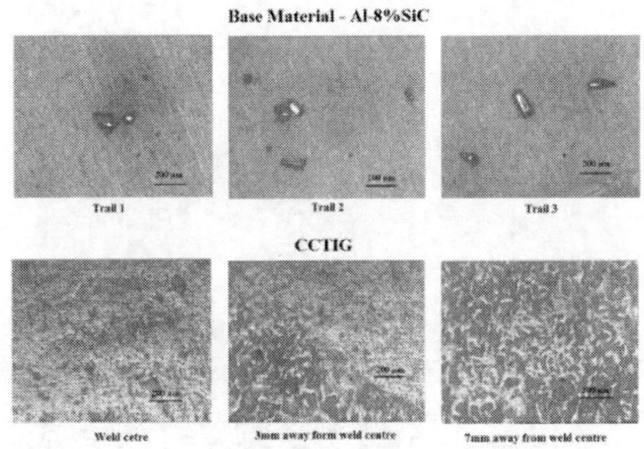

Figure 6: Microstructure of base material (Al – 8% SiC composite), CCTIG welded samples

Figure 6, shows the micro structure of base material, CCTIG welded samples. Figure 7, shows ATIG – Al_2O_3, ATIG – CaO & ATIG – MgO micro structure of weld zone, 3mm away from weld zone & 7mm away from the weld zone otherwise called as heat affected zone.

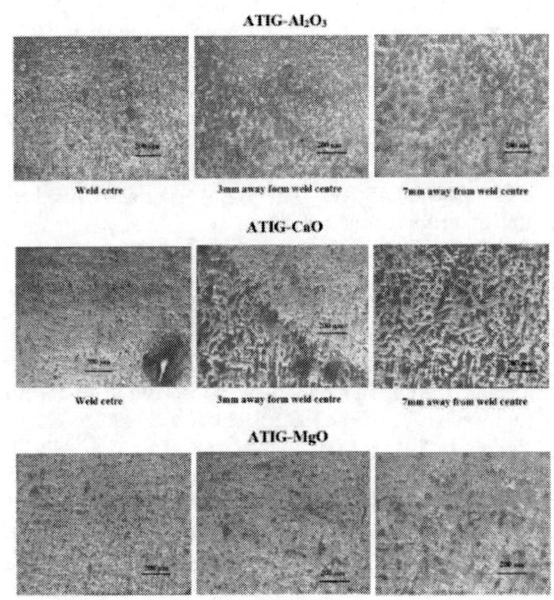

Figure 7: Microstructure of ATIG welded samples with active fluxes such as Al_2O_3, CaO & MgO

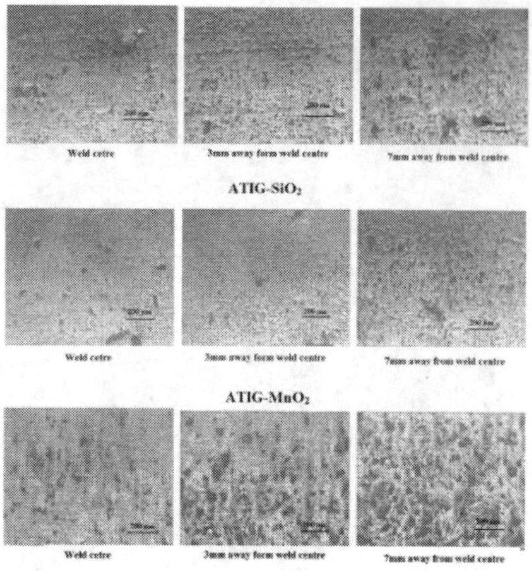

Figure 8: Microstructure of ATIG welded samples with active fluxes such as TiO_2, SiO_2 & MnO_2

Figure 8 shows the micro structure of ATIG – MnO_2, ATIG –SiO_2 & ATIG – TiO_2 weldments. From Figure 6, the following results are observed, very fine grain microstructure is observed in base material and coarse grain size microstructure in weld zone and in HAZ. Figure 7 shows ATIG – CaO & ATIG – MgO is also showed coarse grain structure which is similar to CCTIG welded microstructure. Intermediate size between fine and coarse microstructure is found in weld zone as well as in heat affected zone in ATIG – MnO_2 & ATIG – Al_2O_3. Figure-8 shows ATIG –SiO_2 & ATIG – TiO_2 is having fine grain microstructure in both weld zone and also in heat affected zone. Base material showed very fine grain microstructure.

3.3 Cooling rate – CCTIG and ATIG welded samples

Table 3 shows the cooling rate CCTIG & ATIG weldments of various active fluxes are recorded and tabulated. Highest cooling rate is observed in ATIG-SiO2 followed by ATIG – TiO2 of 295K/s & 290K/s. Lowest cooling rate is observed in CCTIG & ATIG – Al2O3 of 286K/s & 283K/s. Higher cooling rate produces the fine grain microstructure and lower cooling rate produces the coarse grain microstructure in weld zone and heat affected zone. This is due to the temperature gradient observed in weld zone and between weld zone & heat affected zone.

Table 3: CCTIG & ATIG welded samples cooling rate

PCTIG welding condition	TIG	ATIG Al_2O_3	ATIG CaO	ATIG MgO	ATIG SiO_2	ATIG TiO_2	ATIG MnO_2
Rate of cooling (K/s)	286	283	287	286	295	290	287

4. Conclusion

Correlation between micro hardness, microstructure and cooling rate of Constant Current TIG welding and Activated TIG welding on Al-SiC composite has been studied and validated.

Micro hardness are observed in three different locations of weld surface at 1mm, 2mm & 3mm locations below the weld surface and it also observed along the weld zone to heat affected zone upto 12mm for the plotting the micro hardness value around the weldment.

Microstructure observed in weld centre of the weldment. The interaction of weld zone & heat affected zone occurs 3mm away from the weld centre line and heat affected zone formation is 7mm away from the weld centre line.

Minimum micro hardness value is found in ATIG – MnO_2, ATIG – CaO & ATIG – MgO due to intermediate micro structure between coarse and fine grain formation in heat affected zone. ATIG – Al_2O_3 weld zone & heat affected zone and the heat affected zone of ATIG – MnO_2, ATIG – CaO & ATIG – MgO shows coarse microstructure, resulting in reduced micro hardness. ATIG welding on Al-SiC composite shows fine grain weld microstructure resulting in higher micro hardness on ATIG – SiO_2 & ATIG – TiO_2.

Cooling rate for the different CCTIG & ATIG welding are recorded and correlation between the micro structures are studied. Coarse micro structure in weld zone and heat affected zone have least cooling rate. Fine micro structure in weld zone should have higher cooling rate. It is also inferred that heat affected zone directly depends on temperature gradient between the weld centre and weldment's heat affected zone.

5. Acknowledgements

This research work was funded by Management of SASTRA University and the authors gratefully acknowledge and thank the Vice Chancellor, SASTRA University, Thanjavur, India for the support and encouragement.

References

1. LEI Yu-cheng, YUAN Wei-jin, CHEN Xi-zhang, ZHU Fei, CHENG Xiao-nong, In-situ weld-alloying plasma arc welding of SiC_p/Al MMC, Transactions of Nonferrous Metals Society of China, 17, 2007, page 313-317.
2. A. Urena, M.D. Escalera, L. Gil, Influence of interface reactions on fracture mechanisms in TIG arc-welded aluminium matrix composites, Composites Science and Technology, 60, 2000, page 613-622. (PII: S0266-3538(99)00168-2)
3. Selvi Dev, A. Archibald Stuart, R.C. Ravi Dev Kumaar, B.S. Murty, K. Prasad Rao, Effect of scandium additions on microstructure and mechanical properties of Al-Zn-Mg alloy welds, Materials Science and Engineering A, 467, 2007, page 132 -138. (doi:10.1016/j.msea.2007.02.080)

4. R.I. Hsieh, Y.-T. Pan, H.-Y. Liou, The Study of Minor Elements and Shielding Gas on Penetration in TIG Welding of Type 304 Stainless Steel, Journal of Materials Engineering and Performance, 8(1), February 1999, page 68-74.
5. Z.D. Zhang, L.M. Liu, Y. Shen, L. Wang, Mechanical properties and microstructure of magnesium alloy gas tungsten arc welded with cadmium chloride flux, Material Characterization, 59, 2008, page 40-46.
6. Ding FAN, Ruihua ZHANG, Yufen Gu, Masao USHIO, Effect of Flux on A-TIG Welding of Mild Steels, Transaction of JWRI, 30, 1, 2001, page 35-40.
7. Hidetoshi Fujii, Toyoyuki Sato, Shanping Lu, Kiyoshi Nogi, Development of an advanced A-TIG (AA-TIG) welding method by control of Marangoni convection, Material Science and Engineering A, 495, 2008, page 296-303. (Doi: 10.1016/j.msea.2007.10.116)
8. Shanping LU, Dianzhong LI, Hidetoshi Fujii, Kiyoshi Nogi, Time Dependant Weld Shape in Ar-O_2 Shielded Stationary GTA Welding, Journal of Material Science and Technology, 23 (5), 2007, page 650-654.
9. Shanping Lu, Hidetoshi Fujii, Kiyoshi Nogi, Marangoni convection and weld shape variations in He-CO_2 shielded gas tungsten arc welding on SUS304 stainless steel, Journal of Material Science, 43, 2008, page 4583-4591. (Doi: 10.1007/s10853-008-2681-3)
10. Shanping Lu, Hidetoshi Fujii, Kiyoshi Nogi, Marangoni convection and weld shape variations in Ar-O_2 and Ar-CO_2 shielded GTA welding, Material Science and Engineering A, 380, 2004, page 290-297. (Doi: 10.1016/j.msea.2004.05.057)
11. Shanping Lu, Hidetoshi Fujii, Kiyoshi Nogi, Weld Shape variation and Electrode Oxidation Behavior under Ar-(Ar-CO2) Double Shielded GTA Welding, Journal of Material Science and Technology, 26 (2), 2010, page 170-176.
12. A. Berthier, P. Paillard, M. Carin, S. Pellerin, F. Valensi, TIG and A-TIG welding experimental investigations and comparison with simulation Part-2 - arc constriction and arc temperature, Volume 17, Number 8, 2012, Science and Technology of Welding and Joining, page 616-621.
13. S. Leconte, P. Paillard, J. Saindrenan, Effect of fluxes containing oxides on tungsten inert gas welding process, Volume 11, Number 1, 2006, Science and Technology of Welding and Joining, page 43-47.
14. Y.L. Xu, Z.B. Dong, Y.H. Wei, C.L. Yang, Marangoni convection and weld shape variation in A-TIG welding process, Volume 48, 2007, Theoretical and Applied Fracture Mechanics, page 178-186.
15. Kuang-Hung Tseng, Chih-Yu Hsu, Performance of activated TIG process in austenitic stainless steel welds, Volume 211, 2011, Journal of Materials Processing Technology, page 503-512.
16. Chunli YANG, Sanbao LIN, Fengyao LIU, Lin Wu, Qingtao ZHANG, Research on the Mechanism of Penetration Increase by Flux in A-TIG Welding, Journal of Material Science and Technology, 19,1, 2003, page 225-227.
17. Sivachidambaram Pichumani, Raghuraman Srinivasan, Effect of Pulsed Current and Activated Flux with TIG Welding on the Microstructure and Mechanical Properties of Al-SiC Composite, Indian Journal of Science and Technology, Vol. 8(27), 71559, October 2015, pp 1-6.

Nonlinear Dynamic Responses of a Honeycomb Sandwich Plate Subject to Transverse Excitations

Dongmei Wang[1], Wei Zhang[2], Minghui Yao[2] and Yinli Liu[1]

[1]College of science, Tianjin University of Science and Technology, 300457 Tianjin, P. R. China
[2]College of Mechanical Engineering, Beijing University of Technology, 100124 Beijing, P. R. China

Abstract. Nonlinear dynamic behaviors of a simply supported honeycomb sandwich plate subjected to the transverse excitations are investigated in this paper. Based on the classical thin plate theory and Von Karman large deformation theory, the governing equation of motion for the honeycomb sandwich plate is established by using the Hamilton principle. The nonlinear governing partial differential equation is discretized to the ordinary differential equations by differential quadrature method and then solved by Runge-Kutta-Fehlberg method. Based on the numerical simulations, combined with nonlinear dynamic theory, the influences of the frequency and amplitude of the transverse excitation are investigated respectively by using the bifurcation diagrams, Poincare maps and phase portraits. The results exhibit the existence of the period-1, period-2 and chaotic responses with the variation of the excitations, which demonstrate that those motions appear alternately.

1 Introduction

As a kind of special composite structural material, sandwich beam and plates are widely used in a variety of aircraft structures and satellite launch vehicles. This is due to their many advantages such as low density, high strength ratio, high stiffness and ease of manufacture and repair compared with other structural materials. Therefore, sandwich structures have been received great attention, particularly in the recent decades.

A review of literature indicates that many analysis of sandwich plates, involving static, free vibration and stability analyzes, has been carried out. Using several plate models, Zenkour [1-2] performed a comprehensive analysis of functionally grade sandwich plates, including buckling, free vibration and stability analyzes. Li et al. [3] applied the three dimensional Ritz method based on the use of the Chebyshev polynomials for the free vibration analysis of functionally graded material sandwich plates. Dozio[4] studied the free vibration behavior of sandwich plates with FGM core via variable-kinematic 2-D Ritz models using Chebyshev polynomials. Neves et al.[5] combined quasi-3D higher-order shear deformation theories with meshless techniques in order to cope with bending, free vibration and stability analysis of FGM isotropic and sandwich plates. Fiorenzo [6,7] investigated free vibration and thermal stability of functionally graded sandwich plates and sandwich plates with anisotropic face sheets in thermal environment by using Hierarchical Trigonometric Ritz Formulation. Wang and Yuan [8] studied the effects of boundary conditions and different types loading on the displacement and stress distributions of the soft-core sandwich panels by using differential quadrature method.

The aforementioned research works provide the foundation and the guidance for us to study a variety of properties of the honeycomb sandwich plates. However, there is few research focused on the global bifurcations and chaotic dynamics of sandwich plates. We have found two papers on bifurcations and chaotic dynamics of sandwich plates. J.H.Zhang and W. zhang [9] investigated the global bifurcations and multi-pulse chaotic dynamics of a simply supported honeycomb sandwich rectangular plate under combined parametric and transverse excitations, they found that multi-pulse chaotic motions can occur in the honeycomb sandwich rectangular plate. Applying the Galerkin's approach, Zhang et al. [10] investigated nonlinear dynamic behaviors of a simply supported 3D-Kagome truss core sandwich plate subjected to the transverse and the in-plane excitations. The bifurcation diagrams exhibit the existence of period, muti-period and chaotic motions appear alternately.

Usually, the honeycomb sandwich plates can be modeled by nonlinear partial differential equations. As we all know, in most cases it is difficult to obtain closed-form solutions for nonlinear partial differential equations. As a result, people have to apply approximate, numerical or analytical methods to solve partial differential equations which arise in almost all engineering sphere. In the literature mentioned above, the approximate method used by scholars include finite difference method, finite element method, Ritz method, Galerkin method,

https://doi.org/10.1051/matecconf/201815101003

homotopy analysis method, multiquadric collocations method and differential quadrature method.

The differential quadrature (DQ) method, which was introduced by Bellman [11] et al. in 1972, is a highly efficient approach to directly solving a partial differential equation for a finite domain with a set of given boundary conditions. The DQ method can discretize a finite domain by using a set of regularly or irregularly distributed grid points. Using appropriate interpolation polynomials, the unknown functions and their derivatives at an arbitrary point can be represented by a weighted linear combination of the functional values at all the discrete points. The accurate numerical results and little computational effort are obtained by using a smaller number of grid points.

The pioneering work on the application of the DQ method to the structural mechanics was given by Striz et al. [12, 13]. Since the DQ method has lots of merits, such as simple principle, high precision, less computation and easy programming, this method has been used to solve a variety of problems in structural mechanics [14-20]. All these works demonstrated that the application of the DQ method caused the accurate results and less computational effort. However, most of the DQ applications are relevant to the problems of the static or free vibration in engineering fields, few applications are relevant to the problems of the nonlinear dynamics. The author [21] of this paper has applied the DQ method to the analysis of nonlinear dynamics of axially moving viscoelastic beams and the differential quadrature method was compared with other methods, which accuracy is verified.

In this paper, we developed the DQ method to analyze the nonlinear dynamical properties such as bifurcations and chaos of a simply supported honeycomb sandwich rectangular plate under transverse excitation for the first time. It is found from the numerical results that the DQ method can be successfully applied to reveal the complex dynamic phenomena of the honeycomb sandwich rectangular plate.

2 Governing equation of the honeycomb sandwich plate

Consider a simply supported honeycomb sandwich plate, where the edge lengths are a and b and h the thickness is h, respectively. The honeycomb sandwich rectangular plate consists of 3 layers that made of two layers of thin skins and a hexagon honeycomb core. Based on the classical thin plate theory and Von Karman large deformation theory, by using the generalized Hamilton's principle, we have the governing equation of the nonlinear transverse vibration of the honeycomb sandwich plate under transverse load as following:

$$D_{11}w_{,xxxx}-(D_{12}-4D_{66}-D_{21})w_{,yyxx}-D_{22}w_{,yyyy}$$

$$+\left(A_{11}(w_{,x}w_{,xx})+A_{12}(w_{,y}w_{,yx})\right)w_{,x}+\left(A_{21}(w_{,x}w_{,yx})+A_{22}(w_{,y}w_{,yy})\right)w_{,y}$$

$$+\left(\mathrm{A}_{21}\left(\frac{1}{2}w_{,x}^{2}\right)+A_{22}\left(\frac{1}{2}w_{,y}^{2}\right)\right)w_{,yy}+A_{66}(w_{,yy}w_{,x}+w_{,y}w_{,yx})w_{,x}$$

$$+\left(A_{11}\left(\frac{1}{2}w_{,x}^{2}\right)+A_{12}(w_{,y}^{2})\right)w_{,xx}+A_{66}(w_{,xx}w_{,y}+w_{,x}w_{,yx})w_{,y}$$

$$-c\dot{w}+2A_{66}w_{x}w_{y}w_{yx}+f=I_0\ddot{w}\ , \quad \{(x,y)|\,0< x< a,\ 0< y<\}\ (1)$$

Where a dot represents the partial differentiation with respect to time t, and a comma denotes the partial differentiation with respect to a specified coordinate, c is a vibration damping coefficient, $f=F(x,y)\cos\Omega t$ is the transverse excitation that loads the plate

Where

$$(A_{ij},D_{ij})=\sum_{k=1}^{3}\int_{z_k}^{z_{k+1}}Q_{ij}^{k}(1,z^2)dz, \qquad (i,j=1,2,6) \tag{2}$$

Q_{ij}^{k} is the conversion elastic constant of layer k

$$Q_{11}^{k}=\frac{E_1^k}{1-v_{12}^{k}v_{21}^{k}},\ Q_{12}^{k}=\frac{v_{12}E_1^k}{1-v_{12}^{k}v_{21}^{k}},\ Q_{22}^{k}=\frac{E_2^k}{1-v_{12}^{k}v_{21}^{k}},\ Q_{66}^{k}=G_{12}^{k} \tag{3}$$

where $E_1=E_2, v_{12}=v_{21}$.

I_0 is the quality of honeycomb sandwich plates and

$$I_0=\sum_{k=1}^{3}\int_{z_k}^{z_{k+1}}\rho_k dz\ . \tag{4}$$

The boundary conditions are

$$x=0 \text{ or } x=a,\ w=0,\ \frac{\partial^2 w}{\partial x^2}=0\ , \tag{5a}$$

$$y=0 \text{ or } y=b,\ w=0,\ \frac{\partial^2 w}{\partial y^2}=0\ . \tag{5b}$$

3 Differential quadrature analogs

The DQM formulation can be found in literature [22]. For integrity and convenience, this subsection reviews the main idea of the DQM. The idea of the DQM is to directly compute the derivative of a function at any grid point within its bounded domain by estimating a weighted linear sum of values of the function at a small set of points belonging to the domain.

Introduce $M\times N$ unequally spaced sampling points as

$$x_i=\left(\frac{a}{2}\right)[1-\cos\frac{(i-1)\pi}{N-1}],\quad (i=1,2,\cdots,M)\ , \tag{6a}$$

$$y_j=\left(\frac{b}{2}\right)[1-\cos\frac{(j-1)\pi}{N-1}]\ ,\quad (j=1,2,\cdots,N)\ . \tag{6b}$$

Applying the quadrature rules for the derivatives of a function at the sampling points yield

$$w_{,x}(x_i,y_j,t)=\sum_{k=1}^{M}a_{ik}^{(1)}w(x_k,y_j,t)\ , \tag{7a}$$

$$w_{,xx}(x_i,y_j,t)=\sum_{k=1}^{M}a_{ik}^{(2)}w(x_k,y_j,t)\ , \tag{7b}$$

$$w_{,xxxx}(x_i,y_j,t)=\sum_{k=1}^{M}a_{ik}^{(4)}w(x_k,y_j,t)\ , \tag{7c}$$

$$w_{,y}(x_i,y_j,t)=\sum_{z=1}^{N}a_{jz}^{(1)}w(x_i,y_z,t)\ , \tag{7d}$$

$$w_{,yy}(x_i,y_j,t)=\sum_{z=1}^{N}a_{jz}^{(2)}w(x_i,y_z,t)\ , \tag{7e}$$

$$w_{,yyyy}\left(x_i,y_j,t\right)=\sum_{s=1}^{N}a_{js}^{(4)}w\left(x_i,y_s,t\right), \quad (7f)$$

$$w_{,xy}\left(x_i,y_j,t\right)=\sum_{s=1}^{N}a_{js}^{(1)}\sum_{k=1}^{M}a_{ik}^{(1)}w\left(x_k,y_s,t\right), \quad (7g)$$

$$w_{,xxyy}\left(x_i,y_j,t\right)=\sum_{s=1}^{N}a_{js}^{(2)}\sum_{k=1}^{M}a_{ik}^{(2)}w\left(x_k,y_s,t\right). \quad (7h)$$

$$i,k=1,2,\cdots,M;\ j,s=1,2,\cdots,N$$

where $w_{,x}\left(x_i,y_j,t\right)$, $w_{,xx}\left(x_i,y_j,t\right)$, $w_{,xxxx}\left(x_i,y_j,t\right)$, $w_{,y}\left(x_i,y_j,t\right)$, $w_{,yy}\left(x_i,y_j,t\right)$, $w_{,yyyy}\left(x_i,y_j,t\right)$, $w_{,xy}\left(x_i,y_j,t\right)$, $w_{,xxyy}\left(x_i,y_j,t\right)$ are the first, the second and the forth-order partial derivative approximations of x,y at the nth time level at the grid point $\left(x_i,y_j\right)$, respectively, and the DQM weighting coefficients are the expression

$$a_{ij}^{(1)}=\frac{\prod_{k=1,k\neq i}^{N}(x_i-x_k)}{(x_i-x_j)\prod_{k=1,k\neq j}^{N}(x_i-x_j)},\ (i,j=1,2,\cdots,N;j\neq i), \quad (8)$$

$$a_{ij}^{(m)}=m[a_{ii}^{(m-1)}a_{ij}^{(1)}-\frac{a_{ij}^{(m-1)}}{x_i-x_j}],\quad a_{ii}^{(m)}=-\sum_{k=1,k\neq i}^{N}a_{ik}^{(m)}. \quad (9)$$

$$(m=2,3,\cdots;\ i,j=1,2,\cdots N;j\neq i)$$

In this paper, set $a=b=1, M=N$, and the coefficient matrix $\left(a_{ij}^{(m)}\right)$ of the first order, second order and fourth order derivative are expressed in $AB1$, $AB2$, $AB4$, respectively and set the internal node function value vector $\left[w(x_1,y_1,t),w(x_2,y_2,t),\cdots,w(x_{N-2},y_{N-2},t)\right]$ is expressed in W. Modification of the weighting coefficient matrix [22] to implement the boundary condition equations (5), yields

$$\left(a_{ij}^{(2)}\right)=\begin{vmatrix}0 & 0 & \cdots & 0 & 0\\ a_{21}^{(2)} & a_{22}^{(2)} & \cdots & a_{2,N-1}^{(2)} & a_{2,N}^{(2)}\\ \vdots & \vdots & \ddots & \vdots & \vdots\\ a_{N-1,1}^{(2)} & a_{N-1,2}^{(2)} & \cdots & a_{N-1,N-1}^{(2)} & a_{N-1,N}^{(2)}\\ 0 & 0 & \cdots & 0 & 0\end{vmatrix} \quad (10)$$

With $\left(a_{ij}^{(2)}\right)$, the modified weighting coefficient matrix $\left(a_{ij}^{(4)}\right)$ can be computed by using the matrix multiplication method. Substitution of equations (7) into equations (1) yields

$$D_{11}\times(I\otimes AB1\times W)-(D_{11}-4D_{66}-D_{21})\times\left(AB2\otimes I\times(I\otimes AB2\times W)\right)$$
$$-D_{22}\times\left(AB4\otimes I\times W\right)+\left[A_{21}(I\otimes AB1\times W)\bullet\left(AB1\otimes I\times(I\otimes AB1\times W)\right)\right.$$
$$\left.+A_{22}\left(AB1\otimes I\times W\right)\bullet\left(AB2\otimes I\times W\right)\right]\bullet\left(AB1\otimes I\times W\right)+\left[A_{21}\left(\frac{1}{2}\right)\right.$$
$$\left.(I\otimes AB1\times W)\bullet(I\otimes AB1\times W)+A_{22}\left(\frac{1}{2}\right)\left(AB2\otimes I\times W\right)\bullet\left(AB2\otimes I\times W\right)\right]$$
$$\bullet\left(AB2\otimes I\times W\right)+\left[A_{11}(I\otimes AB1\times W)\bullet(I\otimes AB2\times W)+A_{12}\left(AB1\otimes I\times W\right)\right.$$
$$\left.\bullet\left(AB1\otimes I\times(I\otimes AB1\times W)\right)\right]\bullet(I\otimes AB1\times W)+A_{66}\left[\left(AB2\otimes I\times W\right)\right.$$
$$\left.\bullet(I\otimes AB1\times W)+\left(AB1\otimes I\times W\right)\bullet\left(AB1\otimes I\times(I\otimes AB1\times W)\right)\right]\bullet\left(AB1\otimes I\times W\right)$$
$$\left[A_{11}\left(\frac{1}{2}\right)(I\otimes AB1\times W)\bullet(I\otimes AB1\times W)+A_{12}\left(AB2\otimes I\times W\right)\bullet\left(AB2\otimes I\times W\right)\right]$$
$$\bullet\left(AB1\otimes I\times W\right)+\left[A_{11}\left(\frac{1}{2}\right)(I\otimes AB1\times W)\bullet(I\otimes AB1\times W)\right.$$
$$\left.+A_{12}\left(AB2\otimes I\times W\right)\bullet\left(AB2\otimes I\times W\right)\right]\bullet(I\otimes AB2\times W)$$
$$+A_{66}\left[\left(AB1\otimes I\times W\right)\bullet(I\otimes AB2\times W)+(I\otimes AB1\times W)\right.$$
$$\left.\bullet\left(AB1\otimes I\times(I\otimes AB1\times W)\right)\right]\bullet\left(AB1\otimes I\times W\right)-c\dot{W}+2A_{66}\left[(I\otimes AB1\times W)\right.$$
$$\left.\bullet\left(AB1\otimes I\times W\right)\right]\bullet\left(AB1\otimes I\times(I\otimes AB1\times W)\right)+f=I_0\ddot{W} \quad (11)$$

Where I represent a $(N-2)\times(N-2)$ unit matrix, " • "means that two vectors are multiplied by the corresponding elements, "$\times,\otimes$" represent vector product and Kronecker product respectively. Equations (11) includes $(N-2)\times(N-2)$ nonlinear ordinary differential equations with unknown transverse displacements vector W. Given initial conditions, for a set of given parameters E, Ω, F and so on. Then equation (11) can be numerically solved by Runge-Kutta-Fehlberg method, and the transverse displacement $w(x_i,y_i,t)$ at each t can be obtained.

4 Numerical results

In the following, the DQM is applied to the governing equations (1) to investigate the effect of the frequency and the amplitude of the transverse excitation on the nonlinear dynamic behaviors of the honeycomb sandwich plate. The bifurcation diagram is a modern technique used to analyze nonlinear systems. To give the bifurcation diagrams, we will project the Poincare maps onto the displacement axis and the velocity axis. By varying the transverse excitation frequency Ω and the amplitude F, the two types of bifurcation diagrams are respectively presented with all other parameters fixed as shown Figure 1 and Figure 2. In order to illustrate in detail the bifurcation behaviors, the Poincare maps which consists of the displacements w and the velocity $\frac{dw}{dt}$, and the phase portraits which consists of the displacement w and the velocity $\frac{dw}{dt}$ are respectively depicted.

In all present investigations, the center of the plate is chosen to represent the motion of the plate, the initial conditions are

$$w(x,y,0)=0,\ w_t(x,y,0)=0.01 \quad (12)$$

Figure1 presents the bifurcation diagrams for the displacements w via the transverse excitation frequency Ω when the transverse excitation amplitude $F=3000$N.

From Figure 1, we can see that when the frequency Ω is located in the interval [62, 63]Hz, the system movement is complicated chaotic motions. With the increasing of the frequency up to 67Hz, most of the motions are simple periodic motions which interrupted by two small chaotic windows, when $\Omega = 64.475\text{Hz}$, period doubling bifurcation appears, that is the single period motions change into period-2 motions. With the further increase of the frequency of the transverse excitation, the system movement change into chaotic motions again until the last.

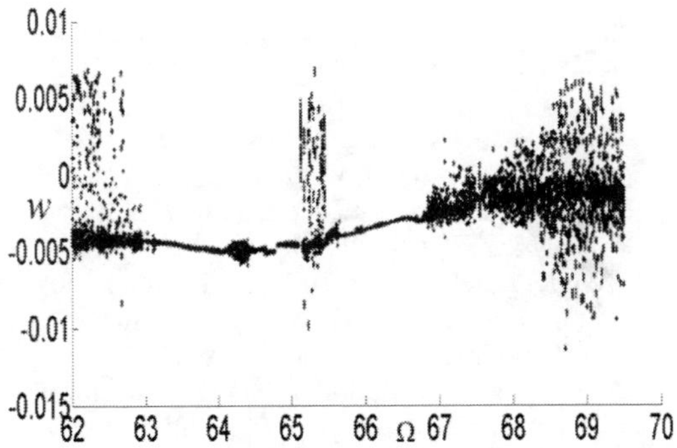

Figure 1. Effect of transverse excitation frequency: bifurcation diagrams for the displacement w via Ω .

Figure2 presents the bifurcation diagrams for the displacements w via the transverse excitation amplitude fluctuation F in $[2500, 2900]\text{N}$, when $\Omega = 62\text{Hz}$. From Figure 2, we can see that chaotic and periodic motions appear alternately when the transverse excitation amplitude F is located in $[2500, 2662]\text{N}$, which most of the motions are large amplitude chaotic motions. Increasing the amplitude of the transverse excitation F , it is observed that a large window periodic motions appear and there is a period doubling bifurcation when $F = 2846\text{N}$. With the further increase of the transverse excitation amplitude from $F = 2850\text{N}$, the system movement change into chaotic motions again until the last.

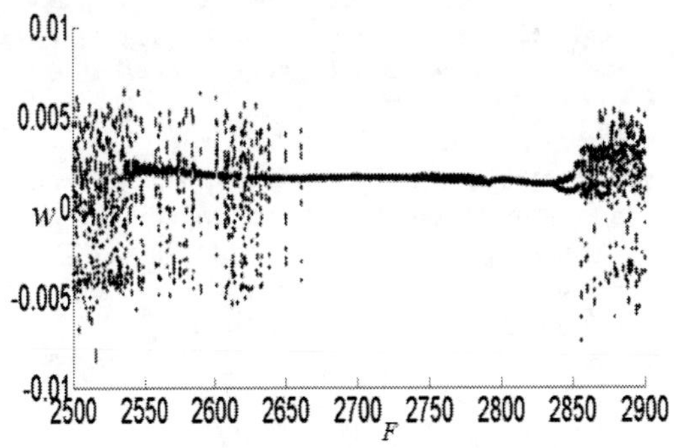

Figure 2.Effect of amplitude of the transverse excitation: bifurcation diagrams for the displacement w via F

Typical periodic motion and chaotic motion are presented in figures 3-5, in terms of (a) phase portrait of w and $\dot{w}$, (b) the Poincare maps of w and $\dot{w}$. They can be easily differentiated. In a projection of a Poincare map, periodic-1 motion is represented by one point, periodic-2 motion is represented by two point and chaos is represented by infinite number of points.

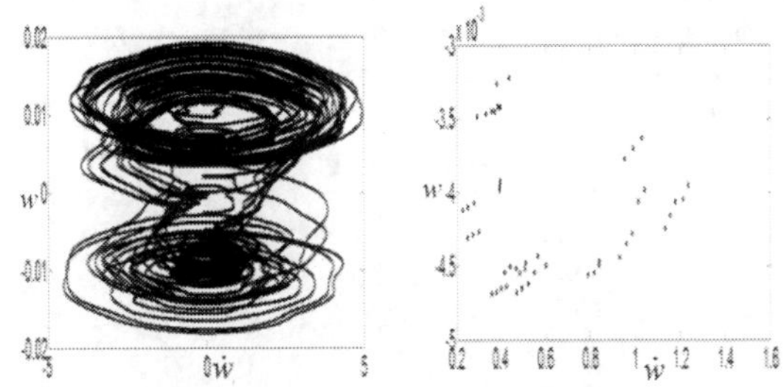

(a) (b)

Figure 3. Chaotic motion appears when $F = 3000\text{N}$, $\Omega = 62.5\text{Hz}$

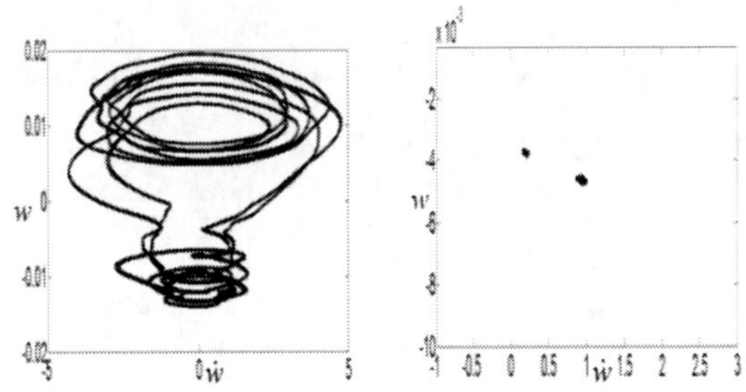

(a) (b)

Figure 4. Period-2 motion appears when $F = 2846\text{N}$, $\Omega = 62\text{Hz}$

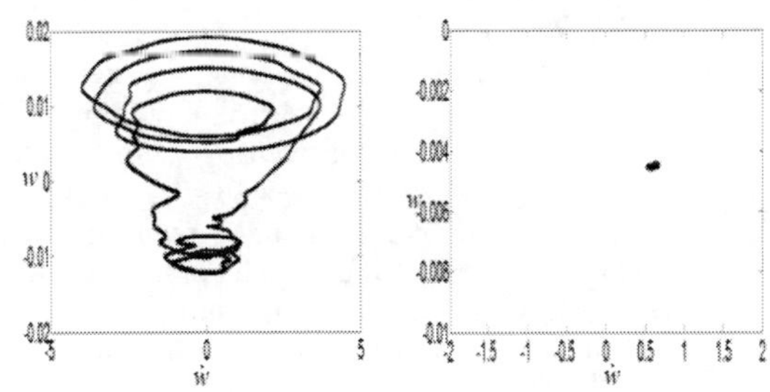

(a) (b)

Figure 5. Period-1 motion appears when $F = 2700\text{N}$, $\Omega = 62\text{Hz}$

5 Summary

This paper studies the bifurcations and chaotic motions for a simply supported honeycomb sandwich rectangular plate under transverse excitation which governing equation is established by using the generalized Hamilton's principle. The DQ method is developed to solve the nonlinear partial-differential governing equation numerically. Based on the numerical solutions, the nonlinear dynamical behaviors are numerically investigated by means of the Poincare map, the phase portraits. Two types of bifurcation diagrams are presented to show the displacement and velocity projections of the Poincare maps changing with the frequency and the amplitude of the transverse excitation. The results show that the chaotic motion and the periodic motion exchange alternately for a sufficiently large fluctuation of frequency and the amplitude of the transverse excitation. In addition, comparing the phase portraits with the Poincare map, the nonlinear dynamical behaviors are identified by the phase portraits are the same as the results obtained using the Poincare map. These results show the DQ method is an efficient method for the nonlinear analysis of honeycomb sandwich plate.

References

1. A. M. Zenkour, A comprehensive analysis of functionally graded sandwich plates:part 2-Buckling and free vibration, Int J Solids Struct. **42**, 5243-5258 (2005).
2. A. M. Zenkour, M. Sobhy, Thermal buckling of various types of FGM sandwich plates, Compos Struct. **93(1)**, 93-112 (2010)
3. Q. Li, V.P. Iu, K.P. Kou, hree-dimensional vibration analysis of functionally graded material sandwich plates, J Sound Vib. **311,** 498-515 (2008)
4. L.Dozio, Natural frequencies of sandwich plates with FGM core via variable-kinematic 2-D Ritz models, Compos Struct. **96** , 561-568 (2013)
5. A.M.A.Neves, A.J.M.Ferreira, E.Carrera, M. Cinefra, C.M.C. Roque, R.M.N.Jorge, et al., Static, free vibration and buckling analysis of analysis of isotropic and sandwich functionally graded plates using a quasi-3D higher-order shear deformation theory and a meshless technique, Compos Part B: Eng. **44(1)**, 657-674 (2013)
6. Fiorenzo A. Fazzolari, Erasmo Carrera, Free vibration analysis of sandwich plates with anisotropic face sheets in thermal environment by using the hierarchical trigonometric Ritz formulation, Composites: Part B **50**, 67-81 (2013)
7. Fiorenzo A. Fazzolari, Natural frequencies and critical temperatures of functionally graded sandwich plates subjected to uniform and non-uniform temperature distributions, Composite Structures **121**, 197-210 (2015)
8. X.W.Wang, Z.X.Yuan, Accurate stress analysis of sandwich panels by differential quadrature method, Applied Mathematical Modeling **43**, 548-565 (2017)
9. J. H .Zhang, and W. Zhang, Multi-pulse chaotic dynamics of non-autonomous nonlinear system for a honeycomb sandwich plate, Acta Mechanica **223**, 1047-1066 (2012)
10. W. Zhang, J. E.Chen., D. X.Cao, L.H. Chen, Nonlinear dynamic responses of a truss core sandwich plate. Composite Structures **108**, 367-386 (2014)
11. R.Bellman, B.G.Kashef, J.Casti, Differential quadrature: a technique for the rapid solution of nonlinear partial differential equations. Journal of Computational Physics **10**, 40-52 (1972)
12. C.W. Bert, S.K. Jang, A.G.Striz, Two new approximate methods for analyzing free vibration of structural components, AIAA Journal **26**, 612-618 (1988)
13. C.W. Bert, S.K. Jang, A.G.Striz, Nonlinear bending analysis of orthotropic rectangular plates by the method of differential quadrature, Computational Mechanics **5**, 217-226 (1989)
14. P.Malekzadeh, G.Karami, Differential quadrature nonlinear analysis of skew composite plates based on FSDT, Engineering Structures **28**, 1307-1318 (2006)
15. A.Alibeigloo, R.Madoliat, Static analysis of cross-ply laminated plates with integrated surface piezoelectric layers using differential quadrature, Composite Structures **88**, 342-353 (2009)
16. J.E.Jam, S. Maleki, A.Andakhshideh, Non-linear bending analysis of moderately thick fundtionally graded plates using generalized differential quadrature method, International Journal of Aerospace Sciences **1**, 49-56 (2011)
17. P.Li, Y.R.Yang, L.Lu, Instability analysis of two-dimensional thin panels in subsonic flow with differential quadrature method, Journal of Dynamics and Control **10**, 11-14 (2012)
18. Y.F.Zhou, Z.M.Wang, Application of the differential quadrature method to free vibration of viscoelastic thin plate with linear thickness variation, Meccanica **49**, 2817–2828 (2014)
19. B.Laxmi, S.Chakraverty, Application of Differential Quadrature method in free vibration analysis of nanobeams based on various nonlocal theories, Computers and Mathematics with Applications **69**, 1444-1462 (2015)
20. X.W.Wang, Z.X.Yuan, Accurate stress analysis of sandwich panels by the differential quadrature method, Applied Mathematical Modelling **43**, 548-565, (2017)
21. W. Zhang, D.M.Wang, M.H.Yao, Using fourier differential quadrature method to analyze transverse nonlinear vibrations of an axially accelerating viscoelastic beam, Nonlinear Dyn **78**, 839-856 (2014)
22. C.Shu, *Differential Quadrature and its Application in Engineering.* (Springer, Berlin 2000)

Material Property and Stress Field near Notch Tip in Steel Plate under High Electric Current

Thomas Jin-Chee Liu , Ji-Fu Tseng, Chun-Der Cheng

Department of Mechanical Engineering, Ming Chi University of Technology, Taishan, New Taipei City, Taiwan

Abstract. Under high electric current, the material property and stress field near the notch tip in the stainless steel plate have been investigated. Using the numerical and experimental methods, the electric current density, temperature, stress, grain image and micro hardness are obtained. Also, the melting and heat-affected zones at the notch tip are investigated. At the tip region, the hardness value improves and the compressive residual stress occurs. This result can reduce the stress level and prevent from the crack initiation when the plate is subjected to the tensile load.

1 Introduction

As shown in Fig. 1, the electro-thermo-structural coupled-field analyses of the fractured metal structure have been reported in the past references [1-6]. This problem concerns the crack detection and crack arrest [6]. Under low electric current, there is a hot spot (shown in Fig. 1) at the crack tip and it can be detected by the infrared camera. However, under high electric current, the hot spot can melt and a crack tip hole can be made. The hole can stop the crack propagation.

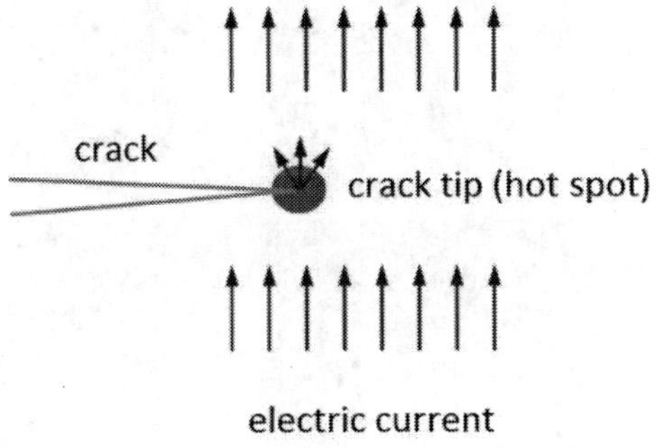

Figure. 1. Coupled-field and hot spot at crack tip.

This paper studies the electro-thermo-structural coupled-field problem of the steel plate with a notch. Using the numerical and experimental methods, the electric current density, temperature, stress, grain image and micro hardness will be obtained. Similar to the crack problem, the concentrations of the electric current and stress exist at the notch tip. Especially, the material property and stress field near the notch tip will be discussed in this paper.

2 Problem definition

In Fig. 2, it shows the sample and configuration of this study. The sample is a steel plate with a notch. The main dimensions of the plate are L, W, a, α, and thickness e. The electric power equipment provides a current I_0. The current will induce the temperature field and thermal tress/deformation. The sample is made of the SUS 304 stainless steel. The material properties are listed in Table 1. The elasto-plastic stress-strain relation with the tangent modulus E_T=0.4E is considered in the analysis.

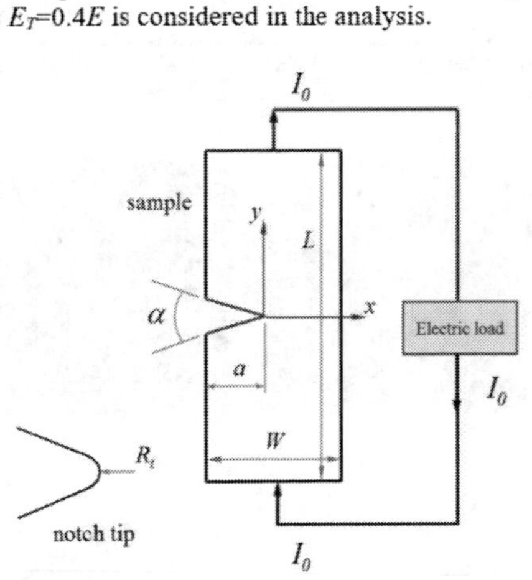

Figure. 2. Problem definition.

3 Experimental and numerical analyses

Fig. 3 shows the experimental equipment of this study. The equipment is modified from the resistance spot welding (RSW) machine. The sample (steel plate) is fixed on two electrodes. The electric power supplier

 https://doi.org/10.1051/mateccconf/201815101004

provides the high electric current. The dimensions of the sample are L=0.1 m, W=0.008 m, a=0.004 m, e=0.001 m, α=20°, and R_t=0.0002 m.

The finite element method and the software ANSYS are adopted for the numerical simulation. In Fig. 4, it shows the finite element model of the steel plate. The model is constructed by the SOLID226 element associated with the electro-thermo-structural coupled-field analysis. The following conditions are considered in the finite element analysis:

(1) Electro-thermo-structural coupled-field.
(2) Three-dimensional degrees-of-freedom (DOFs).
(3) Transient heat transfer (initial condition: 25 °C).
(4) Convection on air/steel interface (natural convection coefficient: 25 W/(m^2-K)).
(5) Steady electric current (DC current).

(6) Elasto-plastic material property.

Table 1. Material data of SUS 304 stainless steel [7,8].

Temperature (°C)	Young's modulus E (GPa)	Yielding strength S_y (MPa)	Coefficient of thermal expansion α (1/°C)	Thermal conductivity k (W/m-°C)	Specific heat C_p (J/kg-°C)	Resistivity ρ (Ω-m)
20	-	-	16.3×10^{-6}	16.2	486	7.2×10^{-7}
27	193	290	-	-	-	-
90	-	-	-	-	496	7.8×10^{-7}
93	192	-	-	-	-	-
100	-	-	16.3×10^{-6}	16.2	-	-
149	187	182	-	-	-	-
200	-	-	16.7×10^{-6}	-	532	8.6×10^{-7}
204	183	-	-	-	-	-
260	179	150	-	-	-	-
300	-	-	17.1×10^{-6}	-	-	-
316	177	-	-	-	-	-
320				-	557	9.5×10^{-7}
371	172	134	-	-	-	-
400	-	-	17.6×10^{-6}	-	-	-
427	168	-	-	-	-	-
430	-	-	-	-	574	10.2×10^{-7}
482	160	125	-	-	-	-
500	-	-	18.0×10^{-6}	21.5	-	-
538	158	-	-	-	-	-
540	-	-	-	-	588	10.8×10^{-7}
593	150	113	-	-	-	-
600	-	-	18.3×10^{-6}	-	-	-
649	145	-	-	-	-	-
650	-	-	-	-	599	11.4×10^{-7}
700	-	-	18.2×10^{-6}	-	-	-
704	141	96	-	-	-	-
760	134	-	-	-	620	11.8×10^{-7}
800	-	-	20.0×10^{-6}	-	-	-
816	125	68	-	-	-	-
870	-	-	-	-	645	12.5×10^{-7}

Poisson's ratio ν= 0.29, density β= 8000 kg/m^3, melting point = 1400 °C.

Figure. 3. Experimental equipment

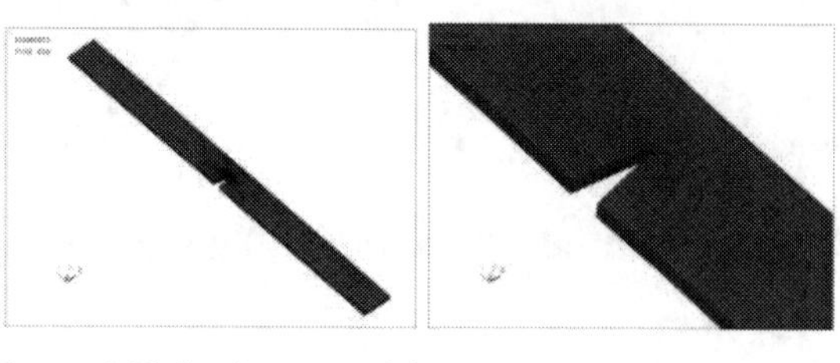

Figure. 4. Finite element model.

4 Results and discussions

As shown in Fig. 3, the steel plate is fixed on two electrodes of the RSW machine. The high electric current I_0=600 A (RMS value) is applied on the steel plate with the operating time t_p. In Fig. 5, the hot spot is seen at the notch tip during the high electric load. The hot spot is due to the electric current density concentration and Joule heating effect at the notch tip.

By controlling t_p, different situations can be seen at the notch tip. In Fig. 6, it represents four situations: (a) surface change, (b) melting/solidification without hole, (c) melting/solidification with hole, (d) burning and breaking. When the electric power is provided enough, the melting zone occurs at the notch tip. Furthermore, the solidification, heat-affected zone (HAZ), shrinking hole or breaking failure can be investigated.

The material property around the notch tip is modified under high electric load and high temperature. The sample in Fig. 6(b) is analyzed by the microscopic investigation and nano- indentation. In Fig. 7, it shows the grain images obtained by the microscope. The grain size in HAZ becomes smaller after high electric load (I_0=600 A, t_p=0.2333 s). Using the nano-indentation method, the hardness of the original material and HAZ are obtained. The hardness values are 2.59 GPa and 3.44 GPa, respectively. At the notch tip region, the hardness improves so that the strength increases.

Figure. 5. Hot spot at notch tip under high current (tp=0.2666 s).

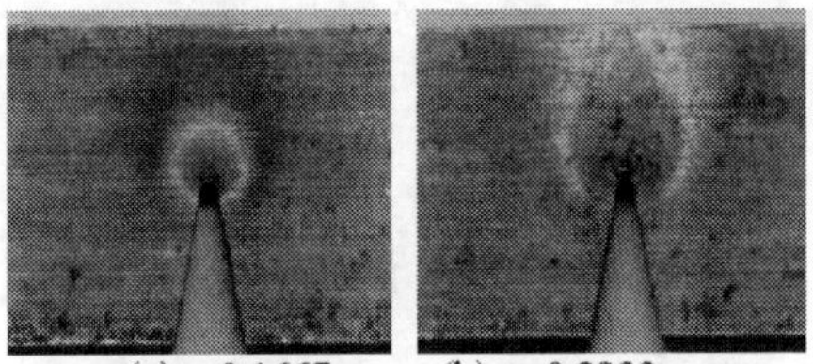

(a) t_p=0.1667 s (b) t_p=0.2333 s

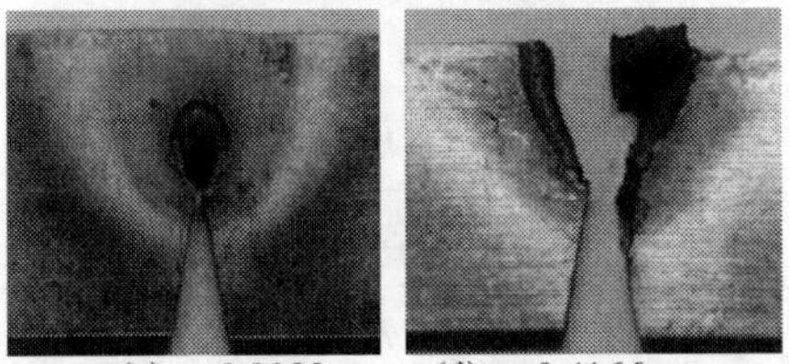

(c) t_p=0.3333 s (d) t_p=0.4166 s

Figure. 6. Different situations at notch tip under high current.

(a)

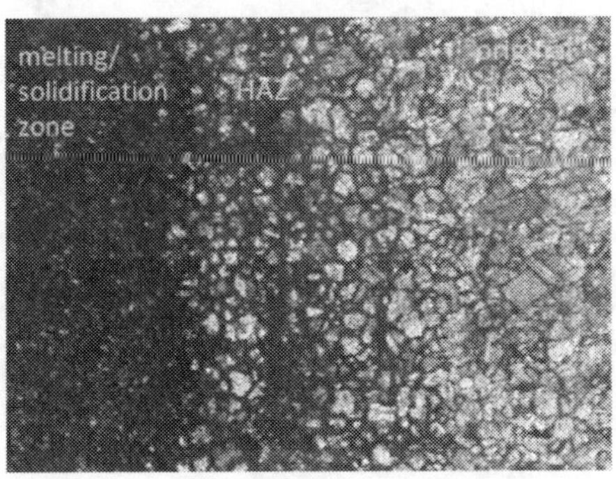

(b)

Figure. 7. Grain images. (a: original steel plate, b: plate under high current) (240X)

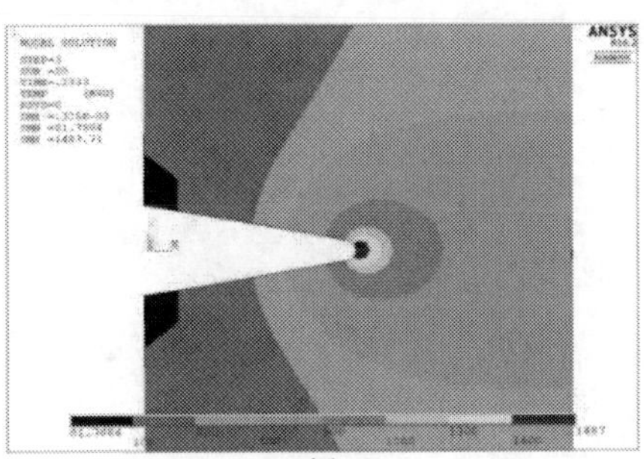

(a)

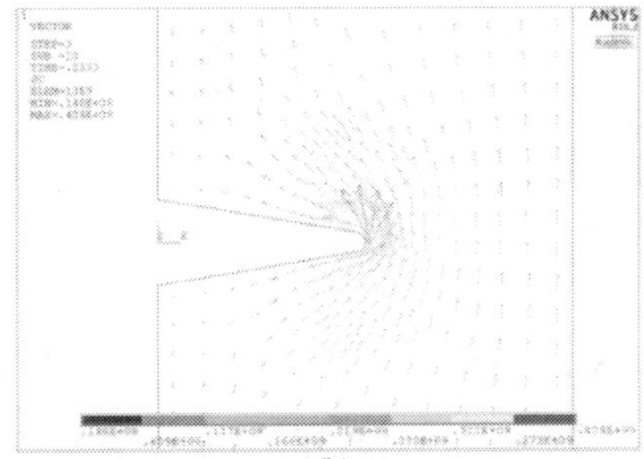

(b)

Figure. 8. (a) temperature field (units: °C), (b) electric current density field (units: A/m^2).

The finite element analysis is used to study the electric current density, temperature and stress fields. Under I_0=600 A and t_p=0.2333 s, the results are shown in Fig. 8. It is investigated that a hot spot exists at the notch tip zone. Also, it shows the electric current density concentration at the notch tip. The temperature contour is identical to the photo in Fig. 6(b).

In addition, the residual stress state is also studied. In the finite element analysis, the cooling process is considered after the electric load is removed. In Fig. 9(a), it shows the residual stress σ_y at t=120 s (cooling time is about 120 s). In the region near the notch tip, the compressive residual stress state is found. Also, the stress history at the notch tip is shown in Fig. 9(b). This result can reduce the tensile stress level and prevent from the crack initiation when the plate is subjected to the tensile load.

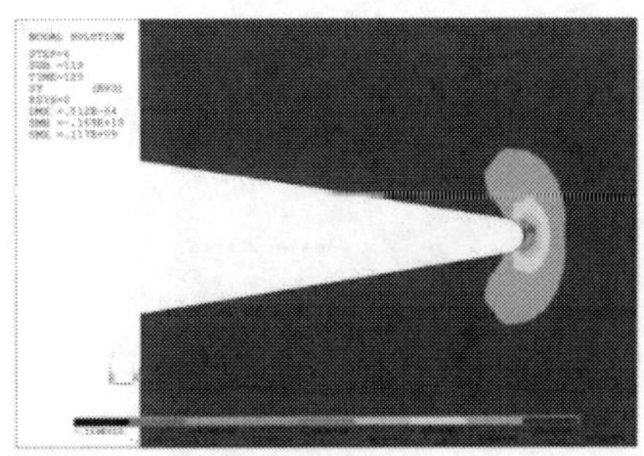

(a)

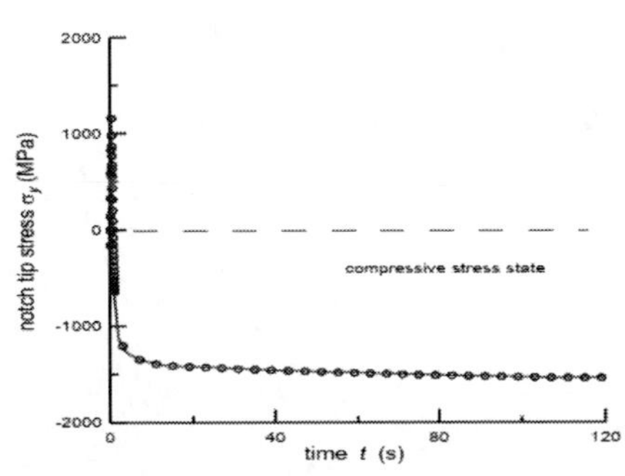

(b)

Figure. 9. (a) Contour of residual stress σ_y (units: Pa), (b) stress history at notch tip.

5 Conclusions

At the notch tip region, the hardness value improves and the compressive residual stress occurs under the high electric current. This result can reduce the stress level and prevent from the crack initiation when the plate is subjected to the tensile load. This conclusion can be applied on the improvement of the structure with notches.

Acknowledgment

The authors would like to thank the Ministry of Science and Technology in Taiwan for the financial support under contract numbers MOST 104-2221-E-131-030 and MOST 105-2221-E-131-014.

References

1. V.Z. Parton and B.A. Kudryavtsev: *Electromagnetoelasticity* (Gordon and Breach, New York 1988).
2. Z. Qin, L. Librescu and D. Hasanyan: J. Therm. Stress Vol. 30 (2007), p. 623
3. Y.M. Fu, X.Z. Bai, G.Y. Qiao, Y.D. Hu and J.Y. Luan: Mater. Sci. Tech. Vol. 17 (2001), p. 1653
4. T.J.C. Liu: Eng. Fract. Mech. Vol. 78 (2011), p. 666
5. T.J.C. Liu: Eng. Fract. Mech. Vol. 123 (2014), p. 2
6. T.J.C. Liu, in: *Encyclopedia of Thermal Stresses*, edited by R.B. Hetnarski, Springer Science + Business Media Dordrecht, Berlin (2014).
7. *Online Materials Information Resource – MatWeb* (MatWeb, LLC., USA 2016). (http://matweb.com/)
8. *High-Temperature Characteristics of Stainless Steels* (American Iron and Steel Institute, USA 2006).

MATEC Web of Conferences 151, 02001 (2018) https://doi.org/10.1051/matecconf/201815102001
ACMAE 2017

Computational modelling of an Organic Rankine Cycle (ORC) waste heat recovery system for an aircraft engine

S. Saadon[1]

[1]Department of Aerospace Engineering, Faculty of Engineering, University Putra Malaysia, Malaysia

Abstract. Escalating fuel prices and carbon dioxide emission are causing new interest in methods to increase the thrust force of an aircraft engine with limitation of fuel consumption. One viable means is the conversion of exhaust engine waste heat to a more useful form of energy or to be used in the aircraft environmental system. A one-dimensional analysis method has been proposed for the organic Rankine cycle (ORC) waste heat recovery system for turbofan engine in this paper. The paper contains two main parts: validation of the numerical model and a performance prediction of turbofan engine integrated to an ORC system. The cycle is compared with industrial waste heat recovery system from Hangzhou Chinen Steam Turbine Power CO., Ltd. The results show that thrust specific fuel consumption (TSFC) of the turbofan engine reach lowest value at 0.91 lbm/lbf.h for 7000 lbf of thrust force. When the system installation weight is applied, the system results in a 2.0% reduction in fuel burn. Hence implementation of ORC system for waste heat recovery to an aircraft engine can bring a great potential to the aviation industry.

1 Introduction

Growing consumption of primary fossil fuels and massive discharge of pollutants are some of the results caused by the world's growing population, and eventually the enlarging energy demand. It is therefore the main concerns that the developing world must face nowadays are the energy shortfall and the environmental destruction. Since 1973, the world energy consumption has been crucially increased and the world energy demands are growing up to 89% starting from 2006 till this year [1]. This affects significantly those industries which waste a huge amount of energy. And for these valid reasons the awareness of the use of the low-grade heat sources has captivated researchers around the world in recent years. To manage this matter, appropriate regulations should be established to further utilize the fossil energy and minimize the misuse energy in a more effective way. Particularly, there is extra effort in the aviation field to reach a higher quality of propulsion system, as the fuel cost is increasing and the future law is becoming more severe. It was recorded that the waste fuel energy from exhaust engine is over 30-40% and only a small fraction of this fuel energy which is roughly 12-25% were converted to useful work [2]. Apart from creating a downturn in fossil fuels market, waste heat recovery will also lead to reduction in greenhouse gases and hence making it the idea of a better future environment more promising.

There are a lot of discussions and researches emerge trying to prove that this waste heat recovery is a practical resource of energy due to its large quantity [3]. An example of steam-based waste heat recovery system applied to several industries is Heat Recovery Steam Generators (HRSGs). However, it is not recommended for a smaller gas turbine engine due to problems of weight as well as supply of water [4]. Hence an Organic Rankine Cycle (ORC) system has been proven to be one of the beneficial exhaust heat recovery technologies due to its small-scale and its undoubtedly potential integration in the next few years' power supply systems. A Rankine cycle is a heat engine thermodynamic cycle which converts the heat from a device for example turbine into mechanical work. In a Rankine cycle, the system applies the heat to rise up the temperature and pressure of an organic fluid. Thus, the name organic Rankine cycle comes from its use of an organic fluid which has a characteristic of a liquid-vapor phase change at a lower temperature than the phase change in water-steam. This allows heat recuperation of Rankine cycle from a lower temperature. Then this heat recovery can be converted into useful work which can be subsequently transforms into electricity.

Concerning the implementation of the ORC technology, various low-grade waste thermal energy industries such as solar energy, biomass energy, waste heat energy and geothermal energy have considered this system [5]. In power plant applications and marine diesel engine, there have been several researches concerning the thermal analysis and design optimization of an ORC using waste heat source [6-8] and recently a study had also been demonstrated experimentally [9]. Even so, to

https://doi.org/10.1051/matecconf/201815102001

integrate the system to an aircraft engine, few factors must be taken into consideration. Principally, the aim of the system for a power plant is to recuperate a great amount of energy from the exhaust engine to supply electricity. However, in an aircraft engine the fundamental point is not to produce power, but to produce thrust and to reduce a considerable amount of fuel consumption. Bronicki et al., [4] also come up with the basis reasons for an ORC to be integrated to an aircraft gas turbine engine sized which marks about 16% of the gas turbine performance's improvement.

Conventionally, a small amount of air at a very high pressure is bled off from an aircraft engine and used to pressurize the cabin by providing air to the environmental control system. Bleed air is fundamentally loss of engine thrust as some air cannot be employed to produce thrust. Therefore, development of bleed less aircraft such as an engine driven electric compressor has started to influence the current technologies. However, this is quite misleading as electric devices need higher electricity power from engine. This is where the ORC system comes as solution. The system draws out waste heat from the turbofan's engine exhaust system and use it to supply electricity to operate the external air compressors to supply flow to the aircraft environmental control system or other aircraft systems. It is a recovery system that intends to increase the overall performance of the engine. As a result, a smaller amount of electricity power is needed from the aircraft engine, hence less fuel consumption, and this lower the engine bleed air and therefore a higher thrust could be achieved.

However, researches on design and performance analysis of an ORC waste heat recovery system integrated to aircraft engine and their possible advantages are still very few. Only one assessment study so far done by C A Perullo et al. [10]. They examined the feasibility and the benefits of an ORC heat recovery system to be used for inflight aircraft power generation. In this study, they utilize the Environmental Design Space (EDS) as the simulation tool [11, 12]. The method applies a refine boiler situated inside the nozzle walls of an aircraft engine to extract heat from the engine exhaust. The organic fluid chosen to evaluate the system was R245fa because of its highest cycle energy performance across multiple operating pressure compare to other three possible fluids.

Nevertheless, because of limited time and resources, they decided to make several modeling assumptions. The researchers decided to use a fixed heat transfer coefficient to search for the off-design conditions and they estimated a pressure loss only in the turbine and although the engine fluid flow progress during the engine running, they assumed a fixed core exhaust flow of 30% that links to the evaporator. Consequently, about 0.9% fuel burn savings is possible but it depends on the whole engine system weight. However, their aim is not to analyze the thermal performance, in terms of how the heat transfer being produced, and the overall performance of the system; hence, a very brief presentation of the model was provided. A more recent study done by Saadon et al., [13] tried to understand clearly the thermodynamic behavior of the ORC system when connected to exhaust turbofan engine. However, their analysis is still very loose in the aim to understand clearly the main cause of such behavior and the impact of the ORC system to the aircraft was not done. Therefore, this study differs from the previous in the following: (i) the detailed study on the heat transfer process of the ORC system conducted and its algorithm; (ii) the analytical study of their thermal performance; (iii) the evaluation of the Thrust Specific Fuel Consumption (TSFC) of the aircraft engine and the fuel burn.

2 Numerical model of ORC integrated to an aircraft exhaust engine

An ORC system integrated with an aircraft exhaust engine is presented in a diagram in Fig. 1 which consists of an evaporator, a turbine, a condenser, a working fluid pump and a regenerator that is used to heat up the organic fluid before getting into the evaporator to achieve a better thermal efficiency of the system. Here, the working fluid is the organic fluid. The Rankine cycle is connected to the aircraft engine at the middle of the low-pressure turbine exit duct and the core nozzle. Fig. 1 below shows the link between the aircraft engine and the ORC system which is based at the evaporator. The system starts at the outlet of the liquid side of the regenerator (station 1). An inlet temperature of the working fluid to the evaporator $T_{EV,WF\_In}(i)$ is assumed and the system is performed.

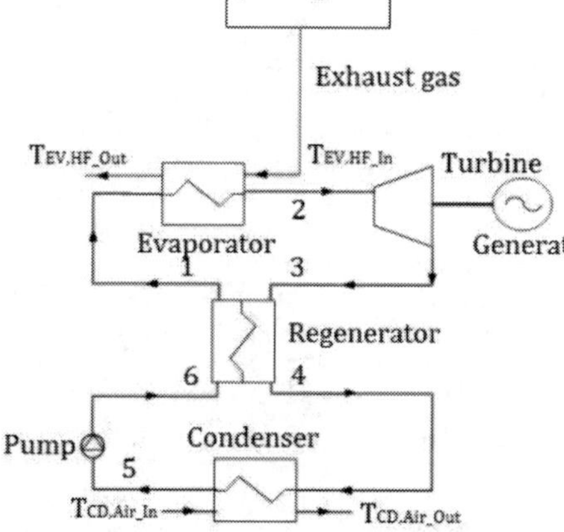

Figure 1. Schematic diagram of the ORC system with regenerator integrated to an aircraft engine. [13]

The ORC system is presented in a T-s (temperature-entropy) diagram as in Fig. 2 below. A shell-tube heat exchanger is chosen as the evaporator as it is the most demanded in many industries and is convenient for higher-pressure applications. It consists of a shell with several tubes inside. The heat is transferred between the two fluids through the tube wall within the shell by entering one fluid inside the tubes, while the other fluid flows outside the tubes.

To model the evaporator, there are three methods of modelling that can be applied to evaluate the heat transfer of the evaporator. They are distributed modelling, zone modelling and single node lump modelling methods J Sun and W Li [8]. The single node lump modelling approach assumes that the temperature difference inside each node is negligible and this implies that the constant specific heat inside is unchanged too. Hence it is normally not possible for this ORC application where liquid-vapor phase change occurs. Meanwhile, with zone modelling, the evaporator is separated into three different zones which are super-heat, two-phase and sub-cool zones. Consequently, the single node lump method can be applied to each zone. However, a vigorous iterative algorithm is required to attain the heat transfer surface area of each zone. Otherwise, a distributed modelling method treats the evaporator as small divisions starting from where the flow enters until towards the outlet. Then eventually, the single node lump method, here it is NTU (Number of Transfer Units) method, can be applied in each division for evaluation of heat and mass transfers. This distributed modelling method is more accurate compared to the first two methods. In this paper, the distributed method is applied to model the evaporator.

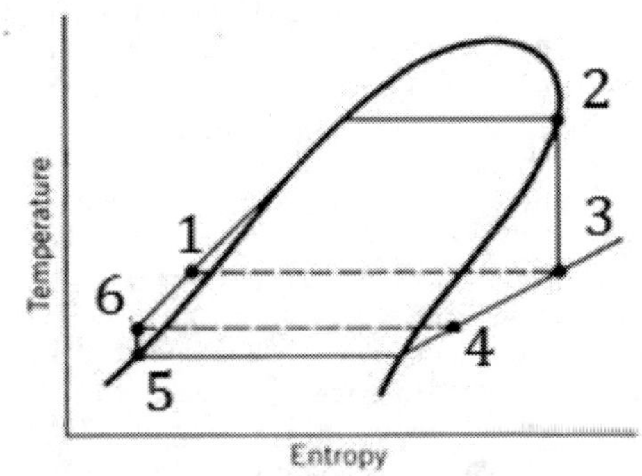

Figure 2. T-s diagram of the ORC system with regenerator.

Fig. 3 below shows the three consecutive parts, noted *i-1, i* and *i+1*. By using the distributed modelling method, it is adequate to assume that the heat capacity of the waste heat engine and organic fluid are roughly invariable in each discrete division. Therefore, to construct a numerical model for each single segment, the NTU method is applied. For each segment *i*, the enthalpy of the organic fluid is supposed to be increased because the heat is absorbed from the exhaust engine. Consequently, the heat transfer rate of each segment is sum up to get the total heat transfer of the evaporator.

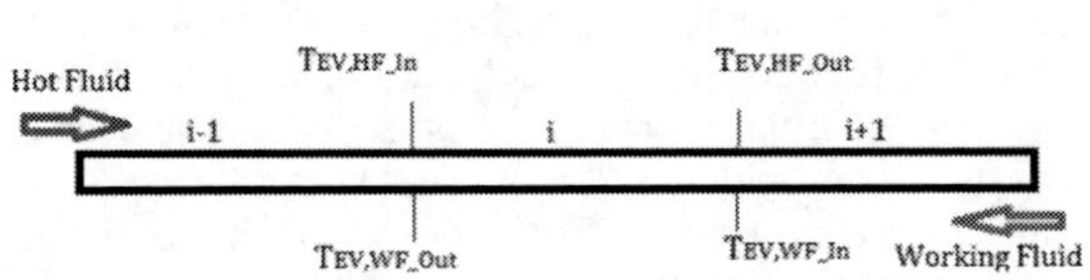

Figure 3. Discrete segments of evaporator. [13]

$$q_{EV,WF}(i) = \dot{m}_{EV,WF}\left[h_{EV,WF\_Out}(i) - h_{EV,WF\_In}(i)\right] \quad (1)$$

According to the Eq. 1, $q_{EV,WF}(i)$, $\dot{m}_{EV,WF}$, $h_{EV,WF\_Out}(i)$ and $h_{EV,WF\_In}(i)$ are heat transfer rate, mass flow rate and inlet and outlet enthalpies respectively of the organic fluid for the section *i*.

Similarly, the heat transfer rate for the exhaust gas, $q_{EV,HF}(i)$ is noted as,

$$q_{EV,HF}(i) = \dot{m}_{EV,HF}\, C_{p,EV,HF}(i)\left[T_{EV,HF\_In}(i) - T_{EV,HF\_Out}(i)\right] \quad (2)$$

Here, $C_{p,EV,HF}(i)$, $\dot{m}_{EV,HF}$, $T_{EV,HF\_In}(i)$ and $T_{EV,HF\_Out}(i)$ are specific heat, mass flow rate and entrance and exit temperatures respectively of the exhaust gas for the section *i*.

The maximum heat transfer rate $q_{EV,MAX}(i)$, throughout the evaporator can be written as,

$$q_{EV,MAX}(i) = C_{EV,MIN}(i)\left[T_{EV,HF\_In}(i) - T_{EV,WF\_In}(i)\right] \quad (3)$$

with

$$C_{EV,MIN}(i) = MIN\left\{\dot{m}_{EV,WF}\, C_{p,EV,WF}(i), \dot{m}_{EV,HF}\, C_{p,EV,HF}(i)\right\} \quad (4)$$

Meanwhile, the effectiveness $\varepsilon_{EV}(i)$ can be defined as [14],

$$\varepsilon_{EV}(i) = \left\{2 \times \left\{\left[1 + C_r(i) + \left(1 + C_r^2(i)\right)^{0.5}\right] \times \frac{1 + \exp\left[-NTU(i) \times \left(1 + C_r^2(i)\right)^{0.5}\right]}{1 - \exp\left[-NTU(i) \times \left(1 + C_r^2(i)\right)^{0.5}\right]}\right\}\right\}^{-1}$$

(5)

$$\text{where} \quad C_{r,EV}(i) = \frac{C_{EV,MIN}(i)}{C_{EV,MAX}(i)} \quad (6)$$

$$C_{EV,MAX}(i) = MAX\left\{\dot{m}_{EV,WF}\, C_{p,EV,WF}(i), \dot{m}_{EV,HF}\, C_{p,EV,HF}(i)\right\} \quad (7)$$

$NTU(i)$ is computed as,

$$NTU(i) = \frac{U(i)A_{EV}(i)}{C_{EV,MIN}(i)} \quad (8)$$

The total heat transfer rate of the evaporator Q_{EV},

$$Q_{EV} = \sum_{i=1}^{N_{EV}} q_{EV}(i) \quad (9)$$

where N_{EV} is the total of evaporator segments.

The pump power consumption in process 5 to 6 is defined by

https://doi.org/10.1051/matecconf/201815102001

$$W_{pump} = \frac{\dot{m}_{WF}\, Cp_{WF}\left(\Pi^{\frac{\gamma-1}{\gamma}} - 1\right)}{\eta_{pump}} \tag{10}$$

where $\Pi = \frac{P_2}{P_1}$, η_{pump} is the efficiency of the pump and γ is the specific heat ratio of the organic fluid.

When the organic fluid exits the pump as a saturated or superheated fluid and flows through the turbine, the power is produced and exits from the turbine as low pressure superheat fluid. Then the turbine power output is obtained from,

$$W_{\exp} = \dot{m}_{WF}\, Cp_{WF}\, \eta_{\exp} T_{\exp\_in}\left(1 - \Pi^{\frac{1-\gamma}{\gamma}}\right) \tag{11}$$

where is $T_{\exp\_in}$ the turbine inlet temperature and $\eta_{\exp}$ is the efficiency of the turbine.

The net power output is then

$$W_{net} = W_{\exp} - W_{pump} \tag{12}$$

And the thermal efficiency of the ORC system is

$$\eta_{net} = \frac{W_{\exp} - W_{pump}}{Q_{EV}} \tag{13}$$

The total of fuel (kerosene) that could be eventually saved is

$$\dot{m}_{fuel} = \frac{Q_{EV}}{HV_{fuel}} \tag{14}$$

where HVfuel is the heating value of kerosene, HVfuel = 43,000 kJ/kg.

As the thermal model of the evaporator depends on the effectiveness of the heat exchanger, and this effectiveness itself depends on the NTU_{EV}, the total heat transfer rate of the evaporator is found numerically by iteration. The procedure is illustrated by the flow chart in Fig. 4. The calculation commences with an initial estimation of NTU_{EV}. The corresponding organic fluid mass flow rate and heat capacity of both fluids are then calculated, which allows determine the new value of NTU_{EV} and $C_{r,EV}$ for each zone. The evaporator effectiveness is then evaluated for each zone. For the first zone, the initial organic fluid mass flow rate is taken as the inlet value. For subsequent zones, the inlet mass flow rate is given by the outlet mass flow rate of the upstream zone. These steps are repeated until the difference between the effectiveness calculated for two consecutive iterations is less than 0.1%. Finally, the heat transfer rate for each zone is calculated and the total heat transfer of the evaporator is attained.

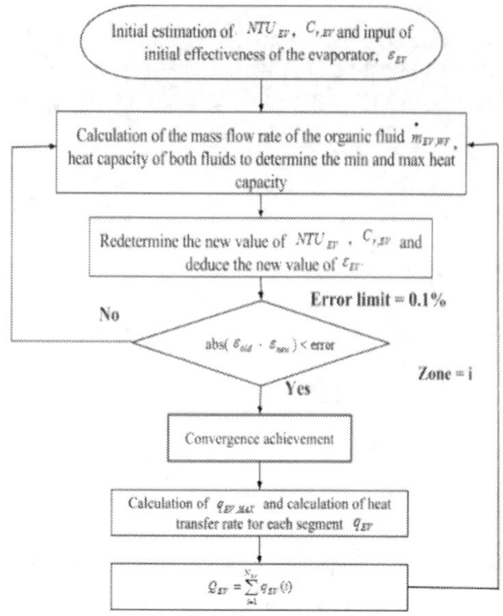

Figure 4. Algorithm of calculation of the total heat transfer rate.

3 Validation of a thermal model

With the intention to validate the numerical model constructed, the ORC model was first compared to the one for industrial waste heat recovery [15] with R123 at cooling water temperature, Tc,in of 293 K and 303 K. The selected waste heat source is the low temperature steam from Hangzhou Chinen Steam Turbine Power CO., Ltd and is depicted in Table 1 below. The specific heat of R123 in the range of the design temperature value is 0.9744 kJ/kg.K, while for kerosene, it is said to be 2.01 kJ/kg.K. The simulation software is independently developed by the researchers through MATLAB. A simulation is then performed by using the mathematical equations defined above, with inlet temperature of R123 at 100 K.

The model was validated by comparing the results of net power output and system thermal efficiency with the previously published results and these results are presented in Fig. 5 and Fig. 6. It is noted that the net power output and thermal efficiency vary with the inlet temperature heat source. It can be observed that there is a slight deviation compared to the results achieved by Jian Song et al. [15], occurred at higher temperature level of the heat source. These may be resulted from the constant inlet organic fluid assumed in the simulation. Basically, the inlet temperature of R123 will vary with the environmental and factory conditions. The relative error of the results obtained compared to works done by Jian et al. [15] for Tc,in = 293 K is 0.43%, while for Tc,in = 303 K, the discrepancies is a bit higher at 10.15%. However, for the aircraft engine which the ORC is going to be applied to in the following section, the inlet temperature of the organic fluid used will be given by the engine data and the value is fixed for the design point. As a whole, the numerical solutions obtained in this study are consistent with those reported in Jian Song et al. [15].

Table 1. Design parameters of ORC system with R123 as the working fluid.

Description	Unit	Value
Mass flow rate of R123	kg/s	21.2
Heat source inlet temperature	K	453
Evaporation temperature	K	350
Turbine inlet pressure	MPa	1.21
Turbine outlet pressure	MPa	0.55
Turbine efficiency	-	0.6
Pump efficiency	-	0.8
Net power output	kW	529
System thermal efficiency	%	12.7

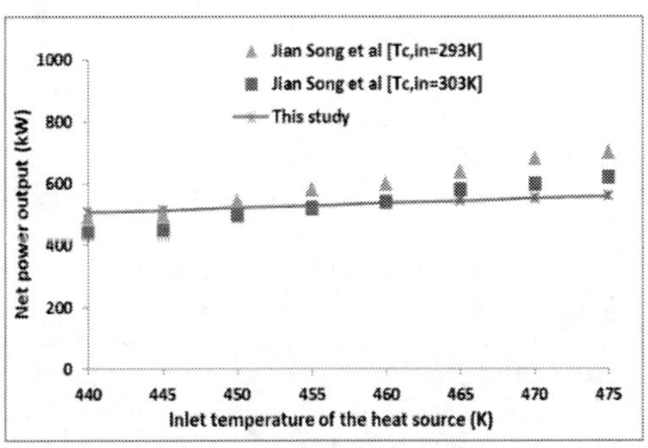

Figure 5. Variations of the net power output of the ORC system with heat source inlet temperature.

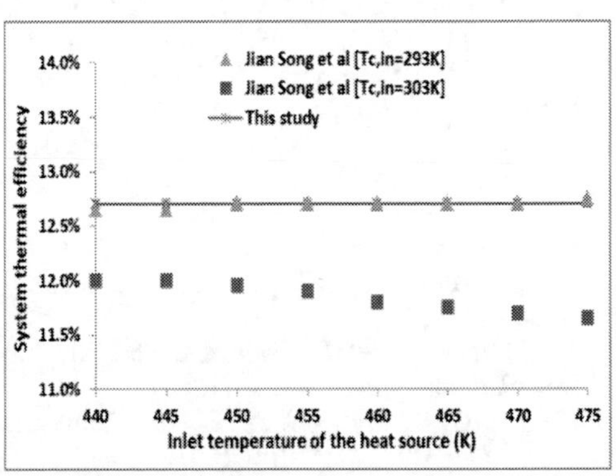

Figure 6. Variations of the system thermal efficiencies of the ORC system with heat source inlet temperature.

4 ORC integrated to an aircraft exhaust engine

This part presents the advantages of ORC system integrated to a CFM56-7B27 turbofan engine on an aircraft size of 737-800. The working fluid chosen is the R245fa. The specific heat of R245fa in the range of the design temperature value is 1.36 kJ/kg.K. Design parameters are listed in Table 2 below [10]. In order to solve this system, the heat transfer flowing out from the exhaust engine is varied until it matches the heat transfer calculated using the heat transfer coefficient and evaporator area from the detailed design. By doing this, it ensures that the evaporator exit temperature set by the thermodynamic limits of R245fa is maintained. The simulation is then executed as before and the results attained are presented in figure below.

Table 2. Main parameter of the ORC system for the turbofan engine

Description	Unit	Value
Mass flow rate of R245fa	kg/s	3.84
Estimated required surface area of evaporator	m^2	23.72
Required heat transfer of evaporator	kW	1105
Exhaust heat temperature	K	843
Inlet temperature of R245fa	K	282
Outlet temperature of R245fa	K	393
Turbine inlet temperature	K	392
Turbine inlet pressure	MPa	1.21
Turbine outlet pressure	MPa	0.55
Turbine efficiency	-	0.87
Pump efficiency	-	0.70

The turbine within the ORC was connected to a simulation of an external compressor in order to evaluate the impact on the engine when engine bleed air was reduced. The result shown below in Fig. 7 is detail common cruise conditions with Mach 0.785 at altitude of 35,000 feet. Even with the small loss in core thrust this reduces the TSFC below the base cycle by an average of 3.1%, defined in Fig. 7 by "Regen + Bleed Savings".

Once the feasibility of the idea of implementing the ORC on an engine basis was confirmed, it was required to estimate the advantages on the combined engine and aircraft system. Fig. 8 described the fuel burn relative to the base cycle and takes into account the vehicle weight throughout the flight. The "Regen + Bleed Savings" scenario show the fuel burn effect only from the TSFC change whereas the "Add 950 lbm per engine" scenario assumes a nominal weight impact of 950 lbm on the detailed ORC system. When the system installation weight is applied, the system results in a 2.0% reduction in fuel burn.

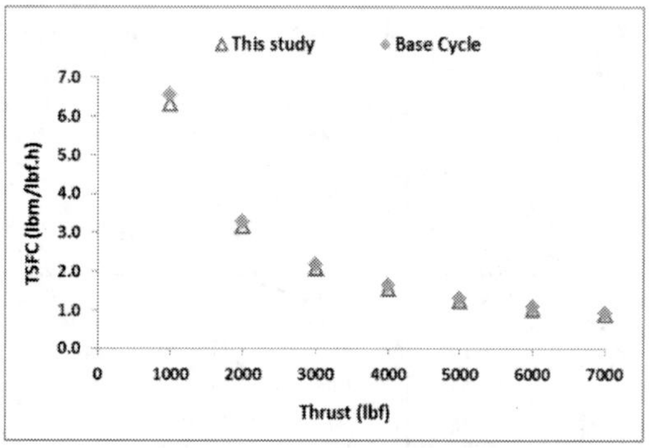

Figure 7. ORC engine with TSFC effects and ORC impact on mission fuel burn

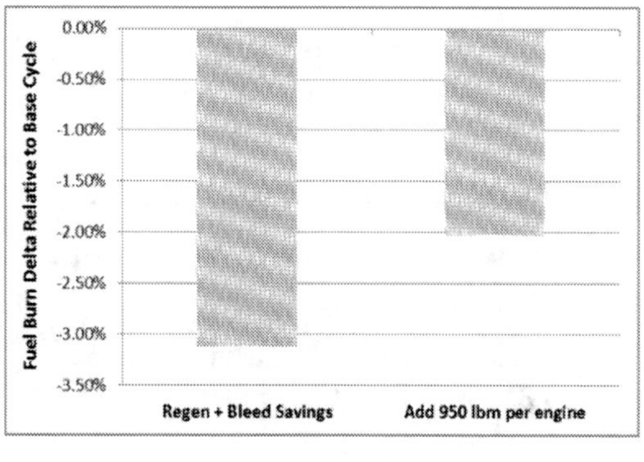

Figure 8. ORC engine with TSFC effects and ORC impact on mission fuel burn

5 Conclusions

This paper presents a detailed study of a one-dimensional analysis method of ORC system implemented in an aircraft engine. A numerical model for the system has been modified from previous authors' work. The model features include an algorithm of calculation of the total heat transfer rate driven by the NTU method. Validation with other researchers' work for an industrial waste heat recovery shows a reasonable agreement with net output power and thermal efficiency although some discrepancies could be found at higher inlet heat source temperature of the evaporator. These discrepancies can partly be attributed to the assumption of lower organic fluid inlet temperature of the evaporator which results in lower mass flow rate of the fluid. At the same time the simulation model needs more parameters to be defined which increases the variabilities in the simulation estimations.

The initial results for the ORC system integrated to a turbofan's exhaust engine provide an idea of the impact of this system on the engine's thrust and its fuel consumption. Parametric optimization shall be conducted in future in order to maximize the total heat transfer rate through the evaporator. Such results will be very useful in future to determine the irreversibility losses through calculation of exergy of each ORC component in order to achieve an optimize system performance and eventually higher thrust power. The presented research proved that an ORC cycle with regenerator gives potential advantages when integrated to an aircraft engine and used to provide electrical power to the aircraft.

References

1. M. Abdolzadeh, M. Sadeqkhani, A. Ahmadi, Energy, **133,** 998-1012, (2017)
2. M. Hasanuzzaman, N.A. Rahim, R. Saidur and S.N. Kazi, , **36,** 233-240, (2011)
3. D. Ziviani, A. Beyene, and M. Venturini, Applied Energy, **121,** 79-95, (2014)
4. L.Y. Bronicki and D.N. Schochet, Proceedings ASME Turbo Expo, **5**, 79-86 (GT2005-68121)
5. B.F. Tchanche, G. Lambrino, A. Frangoudakis and G. Papadakis, Renewable & Sustainable Energy Reviews, **15,** 79-95 (2011)
6. J. Song, Y. Song and C.W. Gu, Energy, **82,** 976-98, (2015)
7. M.H. Yang and R.H. Yeh, Applied Energy, **149,** 1-12, (2015)
8. J. Sun and W. Li, Applied Thermal Energy, **31,** 2032-2041, (2011)
9. W. Pu, C. Yue, D. Han, W. He, X. Liu, Q. Zhang and Y. Chen, Applied Thermal Engineering, **94,** 221-227, (2016)
10. C.A. Perullo, D.N. Mavris and E. Fonseca, Proceedings of ASME Turbo Expo. Turbine Technical Conference and Exposition, (2013)
11. M. Kirby and D. Mavris, 26th International Congress of the Aeronautical Sciences, (2008)
12. J. Schutte, J. Tai and D. Mavris, 48th AIAA/ASME/SAE/ASEE Joint Propulsion Conference & Exhibitition, (2012)
13. S. Saadon and A.R. Abu Talib, IOP Conference Series: Materials Science and Engineering, **152**, Number 1, (2016)
14. J.P. Holman, *Heat Transfer*, sixth ed. McGraw-Hill Book Company, New York, (1986)
15. J. Song, C. Gu and X. Ren, Energy Conversion and Management, 112, 157-165, (2016)

Comparison of Efficiency of Conventional Shaped Circular and Elliptical Shaped Combustor

Dharmahinder Singh Chand[1], Daamanjyot Barara[2], Gautam Ganesh[3], Suraj Anand[4]

[1]*Aeronautical Engineering, Tagore Engineering College, Chennai, Tamil Nadu-600127, India*
[2,3,4] *Undergraduate Student, Department of Mechanical Engineering. SRM University, Chennai, Tamil Nadu-603203, India*

Abstract. There have been concerted efforts towards improving the fuel efficiency of the jet engines in the past, with an aim of reducing the incomplete combustion. The process of combustion in a jet engine takes place in the combustor. A study was conducted for enhancement of air-fuel mixing process by computational analysis of an elliptically shaped combustor for a gas turbine engine. The results of computational analysis of an elliptical shape combustor were compared with a circular shape combustor used in gas turbine engines with a identical cross sectional area. The comparison of the computationally derived parameters of the two combustors i.e. temperature, pressure, and velocity are studied and analyzed. The study intends towards the comparison of the combustion efficiencies of the circular and elliptically shaped combustors. The combustion efficency of elliptical chamber is found to be 98.72% at the same time it was observed 56.26% in case of circular type combustor.

1 Introduction

Since the time aircrafts have taken off, the aviation industry has made strident efforts to upgrade aircraft technology to improve its efficiency. This has led to major advancements in the field of propulsion, aerodynamics, material technology, fuel conservation, etc. The modern day aircrafts have been built while considering these advancements, and performance-wise with significantly high- efficiency standards. However, issues regarding the environmental concerns of pollution and depletion of fuel resources have called for more efforts to be made, especially in the domain of combustion of propulsion system [1].

The combustor is considered as a main component of aircraft, which is responsible for the propulsion of the aircraft. Once aircraft moves forward, the incoming air passes through the engine of the aircraft. The air enters the engine at high velocity and made to pass through the series of compressor blades. The air from the compressor enters the combustor at a high velocity upto 500 ft/s, which is too high for combustion [2]. In order to decrease the air velocity and raise the static pressure, the incoming air is made to pass through the diffuser, before entering the combustor. After passing through the compressor, the velocity of the air decreases significantly, while increasing its pressure. The diffuser is responsible for the reduction in velocity, which helps to maintain a high pressure of the air entering the combustor. This allows for the smooth operation of the engine and flame stability. Usually air-fuel mixture ranges from a ratio 45:1 to 130:1 [2]. The temperature required for the combustion of the air-fuel mixture lies in the range of 850°C to 1700°C [3]. The pressure loss in the combustor has to be very minimal. The air entering the combustor passes through two regions, where 20% of incoming air passes through the snout, while the other 80% of it passes through the passage between the flame tube body and casing. The air entering the snout, passes through the primary zone. Also the 20% of air over the flame tube body, enters the primary zone through the primary holes of the flame tube body. The primary zone is responsible for anchoring the flame and attaining sufficient amount of temperature and turbulence for complete combustion of the incoming air-fuel mixture [4]. Fuel release occurs at the same time through the flame tube, which results in the formation of a chaotic air-fuel mixture. The turbulence of the flow is generated by the toroidal flow pattern owing to swirl vanes and perforated flare. This recirculation of flow pattern used to allow the continuous combustion of the incoming air and fuel [4]. The temperature of the gases released due to combustion lies between 1800°C to 2000°C [2]. In the intermediate zone, 40% of the total airflow enters through the secondary holes, which facilitates in the dropping of the temperature. This drop in temperature allows to complete combustion of CO and other unburnt hydrocarbons [4]. The combusted gases, then pass through the dilution zone of the combustor. The role of the dilution zone is to allow the air through the tertiary holes in order to meet the wall-cooling's requirements. The temperature of the gases is reduced, as the temperature at which the gases are combusted is too for material to withstand used in nozzle vanes of the

 https://doi.org/10.1051/matecconf/201815102002

turbine [4]. The gases exit from the exhaust, after passing through the turbine, thus propelling the aircraft forward.

Hence, it becomes necessary that the combustor's performance is continuously improved. The combustor's efficiency is also directly related to fuel is consumption. The primary considerations of combustor design are ensuring that pressure loss encountered in the combustor is minimized and a high temperature is achieved without compromising with combustion efficiency. While these considerations need to be realized, it must also be ensured that there is efficient fuel consumption and low pollutant emissions [1]. One way to achieve better combustion efficiency is the enhancement of the mixing process. Better mixing helps in producing a better spatial distribution of the air fuel mixture inside the combustor body [5].

The study of non-conventional combustor shapes differs from the conventional design, in terms of mixing and spreading characteristics [6-10]. A change in the cross-sectional shape of the combustor can lead to strong self-induction and instability of vertical structures, which results in a change in the development of the jet, as it expands into its surroundings [5, 8-12].Co-axial jets are capable of producing turbulence and facilitate the mass-entrainment [6-7, 9-10,13]. Thus, an ellipse is such a shape which has a property to vary the axial flow of air through axis-switching. The mixing and scattering characteristics of the air-fuel mixture will vary for the elliptical combustor. The studies have shown that the jet from the non-circular nozzles produces axial vortices and these vortices are augmented by vortex bending through highly curved flow, thus, increasing the intensity of the small-scale turbulence [6-8, 11]. The incorporation of the non-circular jets has led to higher combustion efficiency, enhancing combustor performance and reducing the emissions and combustor instabilities [9, 11-12]. This had led to an introduction of the elliptical combustor with elliptically-shaped fuel injector nozzle and elliptical holes over the flame tube. The elliptical combustor which has a variable curvature produces different downstream velocities along its perimeter. The computational analysis is carried out for both the models. The efficiency of the two combustors i.e. conventional and elliptical is compared using the fundamental efficiency formula of the combustors by Ganesan and Yahya [14-15].

2 Analytical approach

The tests were performed by modelling the conventional and elliptical combustor models on CAD modelling software. Further, they were analyzed in order to get suitable position for the fuel injector. Combustion simulations were further conducted on the two combustor models to get temperature and pressure profiles, which are to be compared.

2.1 Modelling *of* combustor models

The conventional and the elliptical combustor models were designed using the CAD software, Solidworks 2016. The circular model is axis-symmetric. Its inner diameter is 128 mm, with a thickness of 1.5 mm and is modelled over a length of 975 mm. It consists of an inducer inlet, swirler, and fuel injector. The position of the injector is fixed at the point where there is a drop in the Mach number, as compressed air enters the combustion chamber. Similarly, the elliptical model was designed with the same cross-sectional area of conventional model. The fuel injector for both the models is designed as a solid part, aiming that ANSYS FLUENT recognizes the solid part as a fluid region. The CAD models of both the combustors are shown in Fig. 1, Fig. 2, Fig. 3, and Fig. 4 with their corresponding front and side views.

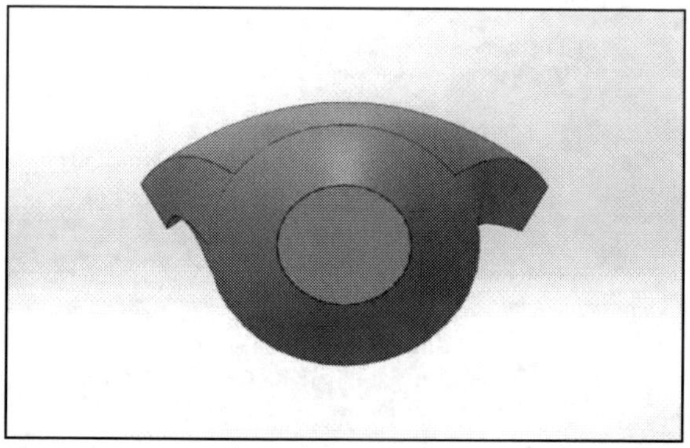

Figure 1. Front View of Conventional Combustion Chamber.

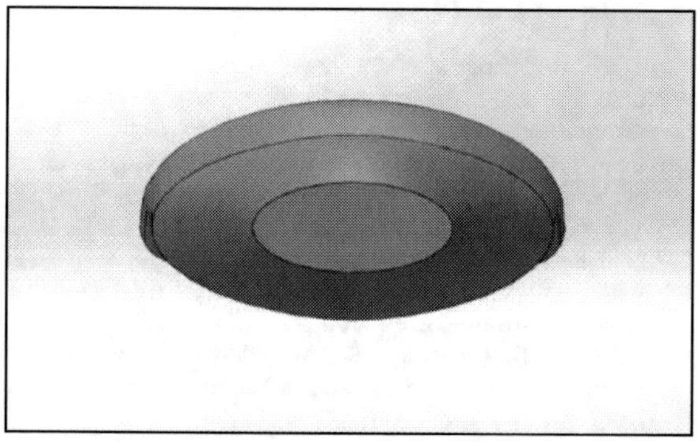

Figure 2. Front View of Elliptical Combustion Chamber.

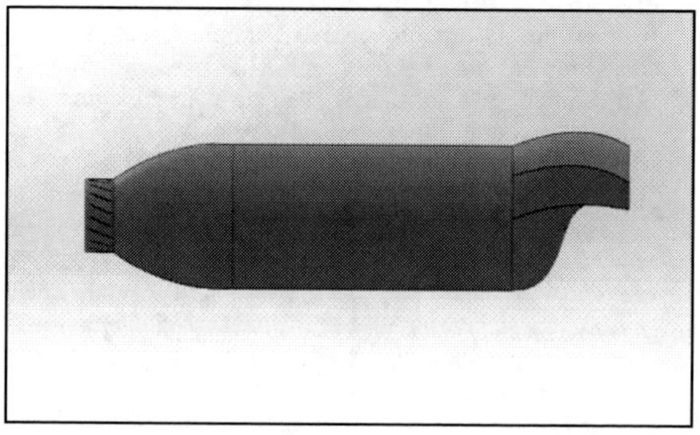

Figure 3. Side View of Conventional Combustion Chamber.

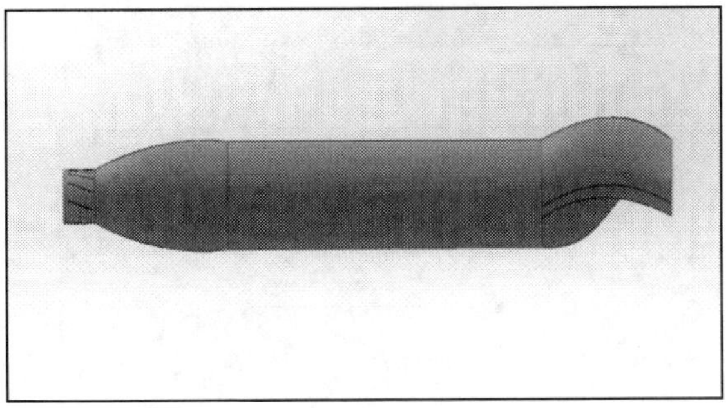

Figure 4. Side View of Elliptical Combustion Chamber.

2.2 Analysis of the combustors

2.2.1 Meshing of the combustor models

The analysis of the combustion models was performed by ANSYS. The models were imported into the software by saving them in the STEP format. The FLUENT module of ANSYS was used to carry out the analysis. The model was meshed in ANSYS meshing software. A tetrahedral mesh was generated, which was later converted to a Polyhedra mesh. The named selections were made and the mesh file was imported to FLUENT for analysis. The default material used in ANSYS, structural steel, was used as the material of the combustor models.The meshed models of combustors are shown in Fig. 5 and Fig. 6.

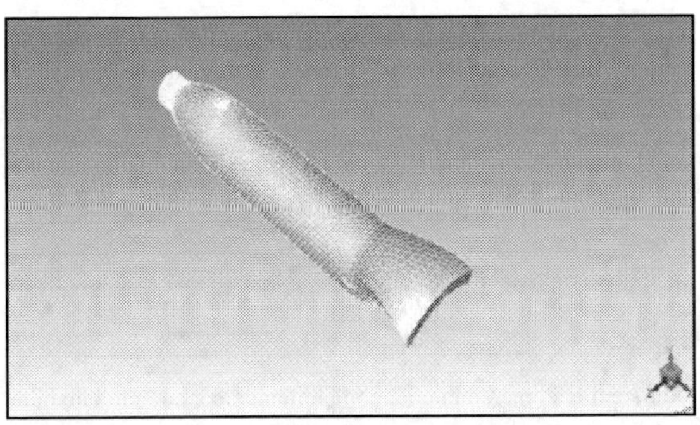

Figure 5. View of the Meshed model of Circular Combustion Chamber.

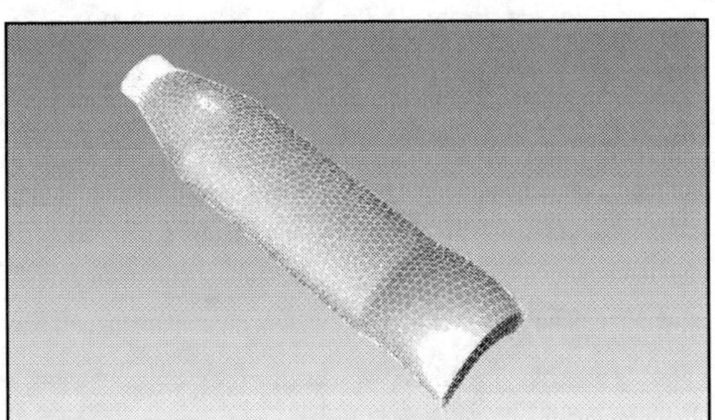

Figure 6. View of the Meshed model of Elliptical Combustion Chamber.

2.2.2 Fuel injector location

The meshed models were imported to ANSYS FLUENT for the combustion simulation. The pressure based solver was used since air flow is in a sonic state. The Energy equation and the k-ε turbulence model were used. The Radiation P-1 model and a non-premixed combustion method were employed. The air inside the compressor is introduced at Mach 0.3. The location where the Mach number drops is used as a station for the fuel injector. At this the point flow is the more stable, due to gradual drop in velocity of the flow. The Fig. 7 illustrates superimposition of the wireframe model of conventional combustor over the graph, to obtain the location of the fuel injector.

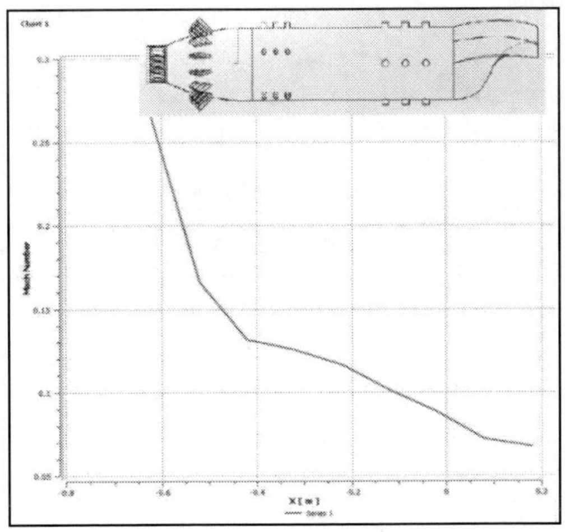

Figure 7. Comparison of the Mach number vs Direction Vector plot with the wireframe model of the Conventional Combustor.

2.2.3 Combustion process

Combustion process simulations were to be performed on the circular and the elliptical models. Combustion being the main aspect of the analysis therefore during the course of work the cooling process was ignored. The Jet-a<g> fuel was used as the fuel for the combustion process. Since the heat addition process in a jet engine takes place at constant pressure, the inlet and outlet pressure conditions were considered invarient as 10 bar. Calculations were carried out for a Mach 0.3 and a stagnation temperature of 288 K. From the calculations, static temperature of the air at the entrance of the combustion chamber was 546.56 K. The mass flow rate of the air was taken as 0.9544 kg/s. While the static temperature of the fuel inlet was at 300 K and the mass flow rate of fuel was taken as 0.01908 kg/s. Using these boundary conditions, the simulations were performed.

3 Profile analysis

The combustion simulations of both the combustor models with the suitable input values were executed and the resulting temperature, pressure, and velocity plots were plotted.

3.1. Temperature profiles

The temperature profiles of the circular combustor chamber and the elliptical combustor chamber were retrieved using CFD Post, a post-processing program. The temperature profiles of the circular and elliptical combustor chamber are shown in Fig. 8 and Fig. 9 respectively.

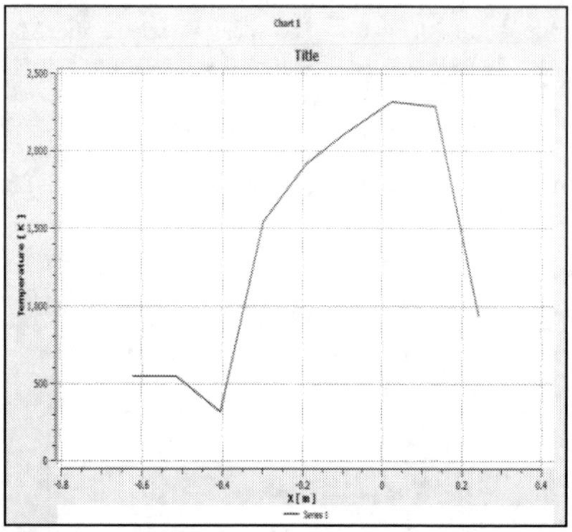

Figure 8. Temperature Profile of Circular Combustion Chamber.

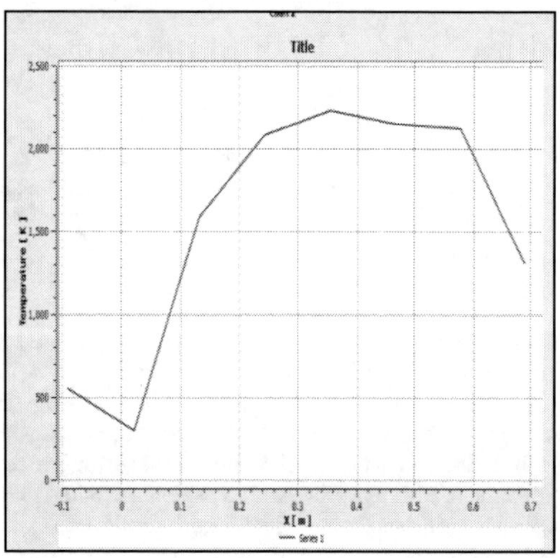

Figure 9. Temperature Profile of Elliptical Combustion Chamber.

Referring to the temperature profiles of both the combustors, it can be inferred that, the temperature at the end of the elliptical combustor is higher than that of the circular combustor. The entrance temperature of the air for both the combustors is same i.e. 546.56 K. The end temperature of the elliptical combustor is 1272 K, while that of the circular combustor is 960 K. The comparatively higher end temperature of the elliptical combustor facilitates higher velocity of the gases at the end of the combustor before entering the turbine.

3.2 Pressure profiles

The pressure profiles of both the combustors are depicted in Fig. 10 and Fig. 11.

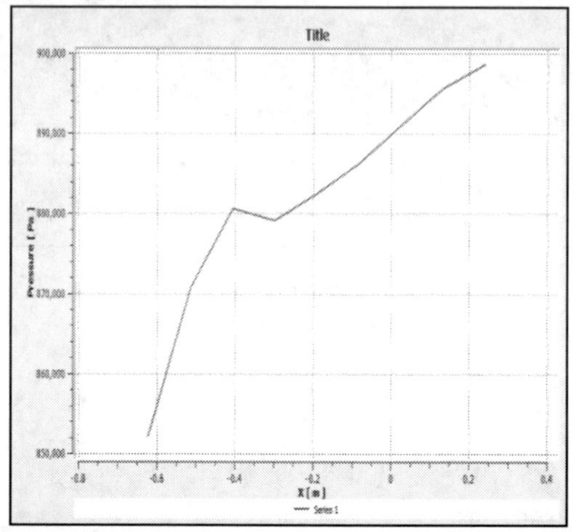

Figure 10. Pressure Profile of Circular Combustion Chamber.

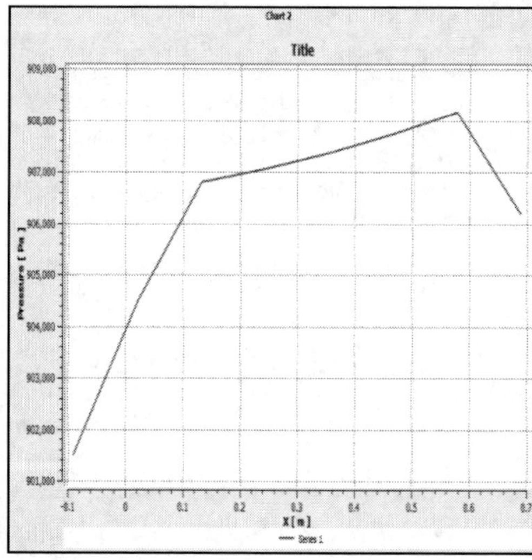

Figure 11. Pressure Profile of Elliptical Combustion Chamber.

One of the requirements of an ideal combustor of a jet engine is to obtain high pressure, without incurring high pressure losses during the process of combustion. It can be inferred from the above profiles that, the pressure achieved in the elliptical combustor is higher than the circular combustor without any occurrence of high pressure drop. There is a need to reduce the pressure loss as the gas turbine engine always operates on the Brayton cycle i.e. constant pressure cycle.

3.3 Velocity profiles

The velocity profiles of both the combustors are illustrated in Fig. 12 and Fig. 13.

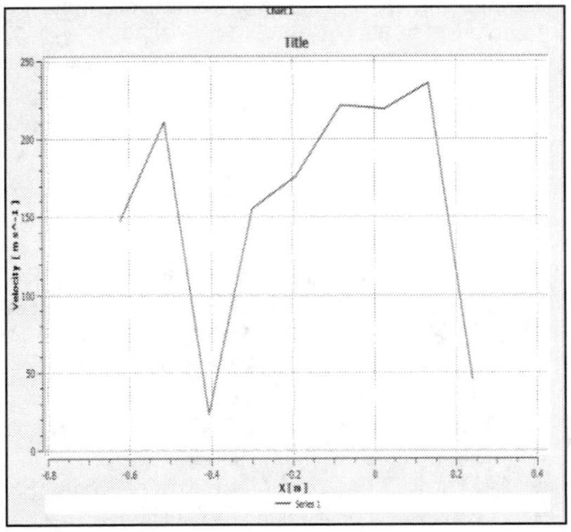

Figure 12. Velocity Profile of Circular Combustion Chamber.

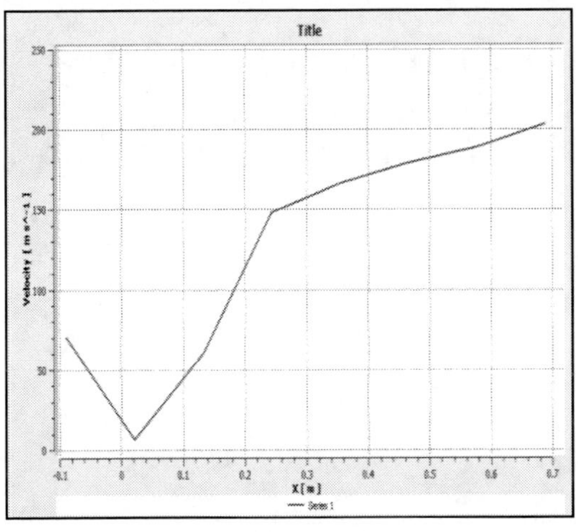

Figure 13. Velocity Profile of Elliptical Combustion Chamber.

The velocity profiles of both the combustion chamber are used to depict the velocity of the gases throughout the combustor along the axial distance. The velocity of both the combustors dip at a point where the fuel injector is placed, as the air and fuel mixture takes place. It can be inferred from the profile that the velocity of the gases at the exit of the elliptical combustor is comparatively, much higher than the circular combustor. The high velocity of the gas is required for producing high thrust and thus propelling the aircraft with a higher thrust.

4 Calculations

The combustor efficiency η_B of the circular combustor and elliptical combustor is calculated by using following formula [11].
According to formula

$$\eta_B = \frac{(\dot{m}_a + \dot{m}_f) * C_p * (T_{03} - T_{02})}{\dot{m}_f * C.V} \quad (1)$$

where,
$\dot{m}_a$ = Mass flow rate of air,
$\dot{m}_f$ = Mass flow rate of fuel,
C_p = Specific heat at constant pressure = 1147 J/kg.K,
C.V= Calorific Value of fuel = $43*10^6$ J/kg,
T_{02} = Stagnation Temperature at the Combustor Inlet,
T_{03} = Stagnation Temperature at the Turbine Inlet.

4.1 Combustion efficiency of circular combustion chamber

The temperature at the inlet of the circular combustor is 546.56 K and at the inlet of the turbine/ exit of the combustor is 960 K by referring to the plot of the temperature curve for a circular combustor as shown in Fig. 6. The mass flow rate of air is 0.9544 kg/s. The mass flow rate of the fuel is 0.01908 kg/s. The combustor efficiency of the circular combustor is given by substituting the values in formula (1),

$$\eta_{Bcircular} = \frac{(0.9544 + 0.01908) * 1147 * (960 - 546.56)}{0.01908 * 43 * 10^6}$$

$$= 0.5626 \text{ or } 56.26\%$$

4.2 Combustion efficiency of elliptical combustion chamber

Similarly, the temperature at the inlet of the elliptical combustor is 546.56K and at the inlet of the turbine/ exit of the combustor is 1272 K, by referring to the plot of temperature curve for an elliptical combustor as shown in Fig. 7. The mass flow rate of air is 0.9544 kg/s. The mass flow rate of the fuel is 0.01908 kg/s. The combustor efficiency of the elliptical combustor is given by substituting the values in formula (1),

$$\eta_{Belliptical} = \frac{(0.9544 + 0.01908) * 1147 * (1272 - 546.56)}{0.01908 * 43 * 10^6}$$

$$= 0.9872 \text{ or } 98.72\%$$

5 Conclusion

It is evident from the calculations and the profiles of combustors that, the variation in temperature, pressure, and velocity are caused due to the change in the shape of the combustors. The improvement in these computationally derived properties of the elliptical combustors is because of the axis-switching phenomenon. The increased velocity and temperature is facilitated by the higher turbulence in the elliptical combustion chamber. The axis switching phenomenon definitely plays a vital role in enhanced mixing of elliptical combustor . Therefore combustion efficiency of elliptical type is more than that of the circular type combustion chamber.

References

1. Wajid.A.Chishty and Manfried Klien, 'Combustion and Emission Issues in Gas Turbines',19th Oct 2009.
2. Rolls-Royce plc, 'The Jet Engine', ISBN 0902121235 (1986).

3. H S Mukunda, 'Understanding Combustion', (1989).
4. A.H. Lefebvre and D. R. Ballal, *Gas turbine combustion*. CRC Press, (2010).
5. D. Wager, 'The Influence of Elliptical Nozzle Holes on Mixing and Combustion in Direct Injection Natural Gas Engines'.
6. E. Gutmark, K. C. Schadow, and K. J. Wilson. 'Subsonic and supersonic combustion using noncircular injectors', Journal of Propulsion and Power, Vol. 7, No. 2 (1991), pp. 240-249.
7. Mohsen Akbarzadeh, Madjid Birouk, 'Liftoff of a Co-Flowing Non-Premixed Turbulent Methane Flame: Effect of the Fuel Nozzle Orifice Geometry'. *Flow, Turbulence and Combustion* 92:4, 903-929, (2014).
8. Miler, R.S., Madnia, C.K. & Givi.P., 'Numerical Simulations of non-circular jets', Computers & Fluids 24, pp1-25, (1995).
9. Husain H.S. & Hussain F., 'Controlled excitation of elliptic jets', Physics of Fluids, 26, pp2763-2766, (1983).
10. Ho, C.M. & Gutmark, E.,'Vortex induction and mass entrainment in a small-Aspect-ratio elliptic jet', Journal of Fluid mechanics,179, pp383-405, (1987).
11. Gutmark, E., Schadow, K.C., Parr, T.P., HansonParr, D.M., & Wilson, K.J., 'Noncircular jets in combustion systems', Experiments in Fluids 7, 1989,pp248-258.
12. E.J.Gutmark, F.F.Grinstein, 'Flow Control with Non Circular Jets', Annual Review of Fluid Mechanics 31(1):239-272, (2003).
13. N L Narasimha Reddy, P Manivannan,K M Kiran Babu, 'The CFD Analysis of Turbulence Characteristics in Combustion Chamber with Non Circular Co-Axial Jets', IOSR-JMCE, Vol.6 , Issue 2, pp 01-10.
14. V. Ganesan, 'Gas Turbines', Tata McGraw Hill Publishing Co., New Delhi, (1999).
15. S. M Yahya,'Fundametals of Compressible Flow with Aircraft and Rocket Propulsion', New Age International Publisher, New Delhi, (1982).

Experimental and Numerical Design and Optimization of a Counter-Flow Heat Exchanger

Salman Bahrami[1] and Mohammad Hassan Rahimian[2]

[1]*Mechanical Engineering Department, University of Tehran, Tehran, Iran*
[2]*Mechanical Engineering Department, University of Tehran, Tehran, Iran*

Abstract. A new inexpensive counter-flow heat exchanger has been designed and optimized for a vapor-compression cooling system in this research. The main aim is to experimentally and numerically evaluate the effect of an internal heat exchanger (IHX) adaptation in an automotive air conditioning system. In this new design of IHX, the high-pressure liquid passes through the central channel and the low-pressure vapor flows in several parallel channels in the opposite direction. The experimental set-up has been made up of original components of the air conditioning system of a medium sedan car, specially designed and built to analyze vehicle A/C equipment under real operating conditions. The results show that this compact IHX may achieve up to 10% of the evaporator capacity while low pressure drop will be imposed on this refrigeration cycle. Also, they confirm considerable decrease of compressor power consumption (CPC), which is intensified at higher evaporator air flow. A significant improvement of the coefficient of performance (COP) is achieved with the IHX employment too. The influence of operating conditions has been also discussed in this paper. Finally, numerical analyses have been briefly presented, which bring more details of the flow behavior and heat transfer phenomena, and help to determine the optimal arrangement of channels.

1 Introduction

In cooling/refrigeration industry, liquid-suction or internal heat exchangers (IHXs) are installed in vapor-compression refrigeration systems to exchange energy between relatively cool low-pressure gas and warm high-pressure liquid refrigerants. The main advantages of the IHX installation are: increasing refrigerating effect at the evaporator, sub-cooling liquid refrigerant to ensure liquid phase entrance to the expansion valve, and minimizing the risk of liquid refrigerant entering at the compressor inlet. The possible disadvantages can be: decreasing the refrigerant mass flow rate delivered by the compressor, increasing compressor discharge temperature, and increasing pressure losses at the suction and liquid lines [1, 2].

Using IHX has been studied mostly in CO2 refrigeration systems. It is an important factor to achieve high system performance; therefore some research works focusing on CO2 trans-critical systems with IHX have been reported [3-8].

Mastrullo at el. [9] presented a chart for predicting the effectiveness of IHX installation for 19 different ozone friendly fluids, and concluded that the advisability of the IHX usage can be evaluated as a function of thermodynamic properties. Variations of evaporating and condensation temperatures were respectively in the range of −40 °C/10 °C and 25 °C/50 °C. Klein at el. [10] identified a dimensionless group to correlate performance impacts attributable to the IHXs, and studied the impact of pressure drops through the IHXs on system performance. They concluded that for given operating conditions the advantages are higher for R134a than R407C and R22.

2 IHX design for a cooling system

In the present work, the effects of adopting a new IHX in an R134a automotive A/C system have been analyzed. The evaporator calorimeter tests have been carried out to evaluate the influences of adaptation of the designed IHX on the cycle performance parameters, which are quite important from point of views of system performance optimization, power saving and fuel consumption.

The IHX is of the counter-flow type made up of aluminum extruded channels. In this design, the relatively warm liquid refrigerant, which is leaving the condenser, passes through the central channel. The relatively cold low-pressure vapor, which is exiting the evaporator, flows in several parallel channels in the opposite direction. The inner- and outer diameters of the IHX have been determined based on geometries of inlet and outlet pipes of the refrigeration system connected to the IHX. The thickness of walls has been mainly dictated by manufacturing constraints. Several IHX lengths have been employed and evaluated from 10 cm up to 30 cm.

The manufacturing process of this type of IHX is simpler and less expensive in comparison with most other

IHXs. In addition, it is light and compact; so quite useful in the limited space of automobiles. Although it has been designed for automotive A/C systems, but the same concept can be implemented in other cooling systems in different industries such as aerospace and oil & gas.

3 Experimental investigations

3.1. Experimental set-up

The experimental set-up has been made up of original components from the air conditioning system of a medium sedan car, arranged in such a way to emulate them in the actual vehicle. It has been made up of three fluid circuits; the refrigerant, the evaporator air, and the condenser air. The diagram of the test apparatus can be found in Ref. [2]. Fig. 1 shows the condenser room with the metal frame in front of the condenser, where 12 thermocouples have been installed on it to find the average temperature of the air flow [2].

R134a is used in experimental test bench as the refrigerant. Temperature and pressure of refrigerant are measured with the T type thermocouples and pressure transmitters, respectively. The mass flow rate of refrigerant is measured with a Coriolis type meter with the capacity of 0- 225 kg/h with 1 kg/h precision, installed in the liquid line. Input parameters of evaporator calorimeter test of automobile A/C system are evaporator air flow rate, dry bulb and wet bulb temperatures of evaporator inlet air, expansion valve inlet pressure, and evaporator average pressure. The condenser is a parallel flow type containing 34 multi-channel flat tubes, which is used in this type of car. The installed evaporator on the system is a laminated type. The evaporator air chamber has been set in the dry bulb temperature of 27°C and 50% relative humidity. Sensible and latent thermal loads of evaporator have been provided by a heater and a humidifier installed upstream of the evaporator. The evaporator blower passes the air flow rate of 200, 300, 400 and 500 m^3/h. Average pressure of evaporator has been set in 0.25 MPa gage. Two inlet expansion valve pressures have been considered; 20 and 23 bar gage. The compressor is a reciprocating swash plate one with the capacity of 150 cc/rev.

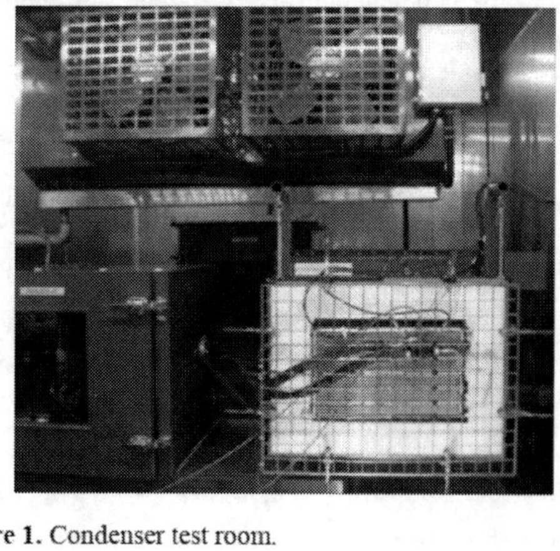

Figure 1. Condenser test room.

3.2. Experimental results

The results indicate that pressure drops of the high-pressure line are lower than the low-pressure line; mainly since the low-pressure flow passes through several parallel channels in contacts with more walls, which causes more friction losses. The pressure drops of the low-pressure vapor over passing through two IHXs of 10 and 30 cm are respectively equal to 2.9% and 6.1% of the inlet pressure, which are suitable in comparison with other IHXs. The capacity of IHX is equal to multiplication of refrigerant mass flow rate and difference of inlet and outlet enthalpies, which could be calculated from temperature and pressure of the inlet- and outlet refrigerant. The highest capacity of IHXs belongs to the higher pressure of expansion valve inlet. It is noticeable that in 30 cm IHX, increasing evaporator air flow causes enhancement of IHX capacity. The results show that by IHX length increase from 10 cm to 30 cm, its capacity will improve 690% and 280% in expansion valve inlet pressure of 20 bar gage and 23 bar gage respectively. IHX capacity is usually compared with the capacity of other components. For instant, Fig. 2 shows the IHX/evaporator capacity ratio obtained in different evaporator air flows. In the best test conditions, the 30 cm IHX capacity is 13% of fabric evaporator capacity.

The most significant effects of IHX employment is related to the compressor performance. Using IHX leads to a maximum CPC reduction of 25%. Installation of 10 cm and 30 cm IHXs causes respectively 13% and 16% CPC reduction in average. The aforementioned achievements can lead to significant automobile net engine power enhancement, as well as fuel economy improvement. Those CPC reductions result the cycle COP increase by 5.4% and 7.7% in average respectively.

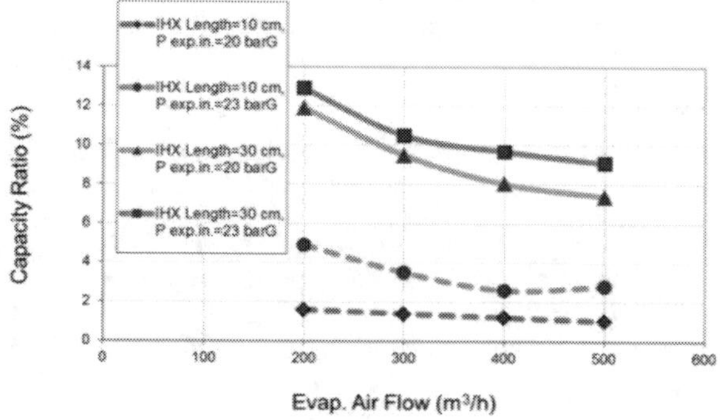

Figure 2. Capacity ratio versus evaporator air flow.

4 Numerical analyses

First, structural analyses were carried out to ensure proper arrangement of fluids in central and outer channels from the maximum stress perspective. Then CFD analyses were executed to learn more about the fluidic/thermal phenomena happening in the IHX. Also several numbers of outer parallel channels were investigated.

4.1 Structural analyses

Static finite element analyses were carried out for two possible configurations of the liquid- and vapor counter flows. Fig. 3 shows the results for the appropriate configuration that poses a lower maximum stress. In this configuration the stress concentration is around the internal channel that contains the high-pressure liquid. The maximum von-Mises stress (6.9 MPa) indicates a proper safe margin, about 6%, considering the properties of the specific Aluminum alloy used. The other configuration (i.e. vapor flows in the central channel) may impose about a 5-times higher maximum stress into the component.

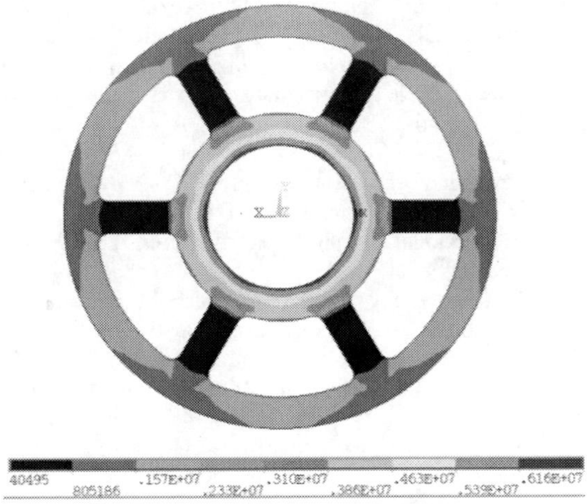

Figure 3. Stress contours (in Pa).

4.2 CFD analyses

Ansys CFX software was used to solve the equations in three different computational domains (i.e. vapor, liquid, and solid), and to obtain velocity, pressure, and temperature contours through them. Steady state condition and second order upwind approach were applied for momentum and energy equations. Boundary conditions were defined according to the experiments, and the results were validated at the same operating conditions. Mass flow inlet boundary conditions for both fluids indicate that flows are quite turbulent; so a k-ε turbulent model was employed according to previous researches and the computational budget. Due to similarity of parallel channels only one channel was simulated with the proportional slice of the central channel, using rotational periodic boundary condition. Proper initializations were considered to obtain the convergence. Mesh independency was studied as well.

The results indicate that pressure drops are well-consistent with experimental investigations, especially for the low-pressure vapor. Heat transfer parameters are also well predicted according to comparison of average outlet temperatures for both fluids. For instance, Fig. 4 shows the temperature contours in the cross section of the IHX with six parallel channels. The big temperature difference between the cores of two fluids (i.e. 63°C) is obvious in this figure, which produces big temperature gradients through the IHX. In addition, producing relatively hot boundary layers inside the outer channels illustrates the proper heat transfer mechanism through the hot walls into the cold channels. In this figure the relatively cold vapor leaves the IHX towards the compressor. The average superheat degree of refrigerant is 7.4°C, which is quite good from the compressor performance perspective. Fig. 5 shows surface heat transfer coefficient contours along the central channel for the aforementioned IHX. These contours are similar to heat flux contours. The parts of walls close to the liquid entrance have lower heat fluxes due to transferring the heat with the vapor that is much warmer when it is leaving the IHX. The results indicate that heat transfer coefficients are less than 50% in outer channels.

The number of parallel channels was investigated numerically. All walls thicknesses and inner- and outer diameters remained constant. From the structural perspective, there is no significant change in the stress contours. From the fluidic point of view, increasing the number of channels reduces the inlet flow area, so in a constant mass flow rate the vapor would enter with a higher velocity. Although it brings a higher convection coefficient, but increases pressure drops as well. While the velocity and heat transfer area are unchanged in the liquid side, increasing the number of parallel channels expands the heat transfer surfaces in the vapor side, and consequently increases the heat absorption potential. It should be noted that the Reynold's number of liquid flow is about up to three times higher than the vapor one. Considering all the aforementioned points and trade-off, based on the numerical results the number of five channels has been identified as the optimum number of channels for the current application.

5 Conclusion and future works

The experimental- and numerical results show that using this proposed IHX can bring significant positive impacts on the performance of car refrigeration cycles. Also, using the studied heat exchanger would impose low pressure drops on the cycle.

Improvement of the CPC is one of the most important achievements of IHX installation. It can increase automobile net engine power, improve its efficiency, and bring economy of fuel. Significant COP enhancement would be expected in the refrigeration cycle as well.

As future works, different mechanisms will be considered to enhance the IHX heat transfer efficiency. For instance adding inner and outer grooves and fines will be studied in order to increase heat transfer surfaces and thermal coefficients. All geometric parameters can be optimized in the next step in order to improve IHX capacity while pressure drops do not increase significantly. Finally, using other refrigerants in other cooling applications will be studied, such as R407C and R22.

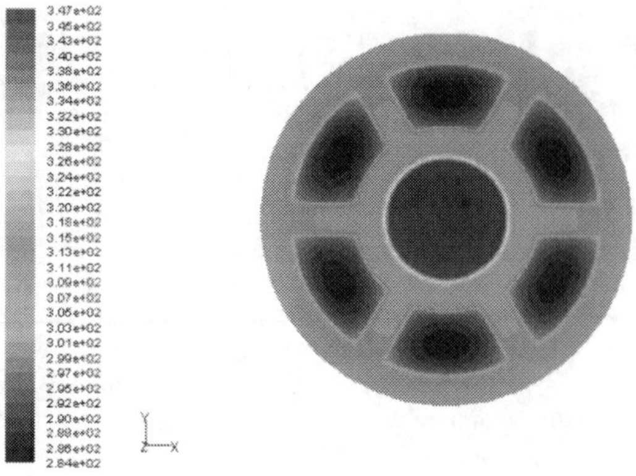

Figure 4. Temperature contours of liquid inlet and gas outlet and IHX walls (in Kelvin).

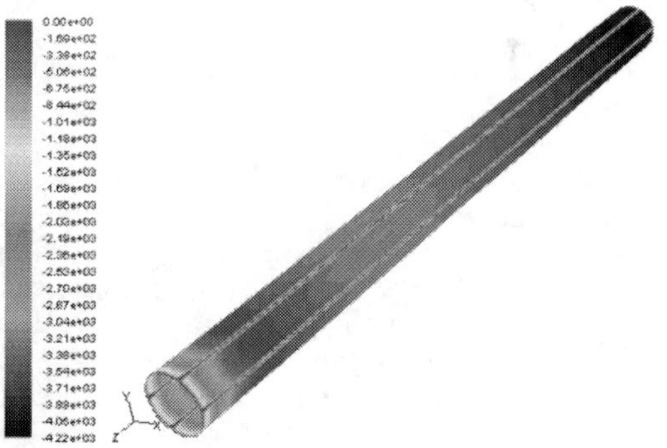

Figure 5. Heat transfer coefficient contours along the central channel (in w/m^{2}K).

Acknowledgements

The authors would like to thank Sardsaz Khodro co. for carrying out the experiments.

References

1. J. Navarro, R. Cabello, E. Torrela, Experimental evaluation of the internal heat exchanger influence on a vapor compression plant energy efficiency working with R22, R134a and R407C, Journal of energy, **30**, 621-636 (2005)
2. S. Bahrami, H. Mohammadbeigi, M.H. Sabour, Experimental analysis of internal heat exchanger for automotive A/C system, *Proceeding of ASME conference, ESDA 2010*, 24485 (2010)
3. Y. Chen, J. Gu, The optimum high pressure for CO2 transcritical refrigeration systems with internal heat exchangers, International Journal of Refrigeration, **28**, 1238-1249 (2005)
4. C. Aprea, A. Maiorino: An experimental evaluation of the transcritical CO2 refrigerator performances using an internal heat exchanger. International Journal of Refrigeration, **31**, 1006-1011 (2008)
5. Y.B. Tao, Y.L. He, Z.G. Wu, Experimental study on the performance of CO2 residential air-conditioning system with an internal heat exchanger, Energy Conversion and Management, **51**, 64-70 (2010)
6. J. Rigola, N. Ablanque, C.D. Perez-Segarra, A. Olive, Numerical simulation and experimental validation of internal heat exchanger influence on CO2 trans-critical cycle performance, International Journal of Refrigeration, **33**, 664-674 (2010)
7. S.G. Kim, Y.J. Kim, G. Lee, M.S. Kim, The performance of a transcritical CO2 cycle with an internal heat exchanger for hot water heating, International Journal of Refrigeration, **28**, 1064-1072 (2005)
8. H. Cho, C. Ryu, Y. Kim, Cooling performance of a variable speed CO2 cycle with an electronic expansion valve and internal heat exchanger, International Journal of Refrigeration, **30**, 664-671 (2007)
9. R. Mastrullo, A.W. Mauro, S. Tino, G.P. Vanoli, A chart for predicting the possible advantage of adopting a suction/liquid heat exchanger in refrigerating system, Journal of Applied Thermal Engineering, **27**, 2443-2448 (2007)
10. S.A. Klein, D.T. Reindl, K. Brownell, Refrigeration system performance using liquid-suction heat exchangers, International Journal of refrigeration, **23**, 588-596 (2000)

Fundamental Study on Operational Parameters of Diaphragmless Shock Tube

Masanori NISHIYAMA[1], Masato TAGUCHI[2] and Masashi KASHITANI[3]

[1,2,3]Department of Aerospace Engineering, National Defense Academy of Japan, 1-10-20 Hashirimizu Yokosuka, Kanagawa, 239-8686, Japan

Abstract. This paper shows influences of initial conditions on a diaphragmless shock tube operation. This facility consists of a driver tube, a driven tube and a damp tank. The driver tube has a circular cross section with diameter of 150 mm and the driven tube, a rectangular cross section (60 mm x 150 mm). The shock tube is operated by using a quick-opening pneumatic piston instead of a diaphragm. For the operation, pressure inside the pneumatic piston (piston pressure) is quickly released by opening a solenoid valve. In this paper, the initial piston pressure was chosen as a parameter to investigate effects on test flow conditions created by the shock tube. As a result, it was found that when the piston pressure at initial condition is large, piston pressure decreased more rapidly than that obtained for a small piston pressure condition, regardless of the pressure ratio of driver and the driven tube. In the condition of a constant initial operational pressure ratio and a different piston pressure, the shock Mach number was almost constant.

1 Introduction

It is well known that a shock tube is used for various researches including high-speed aerodynamics [1, 2]. The shock tube of diaphragmless type is considered as the attractive aerodynamic devices because test flow is not contaminated by the minute fragments of the film. The quick-opening pneumatic piston was installed for diaphragmless operation of our shock tube. This shock tube system has been used to investigate aerodynamic characteristics of the transonic airfoil flows, and to develop the progress flow visualization technique for the past few years. For example, developing the focusing schlieren system to visualize slice image of transonic airfoil flows etc [3-5].

With the diaphragmless system, it is comparatively easy to adjust the Mach number and Reynolds number of test flow. Repeatability of the parameters is also better than conventional diaphragm-type shock tube, because the diaphragmless system remove uncertainties due to rupture of diaphragm. Test flow condition of shock tube is theoretically determined by differential pressure between high pressure driver section and low pressure driven section, and then arbitrary test flow conditions can be obtained by appropriate control of the initial pressure. The Shock Mach number M_s is determined by the pressure ratio between the high and low pressure section p_4/p_1. According to the theory, it is well understood that larger p_4/p_1 creates higher M_s [6]. However, in diaphragmless shock tube, M_s becomes smaller than theoretical value because the piston opens in a finite time. Thus, opening behavior of the piston can induce an uncertainty of the test flow parameters.

Purpose of this study is to understand correlation between pressure inside of the pneumatic piston (piston pressure, p_P) and shock Mach number M_s. The diaphragmless shock tube is operated with several piston pressure to measure the shock Mach number and influences of the piston pressure on the shock Mach number is assessed. Furthermore, opening behavior of the piston is evaluated by monitoring history of piston pressure while the shock tube is operated.

2 Experimental setup

2.1. Diaphramgless shock tube

Dimension of the diaphragmless shock tube is shown in Fig.1. This facility consists of a driver tube, a driven tube and a damp tank. The driver tube (i.e. high pressure section, p_4) has a circular cross section with diameter of 150 mm and the driven tube (i.e. low pressure section, p_1), a rectangular cross section (60 mm x 150 mm). The damp tank is installed at edge of the low pressure section. To measure wall pressure, pressure transducers can be flush-mounted at eleven pressure taps which is created on wall of the low pressure section.

When the shock tube is operated, the initial pressure ratio p_4 / p_1 of the high pressure chamber and the low pressure chamber is adjusted. And the partition part of the high pressure chamber and low pressure chamber is

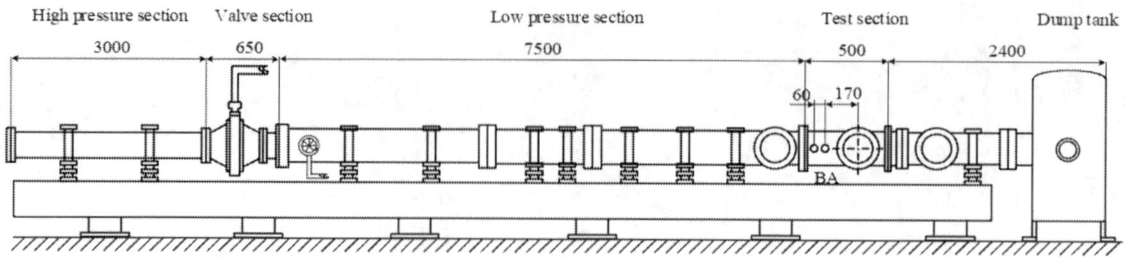

Figure 1. Diaphragmless Shock Tube[3].

quickly opened, immediately after, a primary shock wave is generated and propagates into the low pressure camber. To evaluate the test flow conditions, two PCB pressure transducers (H113A) are installed at tap A and B (see Fig.1). The PCB sensors can detect immediate pressure rise which is induced when the primary shock reaches the locus of the sensors. Shock Mach number M_s is calculated from time difference between the pressure rise at the two pressure sensors by use of a universal counter (Pendulum, CNT-90).

A quick-opening pneumatic piston is installed to partition those two sections. The system of a quick opening valve was designed by Ref. [7]. The schematic diagram of the quick-opining valve of the shock tube is shown in Fig. 2. The valve consists of a piston, a solenoid valve that drives the piston quickly, and a leak section. When the shock tube is operated, pressure inside the piston is quickly released by opening a solenoid valve, and then the piston moves quickly due to pressure reduction in a short time.

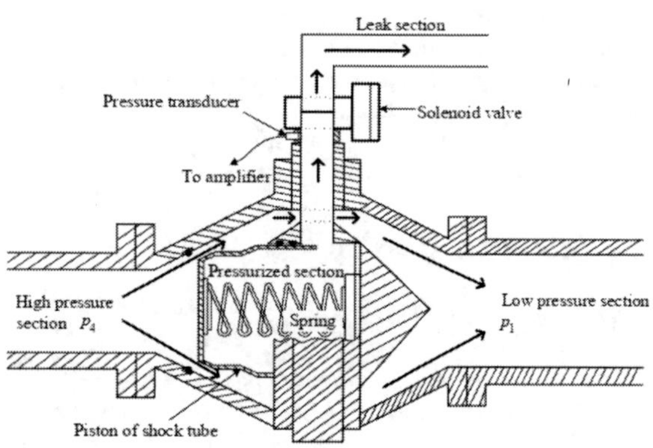

Figure 2. Quick-opening valve.

The pressure transducer is installed at the side wall just upstream of the solenoid valve. A Kulite pressure transducer (XT-190-250A) is used to monitor the pressure of the pneumatic piston p_p. Time response of the piston movement is estimated from history of the pressure reduction due to the shock tube operation. The data of the pressure histories is measured by the digital oscilloscope (Tektronix, TDS 3000 series).

2.2. Experimental condition

Table 1 shows the experimental condition in this study. The conditions of three initial pressure ratios $p_4/p_1 = 60$, 100, 150 are set across those two sections. The initial piston pressure p_p as the operational parameter is adjusted from 311 to 673 kPa. Case A shows the initial piston pressure on the measurement of the history p_p, and Case B shows the condition for estimating the shock Mach number in the different initial pressure p_p. Also, the shock Mach numbers measured in the study are shown in the table, and the result is discussed in section 3.2.

Table 1. Experimental condition.

p_4/p_1	Initial p_p	M_s
60	Case A: 475, 670, 630 kPa	1.63±0.04 (1.59 - 1.66)
	Case B : 311 - 663 kPa	
100	Case A: 475, 670, 630 kPa	1.72±0.03 (1.69 - 1.75)
	Case B : 348 - 652 kPa	
150	Case A: 475, 670, 630 kPa	1.78±0.04 (1.74 - 1.82)
	Case A : 350 - 650 kPa	
	Case B : 345 - 673 kPa	

3 Results and discussion

3.1. Piston pressure history

To understand opening behavior of the piston, the history of the piston pressure p_p is measured for initial piston pressures 475, 570 and 630 kPa. The shock tube operation pressure ratio p_4/p_1 is retained at 60, 100 and 150, respectively. The history of the piston pressure is plotted in Fig3 (a) to (c). The figures show the results of Case A. The horizontal axis represents the time from the initial condition, and the vertical axis represents the piston pressure p_p.

 https://doi.org/10.1051/matecconf/201815102004

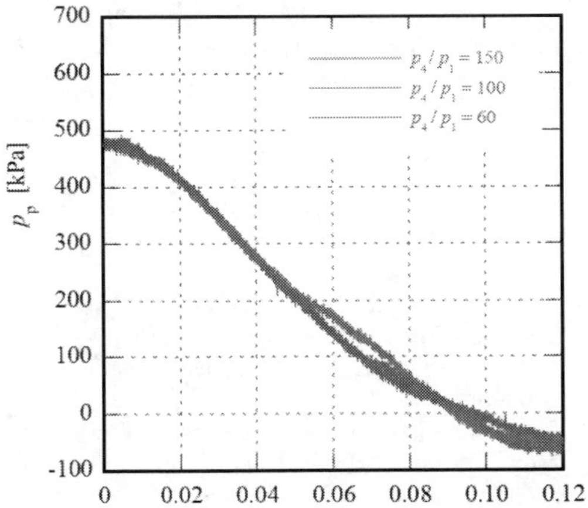

(a) 475 kPa

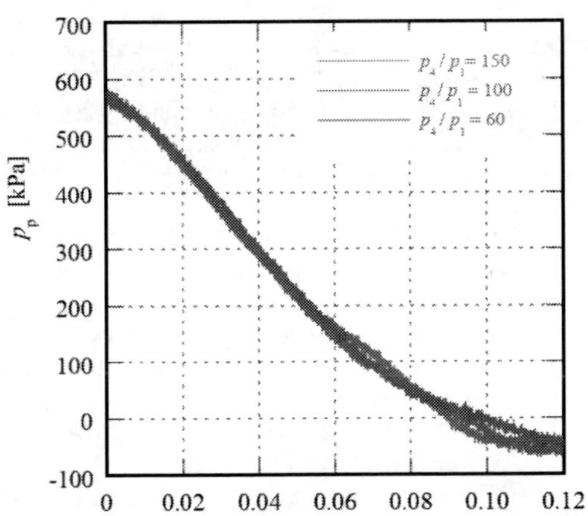

(b) 570 kPa

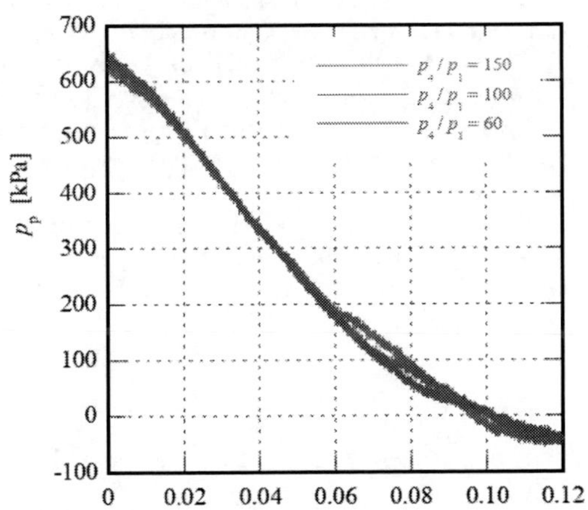

(c) 630 kPa

Figure 3. Piston pressure history.
(Effect of the operational pressure p_4 / p_1)

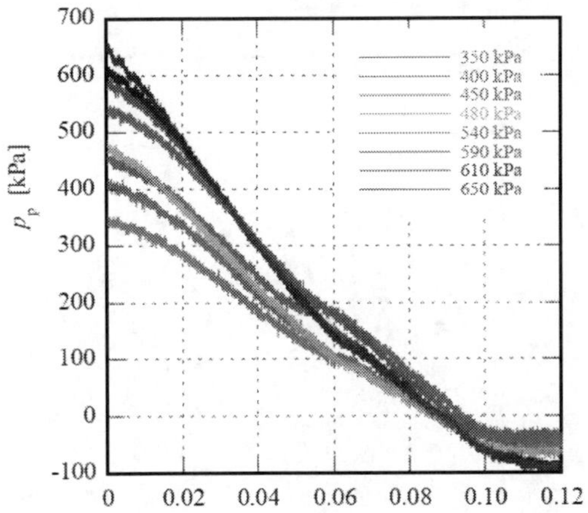

(a) Time history

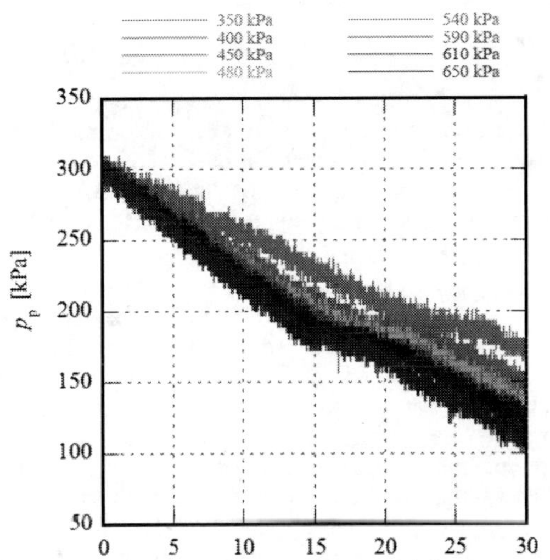

(b) Elapsed time history

Figure 4. Piston pressure history, $p_4 / p_1 = 150$.
(Effect of the initial piston pressure p_p)

In Fig. 3 (a), it is confirmed that the piston pressure decreases from the initial pressure to 0 around 0.09 sec, and it is found that no significant difference is observed in those initial operational pressure ratios p_4 / p_1. Also, it is still not clear under the differences between the three initial piston pressures (see Fig. 3 (b) and (c)). Therefore, the influence of the three initial operational pressure p_4 / p_1 did not greatly affect to the history of measured p_p.

Fig. 4 (a) and (b) show the piston pressure history to understand the influence of the piston pressure p_p. It is the experimental conditions of Case B. The initial piston pressures are adjusted 350 to 650 kPa, the shock tube operational pressure ratio is set to $p_4/p_1 = 150$. Fig. 4 (a) shows the piston pressure history from 0 to 0.12 sec, and Fig. 4 (b) shows the elapsed time from 300 kPa.

https://doi.org/10.1051/matecconf/201815102004

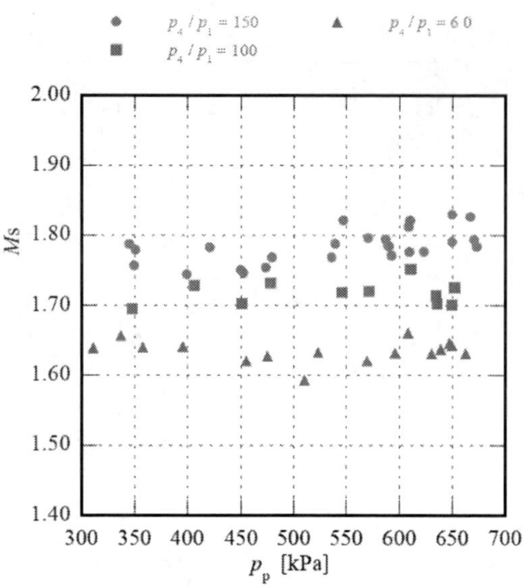

Figure 5. Shock Mach number.

For a large piston pressure at initial condition, piston pressure decreased more rapidly while piston moved than that obtained for a small piston pressure condition. And the all of the initial pressure conditions show 0 around 0.09 sec. It can be confirmed clearly in Fig. 4 (b).

3.2. Shock mach number

The shock Mach number Ms and pressure ratio p_4/p_1, which are measured under several piston pressure conditions p_p, are plotted in Fig. 5. The horizontal axis represents the piston pressure p_p and the vertical axis represents the shock Mach number Ms. Symbols indicate the data obtained at the shock tube operational pressure ratio p_4/p_1 60, 100 and 150 respectively. The tendency shows that higher pressure ratio p_4/p_1 can create the higher shock Mach number. In $p_4/p_1 = 150$, even though the Ms is slightly scattering, but the shock Mach number is almost constant (the detail information is shown in Table 1). The same trend can be seen in $p_4/p_1 = 100$ and 60, and this result indicates that the piston pressure do not have the impact on the determination of Ms. Also, the result shows the same trend as a previous study of a diaphragm-type shock tube [8].

4 Conclusion

This study was performed to understand the effect of an initial parameter for a diaphragmless shock tube which is driven by a pneumatic piston. Pressure inside the piston was employed as the test parameter. A number of tests were conducted under several piston pressure conditions, and shock Mach number and history of piston pressure were measured. The following results were obtained;

1) Pressure ratio p_4/p_1 does not greatly affect the history of piston pressure.
2) For a large piston pressure at initial condition, piston pressure decreased more rapidly while piston moved than that obtained for a small piston pressure condition.
3) The shock Mach number is almost constant, and the piston pressure do not have influence on the determination of Ms.

References

1. William. J. Cook: Shock Tube as a Device for Testing Transonic Airfoils at High Reynolds Numbers, AIAA J., 17-7, (1979) pp.715-721.
2. H. Olivier, T. Reichel and M. Zechner: Airfoil Flow Visualization and Pressure Measurements in High-Reynolds Number Transonic Flow", AIAA J., 41-8, (2003), pp. 1405-1412.
3. M. Kashitani, Y. Yamaguchi, G. Oki, H. Kitano and S. Esashi: Preliminary Study on Diaphragmless Shock Tube for Transonic Airfoil Testing with PDI, AIAA Paper 2010-744, 2010.
4. M. Kashitani, Y. Yamaguchi and H. Kitano, Preliminary Study for Quantitative Measurement of Flow Fields by Focusing Schlieren Method, AIAA Paper 2011-1983, 2011.
5. M. Kashitani, Y. Yamaguchi, D. Miyazaki and G. Oki: A Modified Point-Diffraction Interferometry for Shock Tube Airfoil Testing, AIAA Paper 2012-468, 2012.
6. T. Ikui, K. Matsuo: Mechanics of Shock Wave, Corona Publishing, (1983), pp.149-199 (In Japanese).
7. T. Otomo, T. Kuribayashi. and K. Takayama : The Characteristics of 100 mm × 180 mm Diaphragmless Shock Tube after Improvement, 2002 Shock Wave Symposium, (2003) pp.7-8. (In Japanese).
8. K. Kuraya, S. Tsuchiya: Chemical physics of Shock Wave, Shyokabo Publishing, (1968), pp. 52-56 (In Japanese).

Experimental Research about Shock Wave in a 1+1/2 Counter-Rotating Turbine

Chao Li[1], Huo-Xing Liu[1,2] and Zhi-Hong Zhou[1]

[1]School of Energy and Power Engineering, Beihang University, Beijing, China.
[2] Defense aerothermodynamic laboratory & School of Energy and Power Engineering, Beihang University, Beijing, China.

Abstract. To investigate the internal distribution regularities of shock wave structure in 1+1/2 counter-rotating turbine, numerical simulation and experimental research about the shock wave structure were conducted by using the schlieren apparatus under different working conditions.From the point of the unsteady results, the unsteady effect has few influence on the flow field of high pressure guide vane, but the wake of the high pressure guide leaves periodically sweeps through the front edge of the high pressure blade and there presents strong unsteady effect on flow field of high pressure rotor. Because of periodic influence of external wake and shock wave, the unsteadiness of flow in low pressure rotor is still strong but not that drastic compared to the high pressure rotor. 50% height section of the blade of the three types of blades are extracted respectively to make plane cascades which are conducted blowing experiments in supersonic wind tunnel. The final photograph were analyzed by comparing with the CFD results. Results show that with the increase of expansion ratio, the wave structures in blade channel move toward the exit and the caudal interference between the outer tail wave and is strengthened gradually.

1 Introduction

In 1+1/2 counter-rotating turbine, the high-pressure turbine has one stage, which contains a complete guide vane and a rotor blade. The low-pressure turbine also has one stage but only the rotating blade without guide vanes. The corresponding velocity triangle of the 1+1/2 counter-rotating turbine and the conventional 1+1 counter-rotating turbine are shown in Fig. 1.

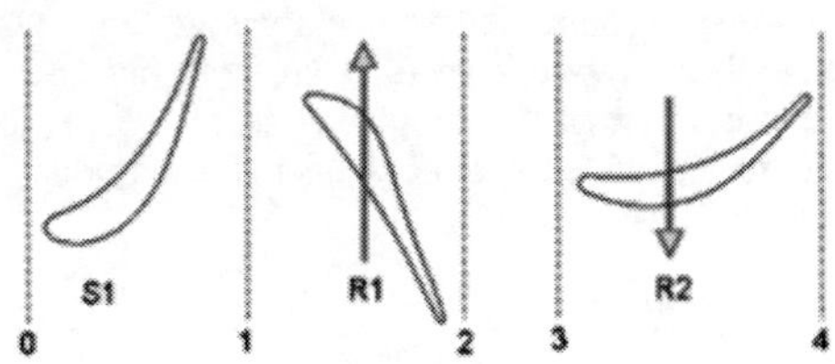

(a) 1+1/2 counter-rotating turbine

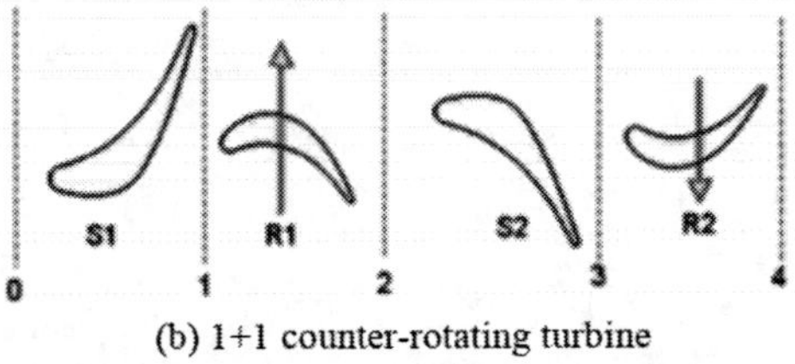

(b) 1+1 counter-rotating turbine

Figure 1. Diagram of blade structure in a 1+1/2 and 1+1 counter-rotating turbine

The adoption of aerodynamic configurations in counter-rotating turbine is one of the most effective techniques to improve the engine performance [1]. The conter rotation of HP and LP turbine can greatly reduce or eliminate the gyroscopic moment in the engine [2]and improve the maneuverability and operability of the aircraft [3]. Moreover, it can reduce the weight of the turbine from the aspect of structure, which makes the engine structure more compact [4]. In 1985, Louis and Mossca of the MIT in the United States conducted resaearch about 1+1/2 and 1/2+1/2 counter-rotating turbine and selected the class load coefficient as the free variable, and the result showed that under the same level of load coefficient, the efficiency of the counter-rotating turbine was higher than that of the conventional turbine [5]. In 1993, Huber from NASA's lewis research center studied counter-rotating turbine in a liquid rocket engine [6] and the results showed that the efficiency of turbine without a second-level turbine guide vane increased by 2%. Lior from Russia used the oblique flow combustion chamber technology in his experiment [7] to provide a certain amount of airflow prewhirl to eliminate the high pressure guide vane and finally he got a 1/2 + 1/2 contra-rotating turbine scheme and found that under the same flow conditions and turbine expansion ratio, the efficiency can be improved by 3% to 5% [8]. In 2000, Haldeman [9] cooperated with P&W in both an experiment and numerical simulation of a 1+1/2 turbine-turbine, the results showed that the low-pressure turbine surface was subjected to larger unsteady stress [10]. The diagram of high pressure in 1+1/2 counter-rotating turbine given by Wang Huishe [11] showed that a set of compressive wave (SSPW) appeared near the 60% axial chord length on the suction surface of the high pressure rotator blade but weakened after its intersecting with the trailing edge (IES). Qi Lei from Beihang university conducted simulation about a small 1 + 1/2 contra-rotating turbine she designed and the results showed that the exit Mach number of the high pressure rotor is high, and complex waves produced in the channel with strong shock wave throughout the channel [12].

In this research, numerical simulation were carried out for a 1+1/2 contra-rotating turbine under design conditions to obtain distribution regulations of the shock wave structure in blade fields, and then schlieren technique was introduced to get the detailed shock wave structure, followed by analyzation of the experimental results.

2 Numerical simulation

2.1 Computational Grid

The object of this calculation is a 1+1/2 counter rotating turbine, in which the number of high pressure guide vanes is 17, the number of high pressure rotor blades is 29, and the number of low pressure rotor blades is 33. Autogrid software is used to mesh the single channel model of the turbine. The grid diagram is shown in Fig. 2.

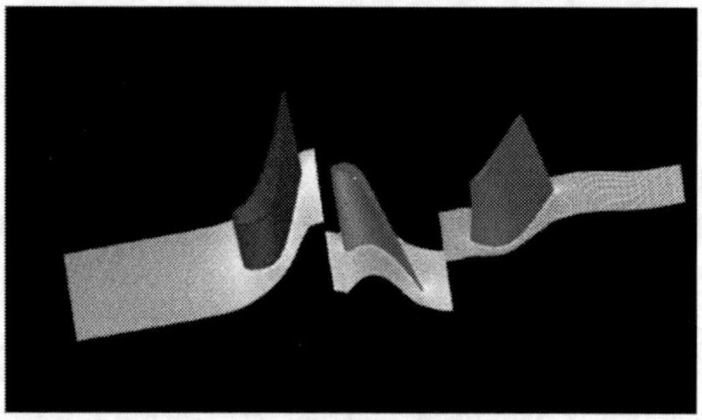

Figure 2. Grid of single channel model

The steady numerical simulation is done by software CFX15.0, and the numerical simulation method is time chasing finite volume method. The calculated refrigerant is set to simulate the gas by changing the specific heat according to the oil gas ratio. When conducting unsteady calculation, set the physical time step of turning the low-pressure rotor through 1 channel as 50 steps, and each step of the virtual time step is 10. Transient Rotor Stator is selected as the dynamic interface type.

2.2 The Calculation Results

Shown in Fig. 3 to 5 is the distribution of isentropic Mach number on surface of blade root, middle and tip (corresponding to 10%, 50%, 90% blade height). It can be seen from the Fig. 3 that the airflow around the surface of tip and middle of the high pressure guide vane is in subsonic flow condition, but supersonic state around the surface of posterior root. The inlet angle of attack is 0 degree and the unsteady results are in good agreement with the steady results at the the front part of the root, middle and tip part of the rotator. The suction peak near the trailing edge of suction surface of unsteady time-average result is located posteriorly compared with that of steady result, and the diffusivity of trailing edge increased. This is because there are various waves in this area, and the flow is complicated. It shows that the unsteady effect has a great influence on the wave flow in area of guide vane.

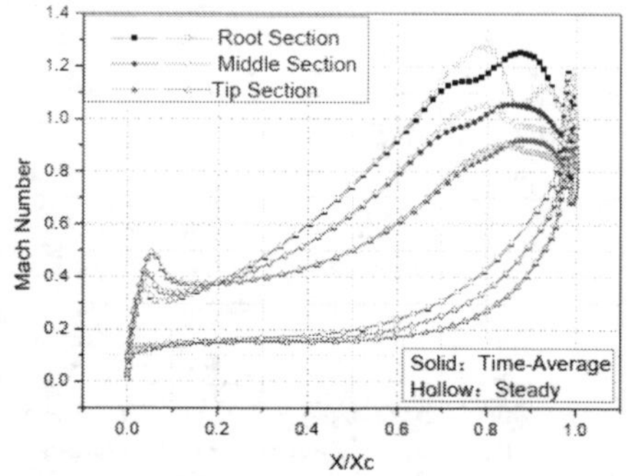

Figure 3. Distribution of Isentropic Mach number in high pressure guide vane surface under steady and unsteady conditions

In order to improve the working capacity of the low pressure rotator blade in 1+1/2 contra rotating turbine, the high pressure rotator blade must have higher outlet Maher number and relative small gas flow angle. Fig. 4 shows that the high pressure rotor blade is in transonic flow state, and the section from trailing edge to the exit of blade is completely in the supersonic state. The suction peak of leading edge also produces supersonic flow due to the large degree of acceleration. It can be seen from the load state at the head of the blade that the inlet flow of the high pressure rotor has large positive angle of attack and strong shock wave structure appears because of the supersonic flow at the trailing edge of the suction side. Besides, the shock intensity decreased gradually from root to tip.

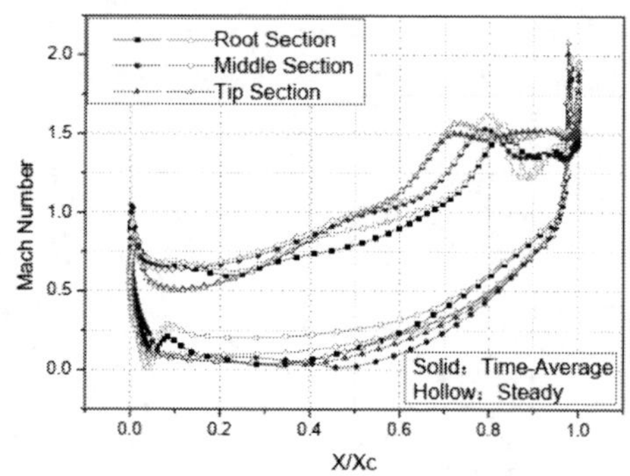

Figure 4. Distribution of Isentropic Mach number in high pressure rotor surface under steady and unsteady conditions

Compared with the high pressure guide vane, the unsteady results are not that agreed with the steady results in high pressure rotator, which is more obvious in the front part of the blade axis. Because of the higher positive angle of attack, the unsteady isentropic Mach number is lower than that of steady result, especially at the middle pressure surface of the blade root and both the suction and pressure surface of middle section of blade.

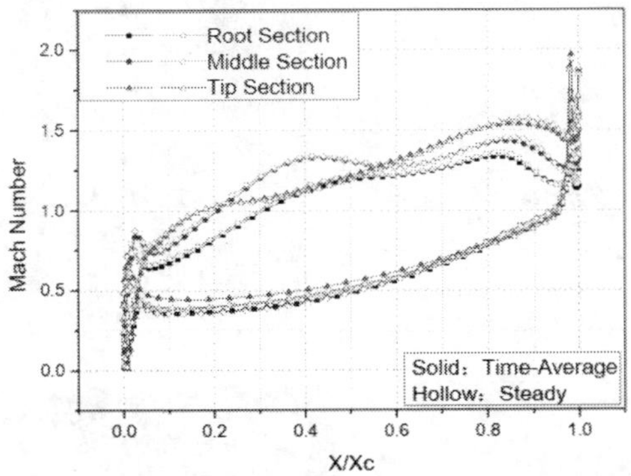

Figure 5. Distribution of Isentropic Mach number in low pressure rotor surface under steady and unsteady conditions

Fig. 5 shows that most regions in the low pressure rotor is in supersonic state, and the distribution of blade surface isentropic Mach number is relatively flat. The load distribution is more reasonable, showing better acceleration performance of airflow in the blade surface. Unlike high pressure blade, the unsteady and steady results of load distribution in low pressure rotor blade are in satisfactory agreement, except slightly upward of isentropic Mach number in the front part of the pressure side of blade tip. It can be seen that most of the areas of the low pressure rotator blade surface are less affected by the unsteady effect.

3 The experimental study

In order to deeply study the dynamic characteristics of the shock wave structure of the designed high pressure guide vane, high pressure and low pressure blade under the supersonic flow condition, 50% height section of the blade of the three types of blades are extracted respectively to make plane cascades shown in Fig. 6, which are conducted blowing experiments in transonic wind tunnel by using the schlieren display technology to record the shock wave structure in three cascade flow passage under the design Angle of attack.

The ratio of inlet total pressure and outlet static pressure was used as the experimental condition variable, and according from the designed outlet Mach number of the turbine, the blow down pressure ratio was set to change from 1.9 to 2.5, and the experimental results were compared with the results of cascade steady numerical calculation.

Figure 6. Plane cascade

3.1 Experimental Equipment

This experiment was conducted on supersonic wind tunnel of Beihang University, schlieren technology was adopted to observe the wave structure in the blade cascade, and the atmospheric pressure and the total inflow pressure were recorded at the experimental steady state. The experiment system is shown as belowing Fig. 7.

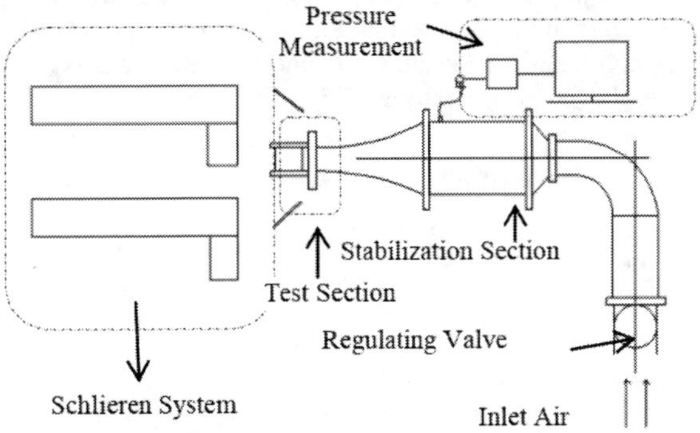

Figure 7. Experiment system

The experimental device consists of seven parts: the air source, pneumatic valve, expansion section, stabilization section, convergence section, test section and gas collecting silencer. The airflow contractes into a rectangular 120mm x 50mm channel after the convergent section, then it flows through the test section, and finally into the atmosphere through the gas collecting silencer.

For the schlieren device of this experiment shown above in Fig. 8, a double reflection schlieren instrument, which is a typical parallel light schlieren instrument, was adopted.

Figure 8. Double reflection schlieren instrument

 https://doi.org/10.1051/matecconf/201815102005

(a) Light source system (b) Main light path

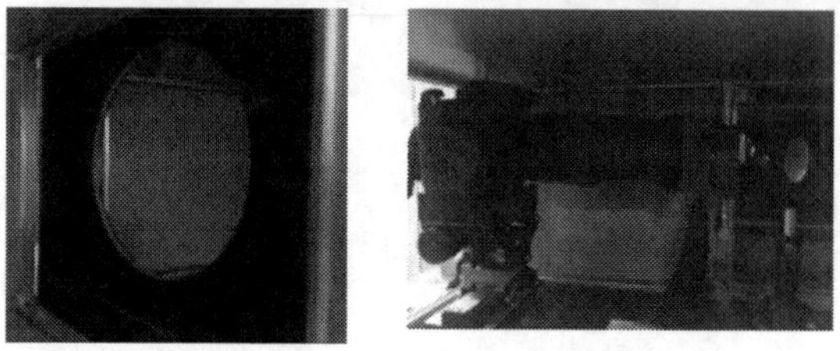

(c) Concave mirror (d) Captured system

Figure 9. Components of schlieren system

The components are shown in the following Fig. 9. The concave reflection mirrior has a diameter of 200mm, a focal length of 2000mm, and the image is taken with a Nikon D810 SLR camera. Camera shutter time is set to 1/1000, and the aperture is F5.6, ISO200.

3.2 Experimental Results

Given below are the numerical simulation results and schlieren test results of the three sets of high pressure guide vanes, high pressure rotator blades and low pressure rotator blades under the blow down ratio of 1.9, 2.1, 2.3 and 2.5 respectively.

It can be seen from the diagram that the related wave structure is obvious, and the internal tail wave, the outer tail wave and the reflected wave of the inner tail wave can be clearly seen. The suction surface of the upper channel has no flow, which has a great influence on the flow inside the channel, so that the two channels of the upper and lower channels lose periodicity. Therefore, the analysis of flow characteristics will mainly focus on the intermediate runner.

Given belowing are simulation and experimental results of high pressure rotor. From the schlieren results in Fig. 10 we can see that with the increase of the pressure ratio, the Mach number of the three sets of the cascade increases gradually, and the structure of the shock wave tends to be more complex. When the Mach number is 1.9,where is no obvious shock at exit of the cascade and when the pressure ratio reaches 2.1, we can see obvious inner tail wave structure and its reflection wave on the adjacent suction surface, while intensity of the outer tail wave is relatively weak.

As the pressure ratio increases, the shock wave structure at the exit of cascade gradually increases, and you can see the position of the inner tail wave hitting on the adjacent blade gradually moves toward the direction of exports, while the reflection wave of tail wave gradually increases. However, the trend of changes of the wave structure is overall weak.

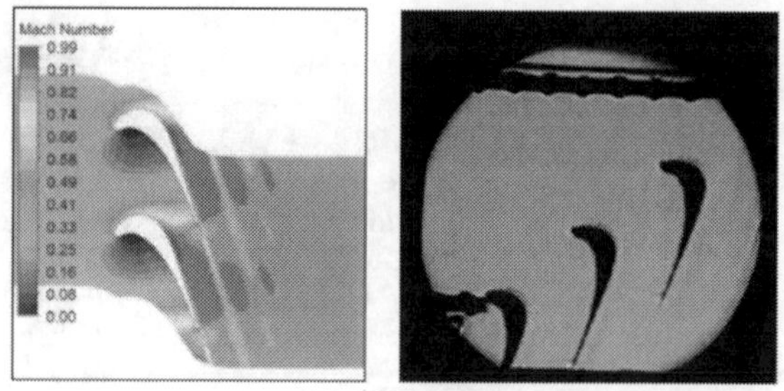

(a) P^*/P_b=1.9

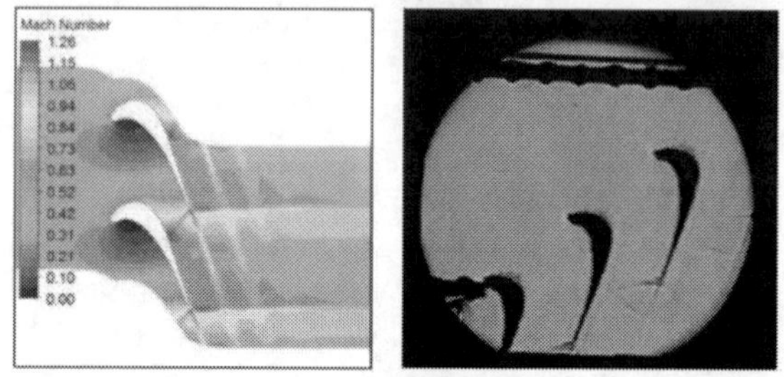

(b) P^*/P_b=2.1

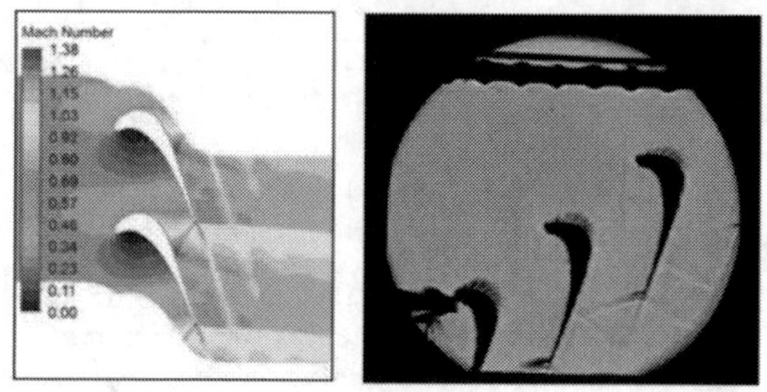

(c) P^*/P_b=2.3

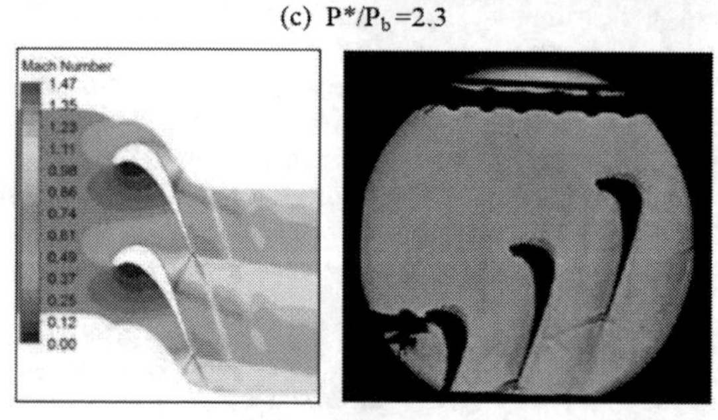

(d) P^*/P_b=2.5

Figure 10. Simulation and Experimental results of high pressure rotor

The numerical simulation results of low pressure rotating blades under different pressure ratios and the corresponding schlieren shock wave structure are shown below in Fig. 11. It can be seen clearly that with the increasing of the pressure ratio, the shock wave structure of the low pressure blade is gradually increasing. When the pressure ratio is 1.9, we can see that the trailing edge shock wave is weak and only an inner tail wave and an external tail wave can be seen without reflection waves. When the pressure ratio is 2.1, the reflection wave of the inner tail wave can be clearly seen, but the intensity is weak. When the pressure ratio reaches 2.3 as shown below, we can see the reflection wave of outer tail wave

B and the reflection wave of inner tail wave A on the adjacent blade clearly. The intensity of internal reflection wave is not obviously weakened after intervening with tail wake F, and second reflection wave is produced after hitting the upstream wake again. The intensity of the outer tail wave of the downstream blade is weakened obviously after the interference with the wake F.

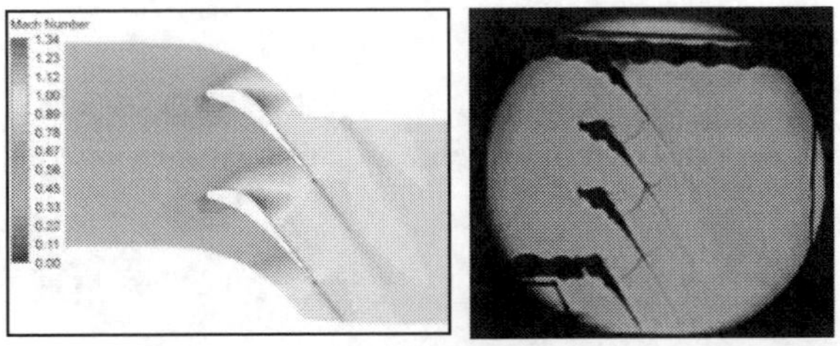

(a) P^*/P_b=1.9

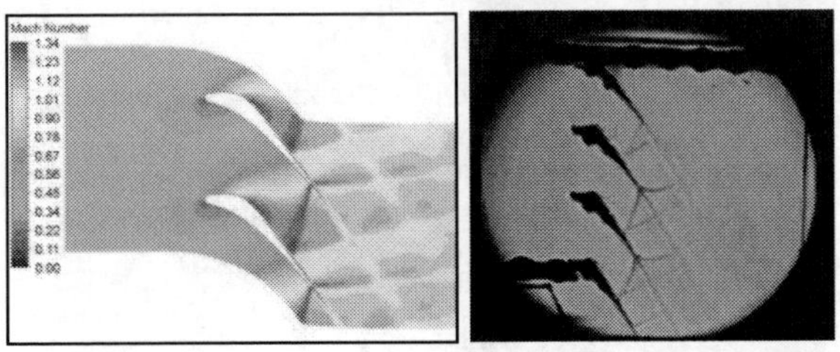

(b) P^*/P_b=2.1

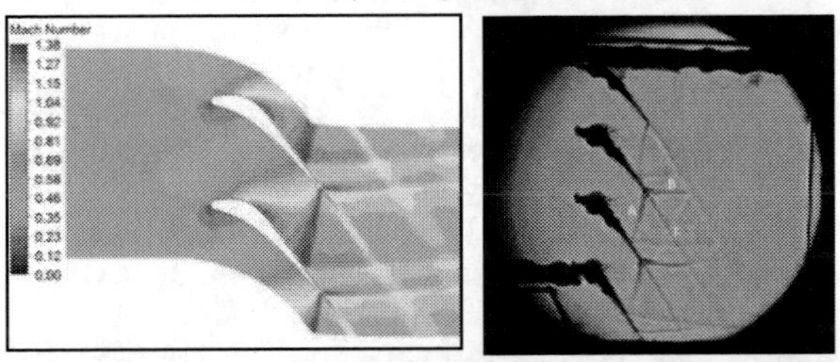

(c) P^*/P_b=2.3

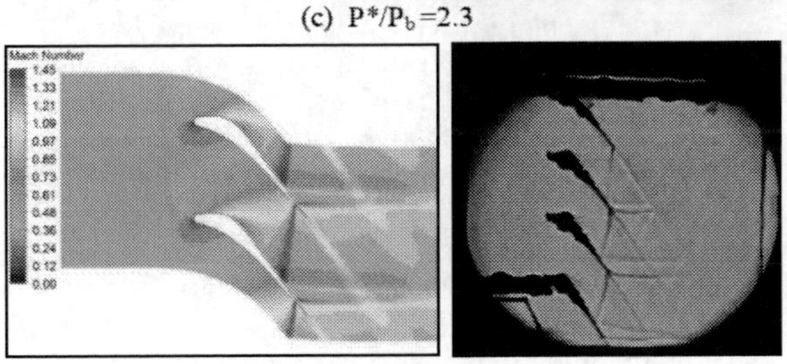

(d) P^*/P_b=2.5

Figure 11. Simulation and Experimental results of low pressure rotor

When the pressure ratio is 2.5, we can clearly see the wave structure moving toward the exit direction, and the wave structure is more complex. The strength of the tail waves and their reflection waves significantly enhances, and multiple reflection waves are produced after the interference between the wake and the primary reflection waves, which brings a great energy loss. The relative position change and intensity enhancement of the outer tail wave and the inner tail wave can be observed obviously by the contrast Fig. 12 of the pressure ratio 2.3 and 2.5.

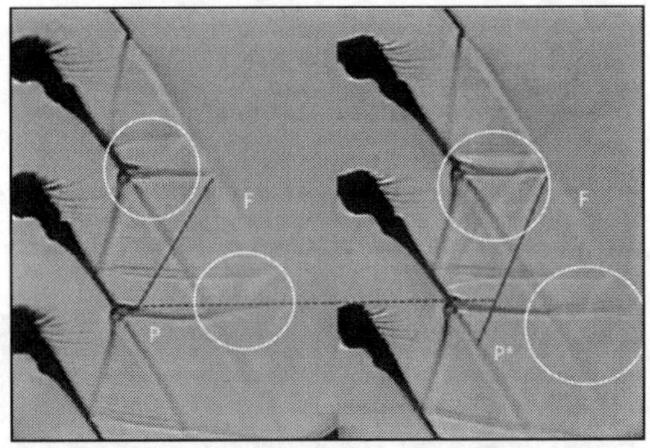

Figure 12. A comparison of wave structures details of LP rotor under pressure ratio 2.3 and 2.5

In the two figures, the positions of intersection points of the reflected wave of the outer tail wave and the wakes are made respectively, and horizontal lines are made to compare the differences of horizontal position between the two points. You can clearly see the intersection points moving toward the exit direction when the pressure ratio increases to 2.5, and the strength of both outer and inner tail wave significantly improves with increase of pressure ratio. At the same time in the white circle position the enhancement of intensity of the reflection wave also can be seen clearly, and when the pressure ratio is 2.5, the stronger reflection wave of the upstream blade mixes together with the reflection wave of the inner tail wave to form a new stronger reflection wave. At this moment, the wave structure is complicated which brings larger shock loss.

Shown in following Fig. 13 are distributions of the shock wave structure in high pressure guide vanes at different pressure ratios. As the blade thickness is small, the stress change caused by laser drilling has a great influence on the density change of the glass plate around the hole. But the wave structure is quite obvious, and primary shock wave from the suction surface and its reflected wave on the adjacent blade can be seen at the pressure ratio of 2.1. Besides, the reflection wave of inner tail wave hitting on the adjacent blade surface reflects again after the interference with the wake. All those waves and reflection waves, as well as the expansion wave system of the trailing edge of the pressure surface, form a complex wave structure system.

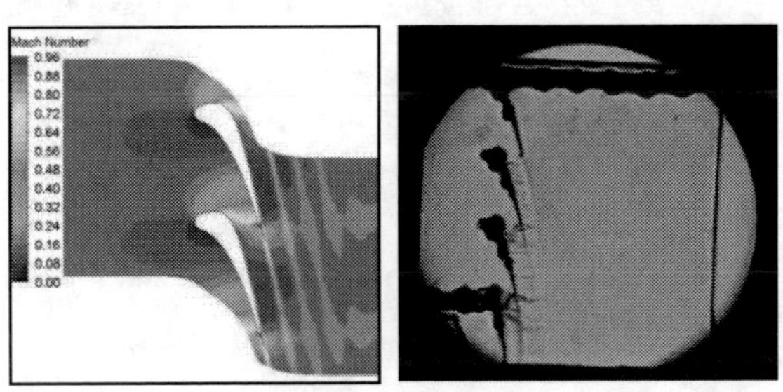

(a) P^*/P_b=1.9

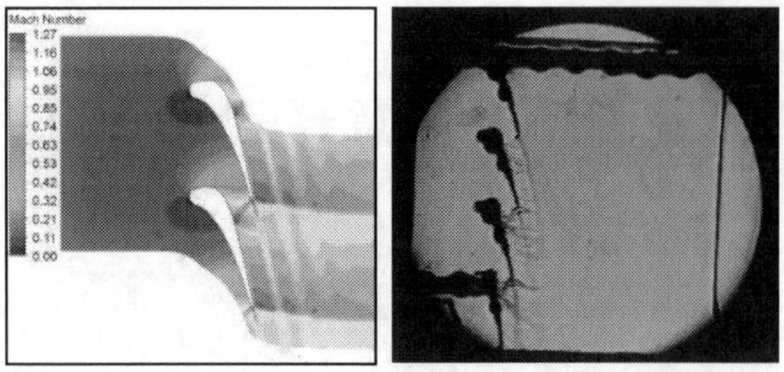

(b) P^*/P_b=2.1

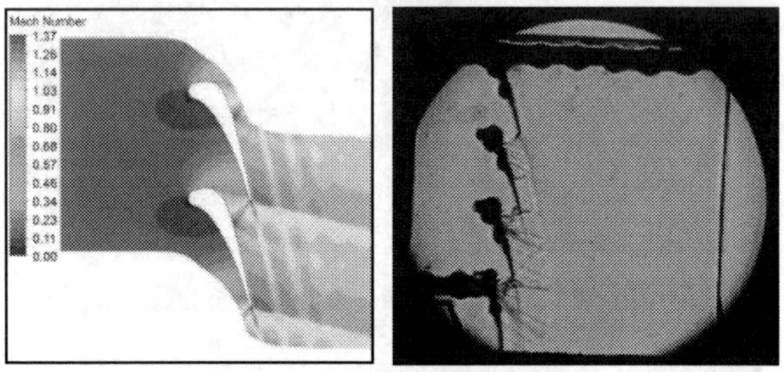

(c) P^*/P_b=2.3

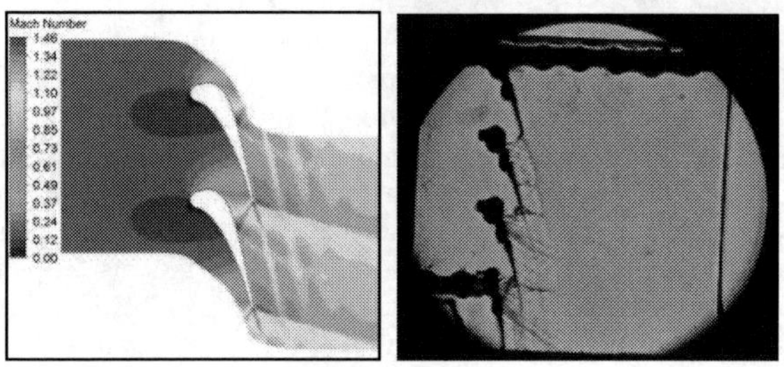

(d) P^*/P_b=2.5

Figure 13. Simulation and Experimental results of high pressure guide vane

4 Conclusion

(1) The unsteady effect has little influence on the flow filed of high pressure guide vane, while the wake of high pressure guide vane periodically sweeps the leading edge of high pressure rotating blade, which brings strong unsteady characteristics and periodic variations of exit Mach number and exit shock wave.

(2) The distribution of pressure in high pressure rotor flow field shows periodic variation along with time, and strong unsteady aerodynamic force appears on the surface of low pressure rotor blade under the periodic effect of not only strong shock wave outside the rotor trailing edge but also strong wake and potential effects. The fluctuations of pressure in the channel of low pressure rotor is extremely volatile.

(3) As the pressure ratio increases, the exit Mach number increases and the wave structure in blade channel gradually moves to the outlet direction. The interference between the outer tail wave and adjacent blade wake is gradually strengthened as well as the reflection wave of internal tail wave, but the overall changing trend of wave structure in rotor channel is weak. On the contrary, the wave structure system in high pressure guide vane changes more severe as the pressure ratio rises, and the primary shock wave and expansion wave significantly enhanced with the increasing of pressure ratio.

References

1. Macioce L.E., Schaefer J.W., Saunders N.T. A Status Report on the Energy Efficient Engine Project[R], NASA-TM-81566, (1980).
2. Koop W. The Integrated High Performance Turbine Engine Technology (IHPTET) Program[A]. ISABE **97**-7175, (1997)
3. Kauser, Fazal B.. An Overview of Gas Turbine Propulsion Technology[A]. *AIAA* Paper[C], AIAA-1994-2828. (1994)
4. Maclin H., Krause F.. Propulsion Technology for Future Commercial Aircraft[A]. *AIAA* Paper[C], AIAA-2003-2544. (2003)
5. Lious J.F.. Axial Flow Contra-Rotating Turbines[A]. *ASME* Paper[C], **85**-GT-218, (1985)
6. Hubber F.W., Branstrom B.R., Finke A.K., et al. Design and Test of a Small Two Stage Counter-Rotating Turbine for Rocket Engine Application[A]. *AIAA* Paper[C], AIAA-**93**-2136, (1993)
7. Lior D.,Priampolsky R. Design of Stator-Less Turbine[R]: Becker System Engineering PPT (2004)
8. Shafer D.G., McNelis N.. Development of a Group Based Mach 4+ Revolutionary Turbine Accelerator Technology Demonstrator (RTATD) for Access to Space[A]. ISOABE[C],(2003)
9. Haldeman C.W., Dunn M.G., Abhari R.S., et al. Experimental and Computational Investigation of the Time-Resolved Pressure Loading on a Vaneless Counter-Rotating Turbine[A]. *ASME* Paper[C], 2000-GT-0445. (2000)
10. Weaver M.M., Manwaring S.R., Abhari R.S., et al. Forcing Function Measurements and Preditions of a Transonic Vaneless Counter-rotating Turbines[A]. *ASME* Paper[C], 2000-GT-375 (2000)
11. H S Wang, H Zhang, Q J Zhao, et al. numerical study of the flow field at the top clearance of rotating rotor in a 1+ 1/2 turbine [J]. Journal of Engineering Physics, **28** (5): 751-754 (2007)
12. L Qi, S N Pan, Livy et al. unsteady flow numerical simulation of 1+1/2 Contra-Rotating turbine [J]. Journal of aerospace power, **25** (8): 1825-1835. (2010)

A time-domain substructure synthesis for finite rotations of flexible mechanical systems

Shulin Chen[1], Li Liu[2] and Zhaoyue Chen[3]

[1,2,3]*School of Aerospace Engineering, Beijing Institute of Technology, Zhongguancun South Street 5, 100081 Beijing, China*

Abstract. This paper presents a floating IBS method for dynamic prediction of structures with finite rotations. The conventional IBS (impulse-based substructuring) method is efficient and accurate in structural transient impact analysis. When facing about finite rotations, the conventional IBS method will generate unreasonable results due to the lack of consideration of geometric nonlinearity. In this paper, the idea of floating frame of reference is introduced. The motions of structures are divided into the moving of floating frame of reference and local vibrations which are described using the idea of IBS method. Two numerical examples validate that the proposed method is available for unconstrained single-body system (free-free rotating beam) and constrained multibody system (a slider-crank system). Meanwhile, to some extent the local elastic vibration degrees of freedom can be reduced by employing interpolation matrixes

1 Introduction

The increase of spacecraft's complexity has rose computational costs of dynamics simulation and design optimization enormously. In order to cut down computing time and improve computational efficiency, dynamic substructuring (DS) techniques emerged.

The DS idea was first accomplished by Hurty in 1960 [1], and now DS have been widely employed in many industrial fields. In general, DS can be classified as: component mode synthesis (CMS), frequency-based substructuring (FBS) methods [2], and impulse-based substructuring (IBS) methods [3]. For more about DS techniques, refer to [4]. In 2010, Rixen first proposed the form of IBS and successfully achieved the implementation of the IBS technique using either numerical IRFs or measured IRFs [5]. After this some further development has been studied [6, 7]. IBS has been proven to be effective in dealing with transient impact problems [3], such as the soft landing process of spacecraft. Dong first achieved the implementation of the IBS technique in dynamic analysis of lunar lander and made some improvement which makes it possible to consider rigid and linear or non-linear elastic connections between substructures using IRFs [8, 9].

However, for the dynamic prediction of complex spacecraft, there may be geometric nonlinearity, such as the rotation of the solar wing about its hinges and the finite rotation of the lunar lander when it lands on the inclined plane. When structures make finite rotation, the conventional IBS method will lead to an unreasonable result. In order to overcome this shortcoming, we propose a floating IBS, which is based on the thought of floating frame of reference. When a body undergoes finite rotation, the actual motion of the system can be divided into the rigid motion of a floating body frame and the linear local elastic vibration with respect to the floating body reference [10]. Then the local elastic vibration can be expressed using the idea of IBS method. To reduce elastic degrees and improve the computational efficiency, an interpolation matrix is introduced. Two numerical examples are given to illustrate the proposed method.

2 Flexible multibody system using floating frame of reference

For simplicity, we take a planar analysis to explain how floating frame of reference works. First we discrete the flexible body by finite element method. Then we define three references, as shown in Fig. 1: the inertial reference (OXY), the floating reference (OfXfYf) and the element reference.

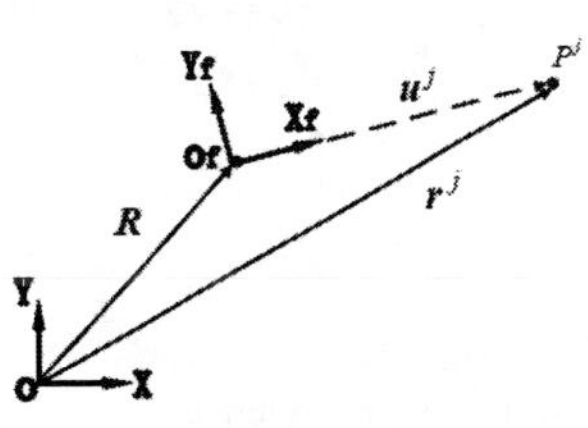

Figure. 1 The illustration of floating frame of reference

The inertial reference is fixed to ground to describe the whole system. The floating reference moves with the flexible body, under which the relative motion of the element node is described. The element reference is defined in finite element discrete process and has a fixed angle with the floating reference [1].

For node j, its displacement can be described in the inertial reference

$$\boldsymbol{r}^j=\begin{bmatrix} x^j & y^j \end{bmatrix}^{\mathrm{T}} \tag{1}$$

and its displacement under the floating reference is

$$\boldsymbol{u}^j=\begin{bmatrix} \bar{x}^j & \bar{y}^j \end{bmatrix}^{\mathrm{T}} \tag{2}$$

Meanwhile, the floating reference's rigid displacement is described as

$$\boldsymbol{R}=\begin{bmatrix} x & y \end{bmatrix}^{\mathrm{T}} \tag{3}$$

For a planar analysis, we have a transform matrix between the floating reference and the inertial reference

$$\boldsymbol{A}=\begin{bmatrix} \cos\theta & -\sin\theta \\ \sin\theta & \cos\theta \end{bmatrix} \tag{4}$$

where θ is rotational angle between floating reference and inertial reference. Then, there is a relationship

$$\boldsymbol{r}^j=\boldsymbol{R}+\boldsymbol{A}\boldsymbol{u}^j \tag{5}$$

As described in finite element frame, $\boldsymbol{u}^j$ can be expressed as

$$\boldsymbol{u}^j=\boldsymbol{D}^j\boldsymbol{S}^j\boldsymbol{B}_1^j\boldsymbol{q} \tag{6}$$

where $\boldsymbol{q}$ is element displacement, $\boldsymbol{S}^j$ is element shape function, $\boldsymbol{B}_1^j$ is a Boolean matrix used to project all generalized displacement to an element, $\boldsymbol{D}^j$ is a transform matrix between the floating reference and the element reference through a rotational angle β^j.

$$\boldsymbol{D}^j=\begin{bmatrix} \cos\beta^j & -\sin\beta^j \\ \sin\beta^j & \cos\beta^j \end{bmatrix} \tag{7}$$

Now we divide element displacement $\boldsymbol{q}$ into initial displacement $\boldsymbol{q}_0$ and elastic displacement $\boldsymbol{q}_f$. To avoid rigid motion, a constrained condition is introduced and there is $\boldsymbol{q}=\boldsymbol{q}_0+\boldsymbol{B}_2\boldsymbol{q}_f$, $\boldsymbol{B}_2$ is a Boolean matrix used to constrain rigid motion. Defining a constant matrix $\boldsymbol{N}^j=\boldsymbol{D}^j\boldsymbol{S}^j\boldsymbol{B}_1^j$, we have

$$\boldsymbol{r}^j=\boldsymbol{R}+\boldsymbol{A}\boldsymbol{u}^j=\boldsymbol{R}+\boldsymbol{A}\boldsymbol{N}^j(\boldsymbol{q}_0+\boldsymbol{B}_2\boldsymbol{q}_f) \tag{8}$$

Taking a derivation of $\boldsymbol{r}^j$ by time, we get

$$\dot{\boldsymbol{r}}^j=\begin{bmatrix} \boldsymbol{I} & \boldsymbol{A}_\theta\boldsymbol{N}^j(\boldsymbol{q}_0+\boldsymbol{B}_2\boldsymbol{q}_f) & \boldsymbol{A}\boldsymbol{N}^j\boldsymbol{B}_2 \end{bmatrix}\dot{\boldsymbol{Q}} \tag{9}$$

where $\dot{\boldsymbol{Q}}$ is the generalized velocity and

$$\dot{A}=\boldsymbol{A}_\theta\dot{\theta},\ \boldsymbol{A}_\theta=\begin{bmatrix} -\sin\theta & -\cos\theta \\ \cos\theta & -\sin\theta \end{bmatrix},\ \dot{\boldsymbol{Q}}=\begin{bmatrix} \dot{\boldsymbol{R}}^{\mathrm{T}} & \dot{\theta} & \dot{\boldsymbol{q}}_f^{\mathrm{T}} \end{bmatrix}^{\mathrm{T}}.$$

Then the kinetic energy of dynamic system can be expressed as

$$T^j=1/2\int \rho^j\dot{\boldsymbol{r}}^{j\mathrm{T}}\dot{\boldsymbol{r}}^j dV^j \tag{10}$$

where ρ is the material density, V is the element volume.

Consider the Lagrange equation

$$\frac{d}{dt}\left(\frac{\partial \boldsymbol{T}^j}{\partial \dot{\boldsymbol{Q}}}\right)^{\mathrm{T}}-\left(\frac{\partial \boldsymbol{T}^j}{\partial \boldsymbol{Q}}\right)^{\mathrm{T}}+\boldsymbol{C}_Q^{\mathrm{T}}\boldsymbol{\lambda}_c=\boldsymbol{F}_g^j \tag{11}$$

where $\boldsymbol{F}_g^j$ is generalized external force, $\boldsymbol{C}_Q$ is the Jacobian matrix of system constraint function, $\boldsymbol{\lambda}_c$ is the Lagrange multiplier for system constraint.

Substituting Eq.(9) and Eq.(10) into Eq.(11), yields Eq.(12), which is the dynamic function under the floating frame of reference. In Eq.(12), ne is the number of element, $\boldsymbol{F}_{out}$ is external force vector. And we define the constants:

$$\boldsymbol{S}_1\equiv\sum_j^{ne}\int_V \rho\boldsymbol{N}^j\mathrm{d}V,\ \boldsymbol{S}_2\equiv\sum_j^{ne}\int_V \rho\boldsymbol{N}^{j\mathrm{T}}\boldsymbol{N}^j\mathrm{d}V,$$

$$\boldsymbol{S}_3\equiv\sum_j^{ne}\int_V \rho\boldsymbol{N}^{j\mathrm{T}}\tilde{\boldsymbol{I}}\boldsymbol{N}^j\mathrm{d}V,\ \boldsymbol{I}_0\equiv\sum_j^{ne}\int_V \rho\boldsymbol{N}^j\boldsymbol{q}_0\mathrm{d}V,$$

$$\boldsymbol{I}_{02}\equiv\sum_j^{ne}\int_V \rho\boldsymbol{q}_0^{\mathrm{T}}\boldsymbol{N}^{j\mathrm{T}}\boldsymbol{N}^j\boldsymbol{q}_0\mathrm{d}V,\ \boldsymbol{I}_{03}\equiv\sum_j^{ne}\int_V \rho\boldsymbol{q}_0^{\mathrm{T}}\boldsymbol{N}^{j\mathrm{T}}\boldsymbol{N}^j\mathrm{d}V,$$

$$\boldsymbol{A}_\theta^T\boldsymbol{A}=\begin{bmatrix} 0 & 1 \\ -1 & 0 \end{bmatrix}\equiv\tilde{\boldsymbol{I}}.$$

$$\begin{bmatrix} \boldsymbol{M}_{RR} & \boldsymbol{A}_\theta\boldsymbol{I}_0+\boldsymbol{A}_\theta\boldsymbol{S}_1\boldsymbol{B}_2\boldsymbol{q}_f & \boldsymbol{A}\boldsymbol{S}_1\boldsymbol{B}_2 \\ & \boldsymbol{I}_{02}+2\boldsymbol{I}_{03}\boldsymbol{B}_2\boldsymbol{q}_f+\boldsymbol{q}_f^{\mathrm{T}}\boldsymbol{B}_2^{\mathrm{T}}\boldsymbol{S}_2\boldsymbol{B}_2\boldsymbol{q}_f & \boldsymbol{I}_{04}+\boldsymbol{q}_f^{\mathrm{T}}\boldsymbol{B}_2^{\mathrm{T}}\boldsymbol{S}_3\boldsymbol{B}_2 \\ \boldsymbol{symmetry} & & \boldsymbol{B}_2^T\boldsymbol{S}_2\boldsymbol{B}_2 \end{bmatrix}\begin{bmatrix} \ddot{\boldsymbol{R}} \\ \ddot{\theta} \\ \ddot{\boldsymbol{q}}_f \end{bmatrix}+\begin{bmatrix} 0 & 0 & 0 \\ 0 & 0 & 0 \\ 0 & 0 & \boldsymbol{K}_{ff} \end{bmatrix}\begin{bmatrix} \boldsymbol{R} \\ \theta \\ \boldsymbol{q}_f \end{bmatrix}+\boldsymbol{C}_Q^{\mathrm{T}}\boldsymbol{\lambda}_c$$
$$=\begin{bmatrix} \dot{\theta}^2\boldsymbol{A}(\boldsymbol{I}_0+\boldsymbol{S}_1\boldsymbol{B}_2\boldsymbol{q}_f)-2\dot{\theta}\boldsymbol{A}_\theta\boldsymbol{S}_1\boldsymbol{B}_2\dot{\boldsymbol{q}}_f \\ -2\dot{\theta}\dot{\boldsymbol{q}}_f^{\mathrm{T}}\boldsymbol{B}_2^{\mathrm{T}}\boldsymbol{S}_2\boldsymbol{B}_2\boldsymbol{q}_f-2\dot{\theta}\dot{\boldsymbol{q}}_f^{\mathrm{T}}\boldsymbol{B}_2^{\mathrm{T}}\boldsymbol{I}_{03}^{\mathrm{T}} \\ \dot{\theta}^2(\boldsymbol{I}_{03}\boldsymbol{B}_2\boldsymbol{I}_f+\boldsymbol{I}_f^{\mathrm{T}}\boldsymbol{B}_2^{\mathrm{T}}\boldsymbol{S}_2\boldsymbol{B}_2\boldsymbol{q}_f)+2\dot{\theta}\boldsymbol{I}_f^{\mathrm{T}}\boldsymbol{B}_2^{\mathrm{T}}\boldsymbol{S}_3\boldsymbol{B}_2\dot{\boldsymbol{q}}_f \end{bmatrix}+\begin{bmatrix} \boldsymbol{I} & \boldsymbol{A}_\theta\boldsymbol{N}^j(\boldsymbol{q}_o+\boldsymbol{B}_2\boldsymbol{q}_f) & \boldsymbol{A}\ \boldsymbol{N}^j\boldsymbol{B}_2 \end{bmatrix}^{\mathrm{T}}\boldsymbol{F}_{out} \tag{12}$$

3 Implementation of ibs algorithm to flexible multibody system

The IBS methods implement the convolution (i.e. the Duhamel's integral) of the IRFs matrix and the applied force vector to represent the dynamic responses of a substructure and use interface compatibility and equilibrium to assemble adjacent substructures. It can be described as

$$\begin{cases} \boldsymbol{u}^{(s)}(t)=\int_0^t \boldsymbol{H}^{(s)}(t-\tau)\left(\boldsymbol{f}^{(s)}(\tau)+\boldsymbol{g}^{(s)}(\tau)\right)\mathrm{d}\tau \\ \boldsymbol{B}\boldsymbol{u}=\sum_{s=1}^{N_s}\boldsymbol{B}^{(s)}\boldsymbol{u}^{(s)}=\boldsymbol{0} \end{cases} \tag{13}$$

where superscript s denotes the serial number of substruc tures, $\boldsymbol{u}^{(s)}$ is the displacement vector of substructure s, $\boldsymbol{H}(t)$ is the impulse response functions (IRFs) matrix, $\boldsymbol{f}(t)$ is the external applied force vector, $\boldsymbol{g}(t)$ is the interf ace force vector, N_s is the number of substructures and $\boldsymbol{B}$ is a signed Boolean matrix identifying the degrees of freed oms (DOFs) on the interfaces. More details refer to [8].

As shown in Eq.(12), $\boldsymbol{q}_f$ is the elastic displacement. Under the floating frame of reference, solving $\boldsymbol{q}_f$ is still a linear elasticity problem. Following the idea of IBS, we use the Duhamel integral

$$\boldsymbol{q}_f=\int_0^t \boldsymbol{H}(t-\tau)\boldsymbol{\lambda}(t)\mathrm{d}\tau \tag{14}$$

where $\boldsymbol{\lambda}(t)$ is the coupling factor of floating reference's rigid motion and local elastic vibration.

First we consider one flexible body system. Then we ignore the system constraint and divide Eq.(12) by floating reference's rigid motion and local elastic vibration

$$\begin{bmatrix} \boldsymbol{M}_{tt} & \boldsymbol{M}_{tf} \\ \boldsymbol{M}_{tf}^T & \boldsymbol{M}_{ff} \end{bmatrix}\begin{bmatrix} \ddot{\boldsymbol{q}}_t \\ \ddot{\boldsymbol{q}}_f \end{bmatrix}+\begin{bmatrix} 0 & 0 \\ 0 & \boldsymbol{K}_{ff} \end{bmatrix}\begin{bmatrix} \boldsymbol{q}_t \\ \boldsymbol{q}_f \end{bmatrix}=\boldsymbol{Q}_v \tag{15}$$

where $\boldsymbol{q}_t=\begin{bmatrix} \boldsymbol{R}^T & \theta \end{bmatrix}^T$ and $\boldsymbol{Q}_v$ stands for right part of Eq.(12).

Write Eq.(15) in a difference scheme

$$\boldsymbol{M}_{n+1}\begin{bmatrix} \ddot{\boldsymbol{q}}_t \\ \ddot{\boldsymbol{q}}_f \end{bmatrix}_{n+1}+\boldsymbol{K}_{n+1}\begin{bmatrix} \boldsymbol{q}_t \\ \boldsymbol{q}_f \end{bmatrix}_{n+1}=\boldsymbol{Q}_{vn+1} \tag{16}$$

where

$$\boldsymbol{q}_{fn+1}=\underset{\sim}{\boldsymbol{q}}+\boldsymbol{H}_1\boldsymbol{\lambda}_{n+1}\mathrm{d}t/2 \tag{17}$$

subscript n means that the variable is at $n\mathrm{d}t$ time, $\mathrm{d}t$ is the time step and

$$\underset{\sim}{\boldsymbol{q}}=\sum_{i=0}^{n-1}\boldsymbol{H}_{n-i}(\boldsymbol{\lambda}_i+\boldsymbol{\lambda}_{i+1})\mathrm{d}t/2+\boldsymbol{H}_1\boldsymbol{\lambda}_n\mathrm{d}t/2 \tag{18}$$

Introduce the Newmark integration strategy

$$\begin{cases} \ddot{\boldsymbol{q}}_{n+1}=\dfrac{1}{\beta\mathrm{d}t^2}\boldsymbol{q}_{n+1}+\ddot{\boldsymbol{Q}}_n^{\mathrm{int}} \\ \dot{\boldsymbol{q}}_{n+1}=\dfrac{\gamma}{\beta\mathrm{d}t}\boldsymbol{q}_{n+1}+\dot{\boldsymbol{Q}}_n^{\mathrm{int}} \end{cases} \tag{19}$$

where

$$\ddot{\boldsymbol{Q}}_n^{\mathrm{int}}=-\frac{1}{\beta\mathrm{d}t^2}\boldsymbol{q}_n-\frac{1}{\beta\mathrm{d}t}\dot{\boldsymbol{q}}_n-(\frac{1}{2\beta}-1)\ddot{\boldsymbol{q}}_n\ ,$$

$$\dot{\boldsymbol{Q}}_n^{\mathrm{int}}=-\boldsymbol{q}_n\frac{\gamma}{\beta\mathrm{d}t}+(1-\frac{\gamma}{\beta})\dot{\boldsymbol{q}}_n+(1-\frac{\gamma}{2\beta})\ddot{\boldsymbol{q}}_n\mathrm{d}t\ .$$

Substituting Eq.(19) and Eq.(17) into Eq.(16), yields

$$\begin{bmatrix} \boldsymbol{M}_{tt} & \boldsymbol{M}_{tf}\boldsymbol{H}_1\mathrm{d}t/2 \\ \boldsymbol{M}_{tf}^T & (\boldsymbol{M}_{ff}+\beta\mathrm{d}t^2\boldsymbol{K}_{ff})\boldsymbol{H}_1\mathrm{d}t/2 \end{bmatrix}\begin{bmatrix} \boldsymbol{q}_t \\ \boldsymbol{\lambda} \end{bmatrix}_{n+1} = (\boldsymbol{Q}_{vn+1}-\boldsymbol{M}_{n+1}\ddot{\boldsymbol{Q}}_n^{\mathrm{int}})\beta\mathrm{d}t^2-\begin{bmatrix} \boldsymbol{M}_{tf}\underset{\sim}{\boldsymbol{q}} \\ (\boldsymbol{M}_{ff}+\beta\mathrm{d}t^2\boldsymbol{K}_{ff})\underset{\sim}{\boldsymbol{q}} \end{bmatrix} \tag{20}$$

Solving Eq.(20) by Newton-Raphson method, we get $\begin{bmatrix} \boldsymbol{q}_t & \boldsymbol{\lambda} \end{bmatrix}_{n+1}^{\mathrm{T}}$. Then $\boldsymbol{q}_{fn+1}$ is obtained by substituting $\boldsymbol{\lambda}_{n+1}$ into Eq.(17).

In order to reduce the elastic degrees, interpolation matrix $\mathbf{T}$ can be introduced

$$\boldsymbol{q}_f=\mathbf{T}\bar{\boldsymbol{q}}_f \tag{21}$$

Eq.(14) now changes into

$$\boldsymbol{q}_f=\mathbf{T}\int_0^t \bar{\boldsymbol{H}}(t-\tau)\bar{\boldsymbol{\lambda}}(t)\mathrm{d}\tau \tag{22}$$

Because of the introduction of $\mathbf{T}$, the dimension of the impulse response function matrix and coupling factor reduces. In this paper we use the Lagrange interpolation function as the interpolation matrix, nevertheless other interpolation function can be used.

For multibody system, the system constraint between adjacent bodies should be considered.

$$\boldsymbol{C}(\boldsymbol{Q},t)=\mathbf{0} \tag{23}$$

where $\boldsymbol{C}$ is the linearly independent constraint equations.

Derivation of Eq.(23) by time once and twice separately yields

$$\boldsymbol{C}_Q\dot{\boldsymbol{Q}}+\boldsymbol{C}_t=\mathbf{0} \tag{24}$$

and

$$\boldsymbol{C}_Q\ddot{\boldsymbol{Q}}+\boldsymbol{C}_{tt}+2\boldsymbol{C}_{Qt}\dot{\boldsymbol{Q}}+\left(\boldsymbol{C}_Q\dot{\boldsymbol{Q}}\right)_Q\dot{\boldsymbol{Q}}=\mathbf{0} \tag{25}$$

where the subscript $\boldsymbol{Q}$ and t separately stand for the partial derivation by generalized displacement vector $\boldsymbol{Q}$ and time t, thus $\boldsymbol{C}_Q$ is the Jacobian matrix of system constraint function.

Then combine with Eq.(15), the constrained system functions can be written as

$$\begin{bmatrix} \boldsymbol{M} & \boldsymbol{C}_Q^{\mathrm{T}} \\ \boldsymbol{C}_Q & 0 \end{bmatrix}\begin{bmatrix} \ddot{\boldsymbol{Q}} \\ \boldsymbol{\lambda}_c \end{bmatrix}=\begin{bmatrix} \boldsymbol{Q}_v-\boldsymbol{K}\boldsymbol{Q} \\ \boldsymbol{\tau} \end{bmatrix} \tag{26}$$

where $\boldsymbol{\tau}=-\boldsymbol{C}_{tt}-2\boldsymbol{C}_{Qt}\dot{\boldsymbol{Q}}-\left(\boldsymbol{C}_Q\dot{\boldsymbol{Q}}\right)_Q\dot{\boldsymbol{Q}}$.

Similar to the above steps substitute Eq.(19) and Eq.(17) into Eq.(26), we get the formula about $\begin{bmatrix} \boldsymbol{q}_t^{\mathrm{T}} & \boldsymbol{\lambda}^{\mathrm{T}} & \boldsymbol{\lambda}_c^{\mathrm{T}} \end{bmatrix}_{n+1}^{\mathrm{T}}$

$$\begin{bmatrix} \boldsymbol{M}_{tt} & \boldsymbol{M}_{tf}\boldsymbol{H}_1\mathrm{d}t/2 & \beta\mathrm{d}t^2\boldsymbol{C}_1^{\mathrm{T}} \\ \boldsymbol{M}_{tf}^T & (\boldsymbol{M}_{ff}+\beta\mathrm{d}t^2\boldsymbol{K}_{ff})\boldsymbol{H}_1 dt/2 & \beta\mathrm{d}t^2\boldsymbol{C}_2^{\mathrm{T}} \\ \beta\mathrm{d}t^2\boldsymbol{C}_1 & \beta\mathrm{d}t^2\boldsymbol{C}_2 & 0 \end{bmatrix}_{n+1}\begin{bmatrix} \boldsymbol{q}_t \\ \boldsymbol{\lambda} \\ \boldsymbol{\lambda}_c \end{bmatrix}_{n+1} = \begin{bmatrix} (\boldsymbol{Q}_{vn+1}-\boldsymbol{M}_{n+1}\ddot{\boldsymbol{Q}}_n^{\mathrm{int}})\beta\mathrm{d}t^2-\boldsymbol{M}_q \\ \beta\mathrm{d}t^2\boldsymbol{\tau}_{n+1} \end{bmatrix} \tag{27}$$

there

$\boldsymbol{C}_Q=\begin{bmatrix} \boldsymbol{C}_1 & \boldsymbol{C}_2 \end{bmatrix}$, $\boldsymbol{M}_q=\begin{bmatrix} (\boldsymbol{M}_{tf}\underset{\sim}{\boldsymbol{q}})^{\mathrm{T}} & ((\boldsymbol{M}_{ff}+\beta\mathrm{d}t^2\boldsymbol{K}_{ff})\underset{\sim}{\boldsymbol{q}})^{\mathrm{T}} \end{bmatrix}^{\mathrm{T}}$.

Employing Newton iterative strategy, the solution of Eq.(27) can be obtained.

4 Numerical examples

To validate the method proposed above, three numerical examples are presented in this section. The first example is a free-free rotating beam which illustrates the IBS method can be use in a flexible rotational system with one body. In this example the Lagrange interpolation function was used to reduce the elastic degrees. In the second example, a slider-crank system with two rotating beams validates this method for flexible multibody system.

4.1 One free-free rotating beam

The simple system is a free-free rotating beam. The whole structure is divided into 41 elements by finite element methods (FEM) and it contains 123 DOFs totally. The

 https://doi.org/10.1051/matecconf/201815102006

one-meter-long beam has a 5mm-radius circular cross section with Young's modulus 70Gpa and density 2.78g/cm3. It was excited by an impact force at the end, as shown in Figure. 2.

Figure. 2 One free-free rotating beam

If geometric nonlinearity is not considered in this example, the system can be described under linear FEM frame, thus

$$M^e\ddot{x}+C^e\dot{x}+K^e x=F^e \tag{28}$$

where M^e, C^e, K^e are separately mass matrix, damper matrix, stiffness matrix and they are all constant matrix. x is element displacement vector and F^e is the external force vector. Using standard Newmark integration strategy shown in Eq.(19), it is easy to obtain the dynamic motion of this system. The motion of the beam during 0 to 0.8s is shown in Figure. 3

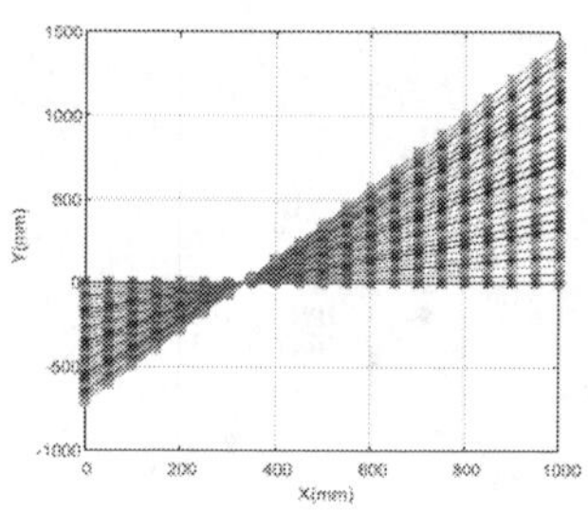

Figure. 3 The configuration of free beam's motion without consideration of geometric nonlinearity

Obviously, the result shown in Figure. 3 is not reasonable. Only displacement in Y direction is excited and the displacement in X direction is nearly zero. As a result, the beam seems to be stretched. The reason for this false phenomenon is the assumption of constant mass and stiffness matrix.

By introducing floating frame of reference and employing Eq.(20), we get a modified result and the motion of the beam during 0 to 0.8s is illustrated in Figure. 4. This result is quite different from the above and is much more reasonable.

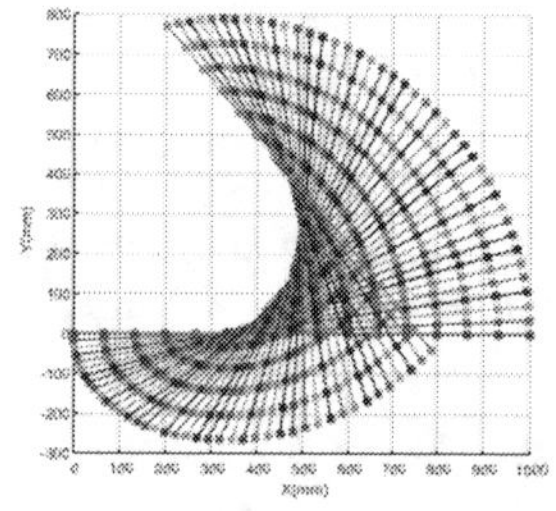

Figure. 4 The configuration of free beam's motion using floating frame of reference

To demonstrate the proposed method, in this example the solution of standard dynamic FE function under floating frame of reference is introduced as a benchmark which is described in Eq.(15).

To implement the IBS algorithm described in Eq.(20), the IRFs matrix $H(t)$ should be obtained. As mentioned above, the motion of this beam can be divided into the moving of floating frame of reference (rigid motion) and local elastic vibration (flexible motion). In this example, the floating frame of reference is set at the position of one end of this beam and we assume that the structure is constrained by a pinned end and a fixed end shown in Figure. 5. Then the IRFs matrix of elastic vibration is achieved through this assumed constrained structure using FE method with Newmark integration strategy.

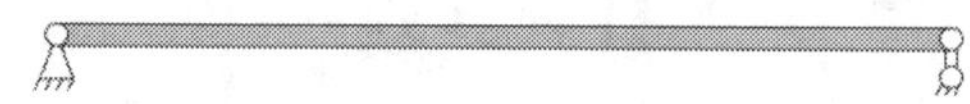

Figure. 5 Constrained structure with a pinned end and a fixed end

The external force applied to this system is

$$F_{ext}=\begin{cases}1000t & 0<t\le 0.01\\ 20-1000t & 0.01<t\le 0.02\\ 0 & 0.02<t\end{cases} \tag{29}$$

The velocity and acceleration of the end point in Y direction is presented in Figure. 6

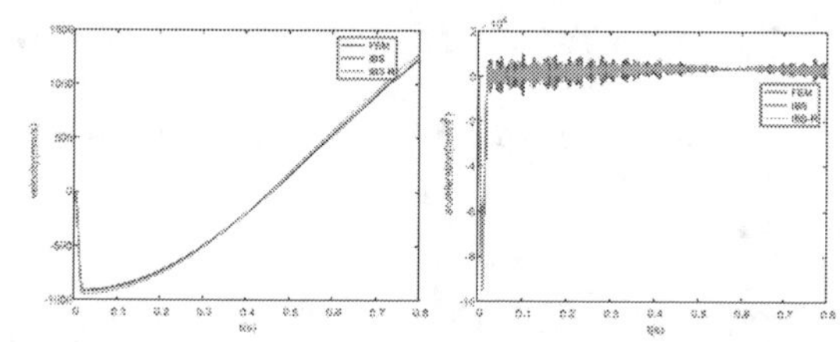

Figure. 6 Results between floating IBS, reduced floating IBS and traditional method

The results of floating IBS meet a great agreement with that of traditional method labeled as FEM in Figure. 6. With reduction of flexible DOFs, there are some errors produced between results of reduced floating IBS and traditional method. To present the accuracy quantitatively, R square(R2) and relative average absolute error(RAAE) are introduced and listed in Table 1. The large R2 and the smaller the RAAE, the more accurate results we obtain.

Table 1 Accuracy of floating IBS and reduced floating IBS compared with traditional method

	floating IBS		reduced floating IBS	
	R2	RAAE	R2	RAAE
V	1.00000	3.09e-05	0.99835	0.03564
A	0.99842	0.02808	0.88014	0.24358

The velocity and acceleration of the end point in frequency domain by Fourier transform is presented in

Figure. 7. Over all, the results are acceptable and this method performs well at low frequency.

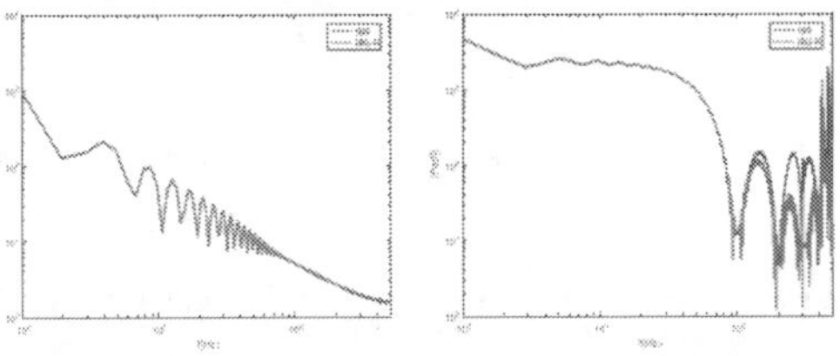

Figure. 7 Results in frequency domain between floating IBS and reduced floating IBS

4.2 A slider-crank system with two rotating beams

To validate the proposed method for multibody system, we introduce a slider-crank system. It consists of two beams and a mass block. They are connected by hinges and shown in Figure. 8.

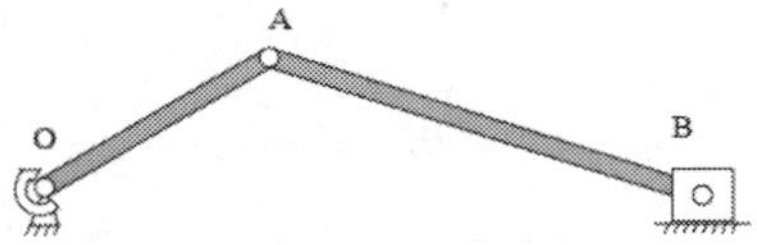

Figure. 8 A slider-crank system

In this system we assume beam OA has an initial angular velocity 25rad/s, mass block B has only one degree of freedom and they are connected by beam AB with hinges. The two beams have a 3mm-radius circular cross section with Young's modulus 200Gpa and density 7.87g/cm3. Beam OA and AB separately have a length of 0.15m and 0.3m. The mass block B has a same mass with beam OA. To make it simple, we ignore the flexibility of beam OA. The finite element model of beam AB is based on Euler-Bernoulli beam theory and the IRFs matrix used in Eq.(27) is obtained by this FE model. When calculating IRFs matrix, beam AB is assumed under the same constrained condition shown in Figure. 5. The system constrained conditions can be described as

$$\begin{cases} \boldsymbol{q}_{oX}^{oA} = 0, \ \boldsymbol{q}_{oY}^{oA} = 0 \\ \boldsymbol{q}_{AX}^{oA} - \boldsymbol{q}_{AX}^{AB} = 0, \ \boldsymbol{q}_{AY}^{oA} - \boldsymbol{q}_{AY}^{AB} = 0 \\ \boldsymbol{q}_{BY}^{oA} = 0 \end{cases} \tag{30}$$

where $\boldsymbol{q}$ is the coordinate in the inertial reference, X and Y stand for the directions of $\boldsymbol{q}$, the superscript and subscript separately stand for the part and the point the coordinates belong to. Using generalized coordinates to express the Eq.(30), we get system constraint equations $\boldsymbol{C}$. Then the dynamic response is obtained by solving Eq.(27).

The angular displacement and angular velocity of beam OA are illustrated in Figure. 9. The results from floating IBS have a great agreement with that of traditional method. It demonstrates that this floating IBS is capable for multibody system with constraint equations.

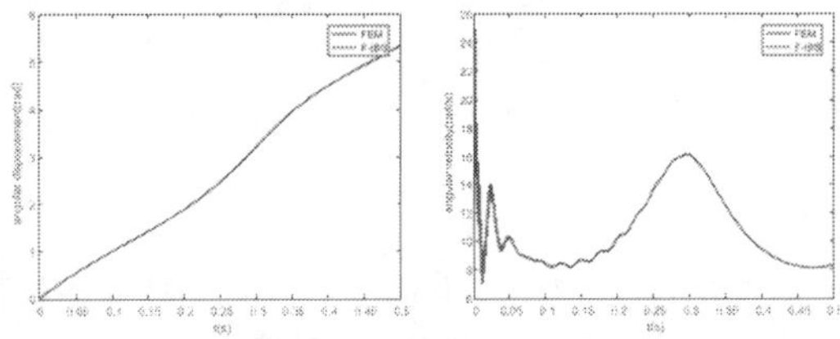

Figure. 9 Results of slider-crank system between floating IBS and traditional method

5 Summary

The proposed method extended the conventional IBS method. The introduction of the idea of floating frame of reference makes it possible to deal with the finite rotations of structures. The numerical examples demonstrate that this method can handle both unconstrained single-body system and constrained multibody system. With the employ of interpolation function, the local elastic DOFs are reduced. To some extent, this reduction of elastic DOFs will lead to some loss of simulating accuracy. In the mass, the results are acceptable and this method performs well at low frequency.

References

1. W. C. Hurty, Journal of the Engineering Mechanics Division **86** (4), 51-70 (1960).
2. B. Jetmundsen, R. L. Bielawa and W. G. Flannelly, Journal of the American Helicopter Society **33** (1), 55-64 (1988).
3. D. J. Rixen and P. L. C. van der Valk, Journal Of Sound And Vibration **332** (26), 7174-7190 (2013).
4. D. de Klerk, D. J. Rixen and S. N. Voormeeren, Aiaa Journal **46** (5), 1169-1181 (2008).
5. D. J. Rixen, Proceedings Of Isma2010 - International Conference on Noise And Vibration Engineering Including Usd2010, 1939-1954 (2010).
6. D. Rixen and N. Haghighat, presented at the 30th IMAC, Jacksonville, FL, United states, (2012).
7. M. V. van der Seijs and D. J. Rixen, Proceedings Of International Conference on Noise And Vibration Engineering (Isma2012) / International Conference on Uncertainty In Structural Dynamics (Usd2012), 3487-3499 (2012).
8. W.-L. Dong, L. Liu, S.-D. Zhou and S.-L. Chen, Aiaa Journal **53** (2), 504-509 (2015).
9. S.-D. Zhou, L. Liu and W.-L. Dong, AIAA Journal **54** (8), 2545-2551 (2016).
10. A. Shabana, *Dynamics of multibody systems*. (Cambridge University Press, 2005).

Analysis of the Problems Existing in Mechanical Structure Design of The Legged Mobile Robot

MingJuan Xie[1], ChunYan Zhang[1] and XingLu Yin[1]

[1] *College of Mechanical Engineering, Shanghai University of Engineering Science ,China*

Abstract. Based on the analysis of the research background and current situation of the legged mobile robots at home and abroad,in this paper we discusses the existing research methods of the legged mobile robot in the mechanical structure design; discusses the mainproblems in the design of the mobile robot, and to focus on the emergence of instability and poor adaptability and obstacle-surmounting performance of the legged mobile robot to find the corresponding solution in the process of movement.

1 Introduction

There are always some special or dangerous environments in nature, so human beings develop and use the mobile robots to engage in some dangerous environment operations. Compared with the wheeled and crawler robot, the legged mobile robot has its unique superior performance [1]: The foothold is the discrete point, which can make the robot choose the optimal support on the ground; with multi degree of freedom and adjustable leg length, the legged mobile robot keeps a stable center of gravity and balance. Therefore, in recent years, many researchers have done much research on the legged mobile robot.

However, the legged mobile robots have their own shortcomings. For example, its walking efficiency is lower and cause rollover easily due to gravity; Encountering extreme terrain will occur foot stuck, so that the entire robot lost "life." The main reason for these problems is the failure of the mechanical structure of the legged mobile robots. Therefore, the reasonable mechanical structure is the key to ensure the sufficient stability and adaptability the full function of the legged mobile robot.

2 Research status of the legged mobile robot

The development of the legged mobile robots begins with bipedal walking robots. In 1968, the British modeled a bipedal mobile robot [2,3], the robot only has two ankles and a hip joint, the operator keeps the balance of the robot through the force feedback, but its stability and obstacle-surmounting performance are poor, so scholars at home and abroad have carried out the research on other legged mobile robot.

2.1 One-legged mobile robot

As shown in figure 2.1, the MIT Robot Lab produced a one-legged robot that showed jump and balance in three-dimensional space in 1983,. In the process of movement, the robot achieved the stability of the jumping robot by connecting the pneumatic jumping leg and the hydraulic cylinder between the links. The mechanical structure is relatively simple and has a certain impact resistance ability [4,5]. A planar jump bow-legged robot developed by Carnegie Mellon University in 1999 shown in figure 2.2. It used the bow spring leg instead of pneumatic jumping leg and optimized the structure design to reduce the quality of robot and improve the robot's adaptive [6].

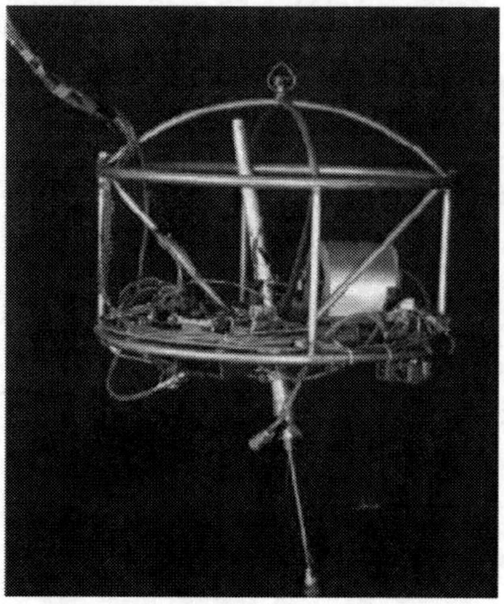

Figure 2.1. Free jumping one-legged robot physical map [4]

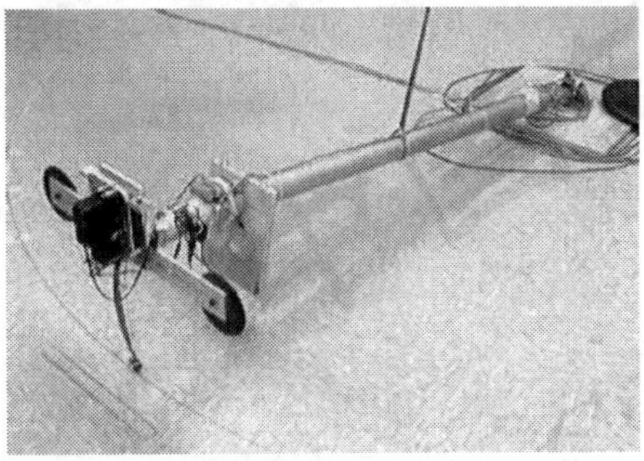

Figure 2.2. Plane jump prototype [4]

The mechanical structure of the one-legged mobile robot is relatively simple. The telescopic jump of leg can be achieved by pneumatic jumping leg or bow spring leg, and the self adaptability of the one-legged robot can be improved by adopting the bow spring leg. These mechanical structures enable the one-legged mobile robot to achieve much mechanical performance, but still have limitations in practical applications.

2.2 Biped mobile robot

In 1973, Kato Ichiro, of Waseda University, in Japan, developed the robot shown in figure 2.3. The robot's main body adopted multi-link mechanism to achieve the robot's static walking. Although it has the most basic components of the biped robot, but the obstacle-surmounting performance and self adaptability could not be satisfied [7].

Figure 2.3. Japan Kato Ichiro bipedal robot WABOT-1 [7]

At present, there are many researches on the biped mobile robot in China. For example, Fan Shuncheng and others have analyzed the stability of the biped robot parallel leg mechanism shown in figure 2.4, each leg of the robot adopted 6-SPS parallel mechanism through the rotation and expansion of each branch and the swing and stretching movements of the legs are completed, which made the robot has good mobility and moves stably [8]. Xu Linsen and others made a study of biped robot walking on water mechanism and promoting mechanism, the propulsion system is the combination of Watt-I mechanism shown in figure 2.5 and the four-bar mechanism, the robot can move stably when the force on the water surface of the robot's sole does not exceed 3N [9].

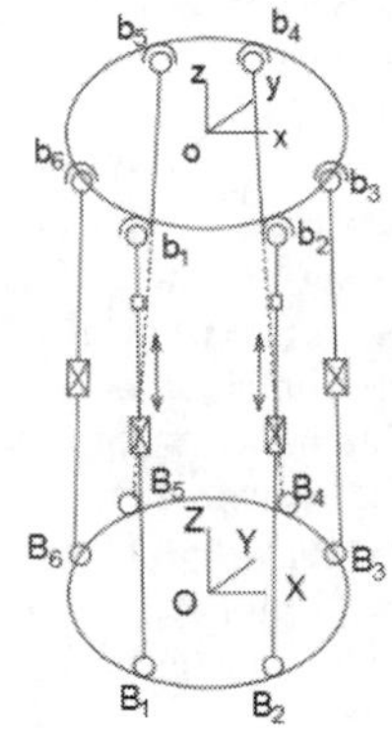

Figure 2.4. Robotic leg parallel mechanism

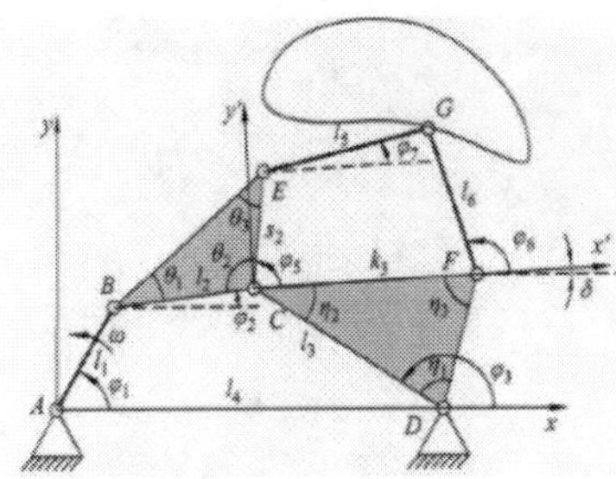

Figure 2.5. Watt-I plane mechanism [9]

Under certain conditions, these institutions can achieve the stability of the biped mobile robot, so it is necessary to consider how to enhance the stability of the robot in a wider range.

2.3 Quadruped mobile robot

The research of the quadruped mobile robot came into develop widely until 1980s.BigDog quadruped robot developed by the United States shown in figure 2.6.The robot was a combination of hydraulic actuator and quadruped mechanism on the whole. Each leg of the quadruped mechanism adopted a linkage mechanism, an elastic mechanism and a spring shock absorber, etc, so as to improve the flexibility and the obstacle-surmounting performance of the robot [10,11].

Figure 2.6. BigDog robot [11]

In 1991, Chen Hao and others in China's Beijing University of Aeronautics and Astronautics studied on quadruped structure of the robot shown in figure 2.7, the robot's body is a pair of cross-linked structure, and the legs can be seen as a series of institutions, the middle of the two links are hinged to each other. The front end is fixed, and the rear end is respectively connected to the left and right bodies through a screw nut, which made the robot's body become wider or narrower, and improve the adaptability and obstacle-surmounting performance of the robot by controlling the nut of the screw [12].Wang Meng and others of Harbin Institute of Technology designed the frog-inspired bionic jumping robot shown in figure 2.8. The hind legs of the robot used the five-bar mechanism as the main body instead of the frog's leg multi-joint series structure, the spring instead of the muscle as the energy storage original, which make sure the stability and the obstacle-surmounting performance in the process of movement [13].

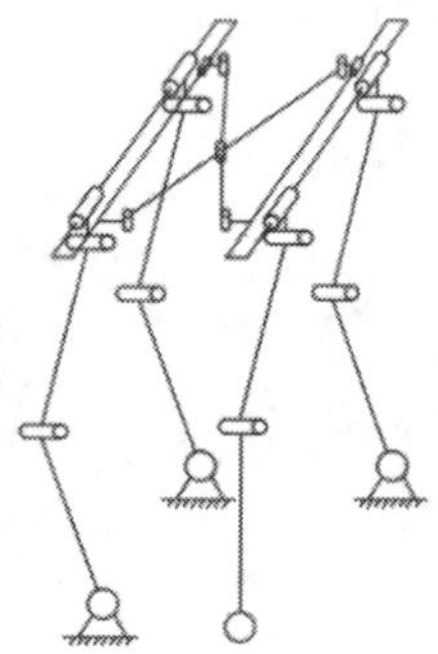

Figure 2.7. Sketch of robot mechanism

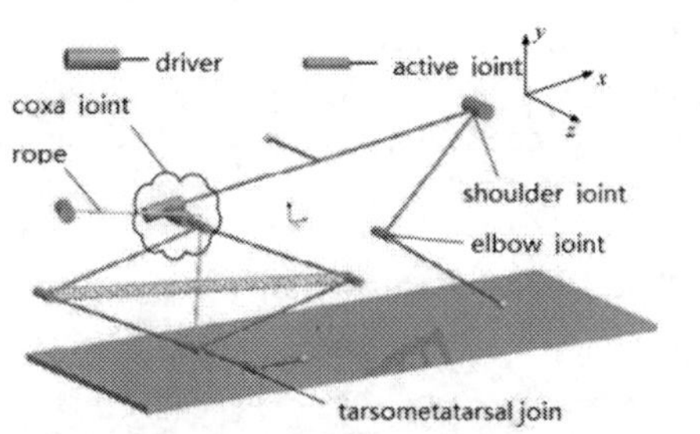

Figure 2.8. The mechanical model of frog [13]

Generally speaking, the quadruped mobile robot is superior to the biped robot in its bearing capacity and stability, and its practicability is relatively high and its technology is relatively mature.

2.4 Six-legged mobile robot

As shown in figure 2.9, Wang Yuefeng and others discussed the mechanical design of pantograph leg mechanism of a six-legged walking machine in 1991. The locomotion mechanism of the robot adopted a telescopic mechanism, and optimized the structure of the mechanism to enhance the stability and enlarge the movement space in the process of movement, but the obstacle-surmounting performance is poor [14]. Six-legged mobile robot has certain stability in the process of movement, but it also has many problems, such as lower flexibility or obstacle-surmounting performance or adaptability.

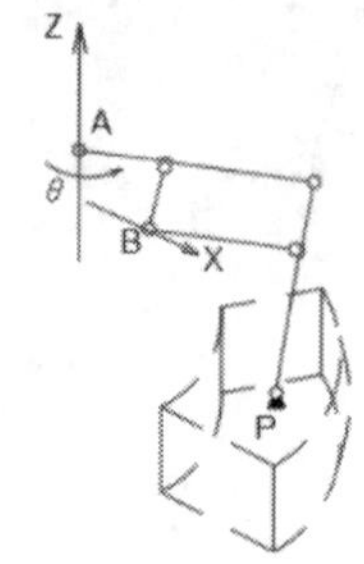

Figure 2.9. Sketch of robot leg scaling mechanism

2.5 Eight-legged mobile robot

Most of the eight-legged mobile robots evolved from bionic robots. At present, there are few researches on eight-legged robots at home and abroad. This paper introduces a new eight-legged gait bionic robot shown in figure 2.10, the robot shares one motor for each pair of feet, the foot movement was caused by the gear driving the disk to rotate, and optimized the leg structure design, which can improve the obstacle-surmounting performance and self adaptability [15]. As shown in figure 2.11, the trajectory of the D point is not only related to the length of each rod, but also to the angle of each member. The disadvantage of this mechanism is that each pair of joints on the foot has a strong coupling, which limits the flexibility of the mechanism and reduces the obstacle-surmounting performance.

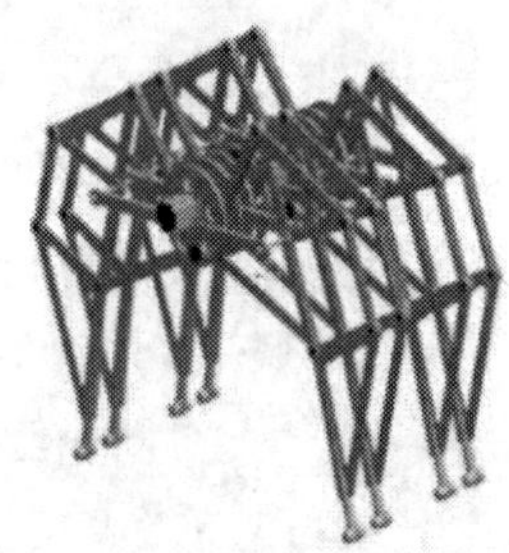

Figure 2.10. The overall model of the eight-legged bionic walking robot [15]

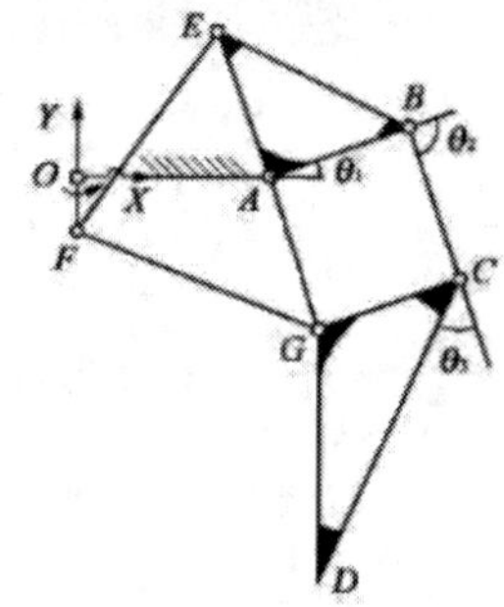

Figure 2.11. Schematic diagram of the legs [15]

From the eight-legged mobile robot, it can be seen that the complexity of the mechanical structure will affect the flexibility and obstacle-surmounting performance of the robot.

To sum up, with the increase in the number of foot, the stability of the legged mobile robot will be improved gradually, but the mechanism has gradually become very complicate, which lead to cause some unfavorable factors, such as volume and weight increase may lead to lower flexibility of the mechanism and poor obstacle-surmounting performance. A single linkage mechanism can achieve some mechanical function, but it may not be as good as the use of the combination with a variety of mechanisms, which is beneficial to improve the overall mechanical performance of the robot. Therefore, on the basis of the existing research about the legged mobile robot in the world, aiming at summarizing the solutions of the problems in the mechanical structure design.

3 The solution to the problems in the mechanism design of the legged mobile robot

Many researches have been done on the key technologies of the legged mobile robots since the 1960s, and some breakthrougsent, hs have been made in some basic theoretical problems, so that the technology of the legged mobile robot has been improved continuously [16]. At preit has already formed some main research methods and theory of the legged mobile robot in the mechanical structure. After applying these research methods to practice, the stability and obstacle-surmounting performance of the robot will be improved, but they still have some problems [1]. Here are some suitable solutions of the problems that exist in the mechanical structure of legged mobile robots:

3.1 In the process of movement,the legged mobile robot causes rollover or instability due to the center of gravity.

In one-legged and biped mobile robots, it is common for a mobile robot to cause rollover during driving because of the center of gravity.

3.1.1 The use of parallel mechanism to solve the problem of rollover or instability in the process of movement.

The parallel mechanism has the characteristics of high rigidity, high precision and easy to control in real time. According to the characteristics of the parallel mechanism, the leg structure of robot can adopt 6-SPS parallel mechanism, which can complete the swing and curve movement of the leg and keep the center of gravity in the stable walking area through the rotation and expansion of each branch chain to solve the problem of rollover. Therefore, parallel mechanism is one of the ways to solve the problem of rollover.

3.1.2 A combination of various mechanisms is adopted to solve the problem of rollover or instability in the process of movement.

As shown in Fig.3.1, the four-bar mechanism, the crank-rocker mechanism and the cam mechanism are used as the locomotion mechanism of the quadruped mobile robot, in which the four-bar mechanism functions as a final action, and the crank rocker mechanism is used to transmit the leg power, Cam mechanism is used to achieve the coordination of the legs of the quadruped mobile robot, thus the combination of these three mechanisms can achieve the stability of the robot [17]. Therefore, in the four-bar mechanism, pantograph mechanism, parallel mechanism and multi joint series mechanism of structure, it can be used with two or more than two simple mechanism to solve the problems of rollover or instability in the process of movement.

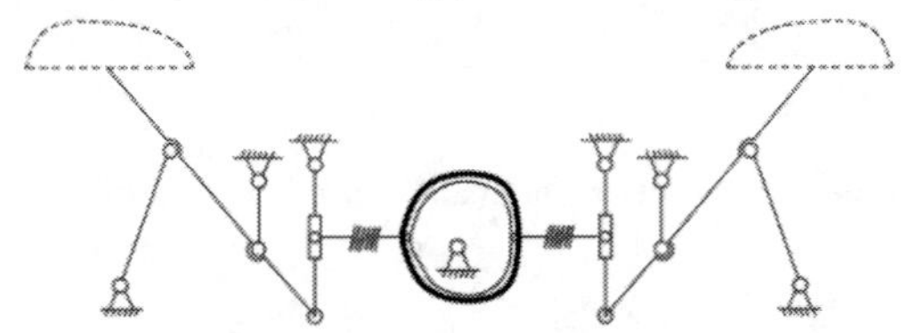

Figure 3.1. Robot half body setback mechanism

3.2 The interference or stuck of the mechanical structures of the legged mobile robot in the process of movement.

In the multi-legged mobile robot, with the increase of the foot, the robot's mechanism becomes complicate. At this time, it is not only necessary to optimize the design of the length of the link and the interrelation in each leg structure, but also to optimize the design of the relationship between the legs. Or it can also solve the interference or stuck problems of the mechanism by using structural singularity in reason.

3.3 The obstacle-surmounting performance and adaptability of the legged mobile robot in the process of movement are low.

Compared with the one-legged mobile robot, due to the complexity of the mechanical structure of the multi-legged mobile robot, the obstacle-surmounting performance is lower than that of one-legged mobile robot. One-legged mobile robot generally uses a similar spring leg structure for obstacle crossing; BigDog robot also has an elastic mechanism and spring shock absorber and other structures to improve the obstacle-surmounting performance of the legged mobile robot. Therefore, an elastic mechanism may be the way to solve the problem of obstacle-surmounting performance.

As shown in Fig.3.2, the spatial pantograph mechanism acts as a branch of the parallel mechanism to control the lifting of the upper platform of the parallel mechanism, so that the mechanism can enlarge or reduce the operating space relatively [18].The most obvious feature of the pantograph mechanism is that it can reduce the space, and it can improve the obstacle-surmounting performance and adaptability of the mobile mechanism and achieve the overall deformation of the mobile mechanism by using the expansion and the shrinkage characteristic of the pantograph mechanism. Therefore, the pantograph mechanism can solve the problem of poor obstacle-surmounting performance and adaptability.

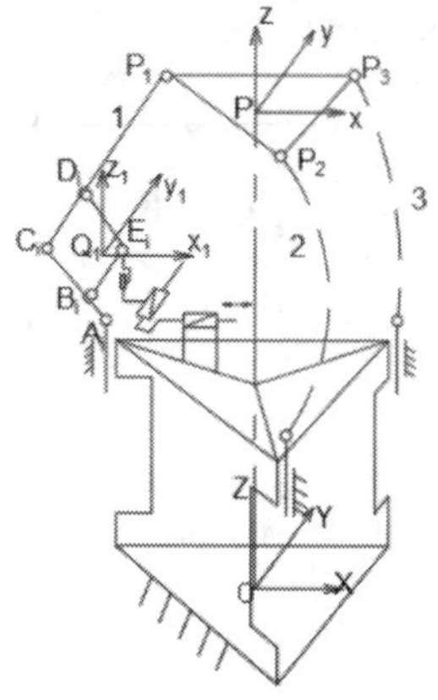

Figure 3.2. A pantograph parallel platform mechanism

4 Conclusions

In summary, based on the analysis of the situation and background of legged mobile robots at home and abroad, this paper analyzes and discusses the various mechanical structures of the legged mobile robots, and put forward the corresponding solution to the main problems in the design of mechanical structure. In the process of movement, the legged mobile robot can adopt the parallel mechanism or the combination of various kinds of institutions to solve the problem of rollover or instability; It can also solve the interference or stuck problems of the mechanism by optimizing the mechanical structure design and using structural singularity in reason; It can also use elastic mechanism or pantograph mechanism to solve the problem of low obstacle-surmounting performance and adaptability.

References

1. Liu Jing, Zhao Xiaoguang, Tan Ming. Legged robots A Review [J]. ROBOT, 2006,28 (1): 81-88.
2. Liu Zhiyuan. Research on Dynamic Walking of Biped Robot [D]. Harbin Institute of Technology, 1991.
3. McGhee.R.B.Robot locomotion[A].In R.Herman, S.Grillner,P.Stein,and D.Smart,editors[C].Plenum Press,1976:237-264.
4. Su Xing. Research on Dynamic Characteristics of a One-legged Hopping Robot Based on Pactuated Leg in Vertical Direction [D]. Harbin Institute of Technology, 2013.
5. Raibert.M.H,Brown.H.B,Jr.Chepponis.M.Experiment in balance with a 3D one legged hopping machine.International J.Robotics Research. 1984(3) : 75-92.
6. Sun Zhichao. Modeling and Simulation of One-legged Hopping Robot Using Multi-body Dynamics[D]. Harbin Institute of Technology, 2013.
7. Chen Xiaogang, Chou Zhongchen, Guan Jialiang. Development status and Prospects of Biped Robot [J]. Mechanical Engineer, 2007 (02): 17-19.
8. Fan Shuncheng, Han Shukui, Gu Huiru. The Stability of Biped Walking Robot Parallel Leg Mechanism[J].Journal of Hebei University of Technology.2004,33. (01) 27-31.
9. Xu Linsen, Mei Tao, Luo Minzhou, Huan Juan, Wei Xianming, Cao Kai. Research on Dynamical Mechanism and Propulsion System of a Biped Robot Walking on Water [J]. ROBOT, 2013,35 (03) 257-262.
10. Raibert M,Blankespoor K,Nelson G,et al.BigDog,the roughterrain quadruped robot //Proceedings of the 17th International Federation of Automation Control.Seoul,2008:10822.
11. Ding Lianghong, Wang Runxiao, Feng Huashan, Li Jun.Brief Analysis of a BigDog Quadruped Robot [J]. China Mechanical Engineering, 2012,23 (05): 505-514.
12. Chen Hao, Ding Xilun. Kinematics analysis of a quadruped robot with variable structure [J]. Journal of University of Science and Technology Beijing, 2012,34 (01): 53-58.
13. Wang Meng. Research on Frog-inspired Biomimetic Jumping Robot [D]. Harbin Institute of Technology.2009.
14. Wang Yuefeng, Gan Jianguo,Gan Dongying. Mechanical Design of Pantograph Legged Mechanism for a Six-legged Walking Robot [J]. Optics and Precision Engineering.1991,03:29-31.
15. Zang Hongbin, Tao Junjie. Developed of the New Eight-legged Walking Bionic Robot [J]. Journal of Mechanical Transmission.2015,39 (09): 181-185.
16. Zhu Daohong. Research of Walking Attitude Tracking Control Method of Biped Robot [D]. Xi'an University of Science And Technology, 2001.
17. Wang Jian, He Xin, Han Wang. Based on cam and crank-rocker four-legged robot gait analysis[J]. Hunan Agricultural Machinery.2011.38 (07): 60-61.
18. Zhao Tieshi, Zhao Yongsheng, Jing Wong, Liu Lechun, Huang Cheng. A Pantograph 6-dof Parallel Platform Mechanism and Its Displacement Analysis [J].ROBOT. 1998,20 (05): 346-351.

A computational study of the aerodynamic performance of a dragonfly forewing in gliding flight

Huixiang Liu[1], Guoyi He[2], Chengyu Ma[1], Qi Wang[3] and Yun Luo[1]

[1] Nanchang Hangkong University School of Aircraft Engineering,student,2017, China
[2] Nanchang Hangkong University School of Aircraft Engineering, associate professor,2017, China
[3] Nanchang Hangkong University School of Aircraft Engineering, professor,2017, China

Abstract. Gliding flight is a common mode of flight for dragonfly, the objective of the current research is to use numerical simulations to explore whether the corrugations have positive effect on aerodynamic performance of the dragonfly wings in gliding flight. In order to compare aerodynamic performance of the dragonfly wing and flat plate, a three-dimensional model of the dragonfly forewing and a three-dimensional flat plate with the same shape of the dragonfly forewing are established. The flow fields around three-dimensional dragonfly forewing and flat plate are simulated for Re=10000 and angles of attack changing from 0° to 25°(with an interval of 5°), numerical simulation indicate that aerodynamic performance of the dragonfly wing is slightly better than the flat plate over the entire range of parameters tested, especially the effect of the corrugateions on the flow is more evident at large angle of attack.

1 Introduction

Dragonflies, as one of the best flying insects in nature, are capable of carrying out dramatic flight manoeuvres. They can hover, accelerate in almost any direction and manoeuvre precisely at high speed to intercept other insects[1], however, gliding flight is also an important flight mode of the dragonflies[2], especially at high temperature, in order to avoid higher heat quantity in body[3], gliding flight is observed more frequently in dragonflies. It is interest that dragonflies could keep only a pair of forewings beat frequency and hindwings are immovable, or keep all of wings stationary in gliding flight.

The incredible flight ability produced by dragonflies is through the flapping of their forewings and hindwings. Unlike typical engineered airfoil, dragonfly wings are not smooth, wing cross-section are highly corrugated. Plated wings vary across the wing span and chord, it has been shown that corrugations could reduce the weight of the wings[4], improving the stability of the wings[5]and enhance the spanwise stiffness in the wings[6]. Howerer, whether the corrugations can improve the aerodynamic performance of the dragonfly wings in gliding flight,there is a difference in the current study.

Experimental measurements of the aerodynamic effect of corrugations were carried out by Kesel[7], Kesel have shown that corrugations plays an important role in the lift production of dragonfly wings,similar results were also found from a computational study corresponding to the same configuration has undertaken by Vargas[2]. On the contrary,early wind tunnel experiments of inset wings conducted by Reese[8]concluded that the corrugations has no aerodynamic significance, a computational study of the aerodynamic effect of corrugations were taken by Sun [9] both concluded that the effect of corrugations were to decrease the lift, the lift of corrugated wings are less than the lift of flat plate. There is still no consensus on whether corrugated wings perform better than flat plate, their research was limited comparing between the aerodynamic performance of different dragonfly wing cross-sections and two-dimensional flat plate. In fact, experimental measurements[7][10]of dragonfly wing cross-sections shown that corrugations vary with the relative span length.

It is this reason that makes further investigation into the aerodynamic aspect of dragonflies' wing worthwhile, in order to provide some perspective for the different performance between the corrugated wings and flat plate, this study focus on whether the corrugations have positive effect on aerodynamic performance of the dragonfly wings in gliding flight.

2 Numerical method

2.1 Geometric models of dragonfly wing

Dragonfly wings are complex in structure,composed of thin cuticular membrane that is supported by a system of veins, moreover, the thickness of thin cuticular membrane are only 4 microns [11], veins can be divided into longitudinal veins and crossveins, It is possible to construct the three-dimensional model of the dragonfly wings according to the actual measured data by

Okamoto[10]and Ren[12]. In order to simplify the model, influence of the irregular shape of the veins, impact of the extremely tiny corrugations, size variation of the wing along the span and chord and other structures like pterostigma, nodus etc. were ignored.The geometry of the dragonfly forewing was established by using the 3D modeling software through reverse engineering. The size of the veins are 0.035mm, veins are shown in Figure 1. The veins are filled with 0.004mm cuticular membrane as in Figure 2, the shape of cuticular membrane is from the forewing of dragonfly in nature. Figure3 shows the three-dimensional model of the dragonfly forewing, the spatial distribution of veins and cuticular membrane forms corrugation.

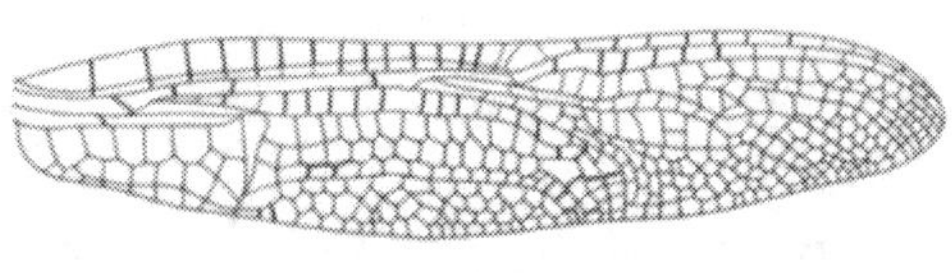

Figure 1. Veins of dragonfly forewing

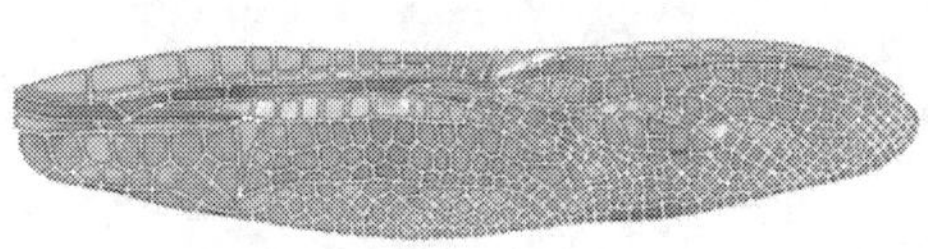

Figure 2. Cuticular membrane

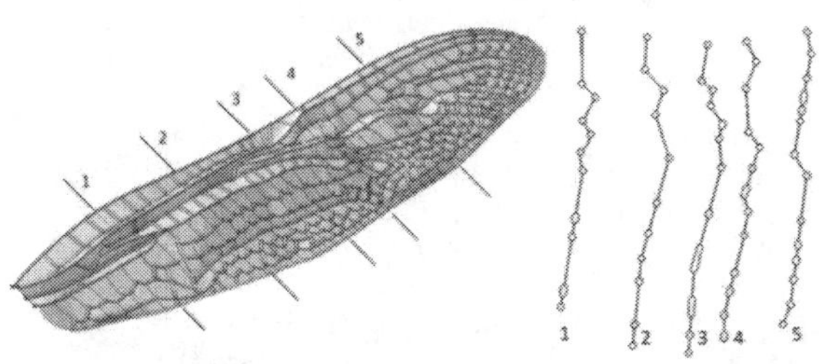

Figure 3. Three-dimensional model of the dragonfly forewing

In order to answer such a question: whether the corrugations have positive effect on aerodynamic performance of the dragonfly wings in gliding flight, a three-dimensional flat plate with the same shape and reference values as the dragonfly forewing was built, and the thickness of the flat plate is 0.035 mm, as shown in Figure 4.

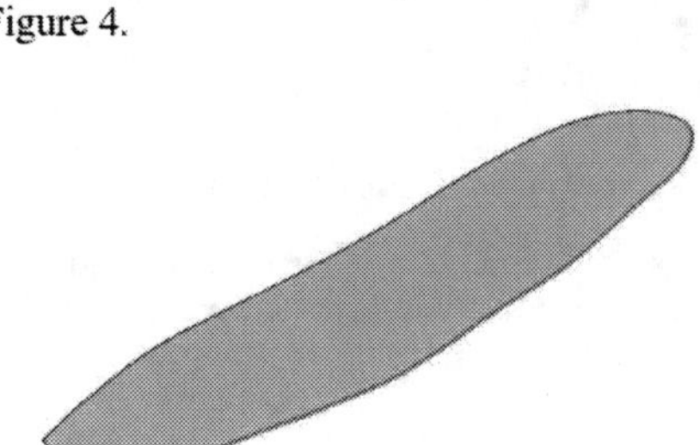

Figure 4. Three-dimensional flat plate

2.2 Governing equations

The incompressible 3D Navier-Stokes equations are follows:

$$\frac{\partial u}{\partial v}+\frac{\partial v}{\partial y}+\frac{\partial w}{\partial y}=0 \tag{1}$$

$$\frac{\partial u}{\partial t}+u\frac{\partial u}{\partial x}+v\frac{\partial u}{\partial y}+w\frac{\partial u}{\partial y}=-\frac{\partial p}{\partial x}+\frac{1}{\mathrm{Re}}\left(\frac{\partial^2 u}{\partial x^2}+\frac{\partial^2 u}{\partial y^2}+\frac{\partial^2 u}{\partial z^2}\right) \tag{2}$$

$$\frac{\partial v}{\partial t}+u\frac{\partial v}{\partial x}+v\frac{\partial v}{\partial y}+w\frac{\partial v}{\partial y}=-\frac{\partial p}{\partial y}+\frac{1}{\mathrm{Re}}\left(\frac{\partial^2 v}{\partial x^2}+\frac{\partial^2 v}{\partial y^2}+\frac{\partial^2 v}{\partial z^2}\right) \tag{3}$$

$$\frac{\partial w}{\partial t}+u\frac{\partial w}{\partial x}+v\frac{\partial w}{\partial y}+w\frac{\partial w}{\partial y}=-\frac{\partial p}{\partial y}+\frac{1}{\mathrm{Re}}\left(\frac{\partial^2 w}{\partial x^2}+\frac{\partial^2 w}{\partial y^2}+\frac{\partial^2 w}{\partial z^2}\right) \tag{4}$$

Where u , v and w are the non-dimensionalized components of velocity along x , y and z directions; P and t is the non-dimensional fluid pressure and time, The equations are non-dimensionalized. corresponds to the Reynolds number defined as:

$$\mathrm{Re}=\frac{\rho Uc}{\mu}=\frac{Uc}{v} \tag{5}$$

Where ρ , U and c denote respectively the density, the free stream velocity and length, and represent the dynamic viscosity coefficient and the kinematic viscosity coefficient respectively.

The key examined are the lift and drag coefficient which are defined as

$$\mathrm{Cl}=\mathrm{F_l}/0.5\rho U^2 c \qquad \mathrm{Cd}=\mathrm{F_d}/0.5\rho U^2 c \tag{6}$$

Where F_l and F_d are the lift and drag of an airfoil respectively.

2.3 Boundary conditions

The three-dimensional domains implemented in the current study consist of six boundary conditions located front, back, left, right, top and bottom of the computational domain, the size of the computational domain is shown in Figure 5, the boundary conditions of left field and the dragonfly forewing are symmetry and wall respectively.

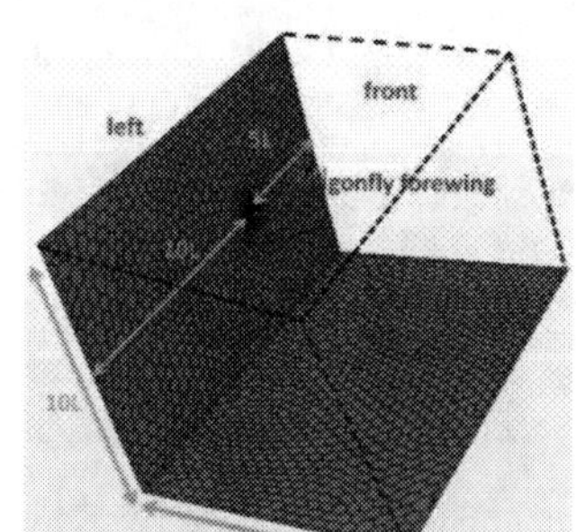

Figure 5. Computational domains

2.4 Validation case

In order to validate that the current numerical solver do not depend on the selection of the grid, we build three type of grids : the first type of the grid is that minimum size of the grid was 0.04 mm, the second type of the grid is that minimum size of the grid was 0.03 mm, the third type of the grid is that minimum size of the grid was 0.02 mm. The results show that the relative error of the lift coefficient is 14.9% in the first and the second type of grid, and the relative error of the second and the third type of gird is 4.7%. The difference in the lift coefficient is no more than 5%, that is an acceptable value. Considering a balance between accuracy and efficiency of the numerical simulation, we chose second type of grid as the solution of grid, as shown in Figure 6.

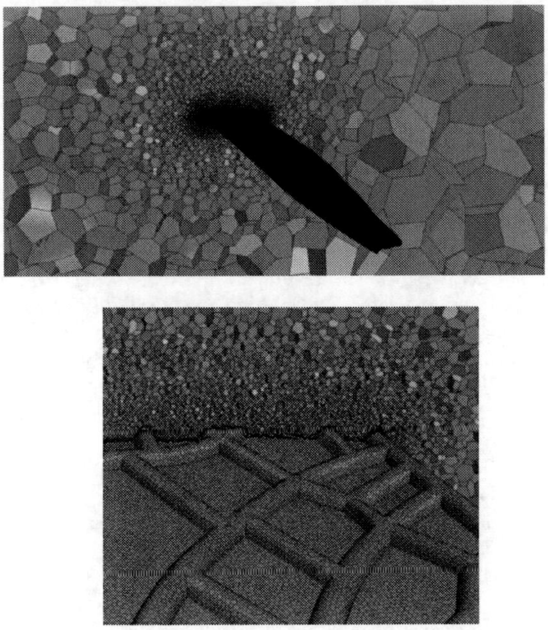

Figure 6. Local showing of computational grid.

Six validation cases were conducted using three-dimensional flat plate at angle of attack of 0° to 25°(with an interval of 5°)with Re=10000, The simulations were performed on second gird for 0° to 25°angle of attack cases respectively.The comparison between the experimental and force coefficient in Figure 7 show indistinctive discrepancies between the two results. The numerical lift coefficient is close to the experimental value at angle of attack of 0° to 10° , the numerical lift coefficient is slightly larger than the experiment value at angle of attack of 15° and 20° , The numerical drag coefficient is close to the experimental value at angle of attack of 0° to 25° , the numerical drag coefficient is slightly smaller than the experiment value at the other angle of attack. It should be noticed that the overall trend of two results is consistent and the maximum relative error is below 10%.

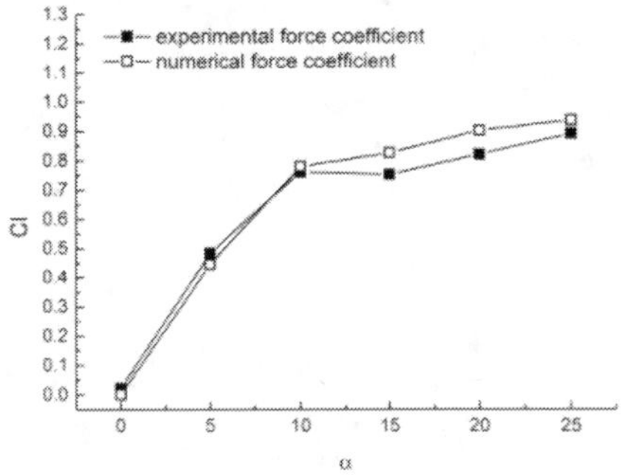

(a)

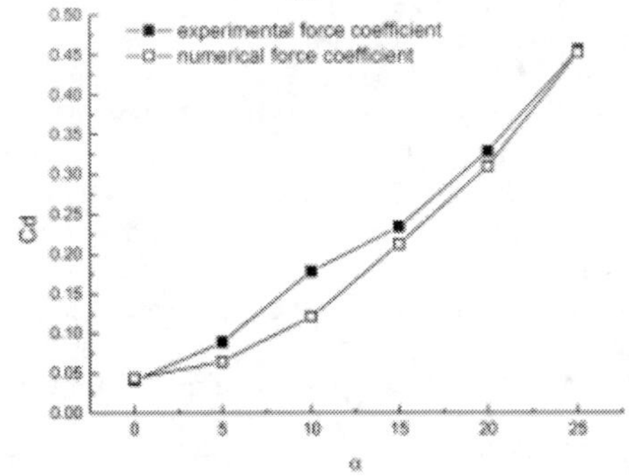

(b)

Figure 7. Comparison between the numerical and the experimental force coefficient of Kesel for Re=10000 and various anglesof attack.(a)lift coefficient.(b)drag coefficient.

3 Results

The flow fields around three-dimensional dragonfly forewing and flat plate are simulated for Re=10000 and angles of attack changing from 0° to 25°(with an interval of 5°), the following sections discuss the force coefficient of three-dimensional dragonfly forewing and flat plate obtained from numerical simulation.The lift and drag coefficient curves for dragonfly forewing and flat plate are shown in Figure 8, it was found from Figure 8 that lift coefficient of dragonfly forewing is slightly higher than the flat plate at all angle of attack,however, drag coefficient of dragonfly forewing is slightly higher than the flat plate at all angle of attack.

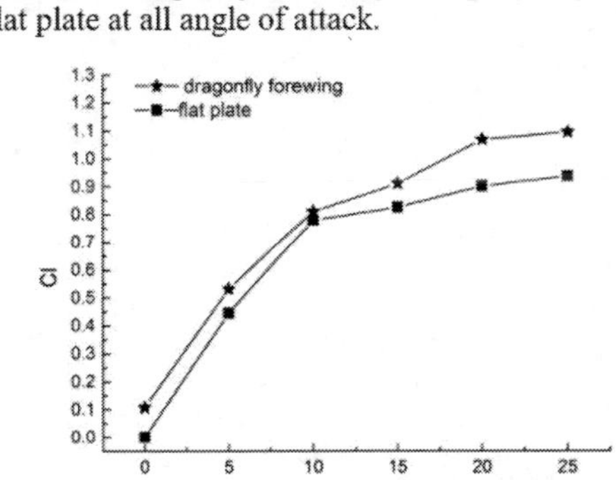

(a)

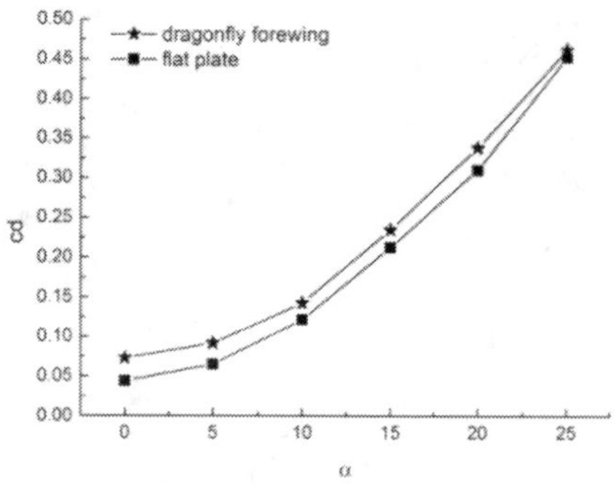

(b)

Figure 8. Aerodynamic performance of dragonfly forewing and flat plate (a)lift coefficient, (b)drag coefficient.

Figure 9(a) and (b) show that there is a trapped vortex in each fold which causes the overall flow to resemble the smooth airfoil. The higher lift coefficient of dragonfly forewing due to these trapped vortexes inside each corrugation, it is important these corrugations delayed flow separation so that the lift coefficient of dragonfly forewing are more higher than flat plate at large angle of attack, however, the higher drag coefficient of dragonfly forewing are produced because of these corrugations.

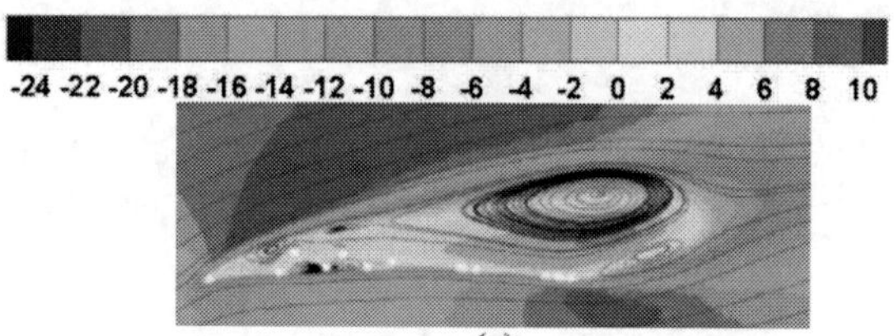

(a)

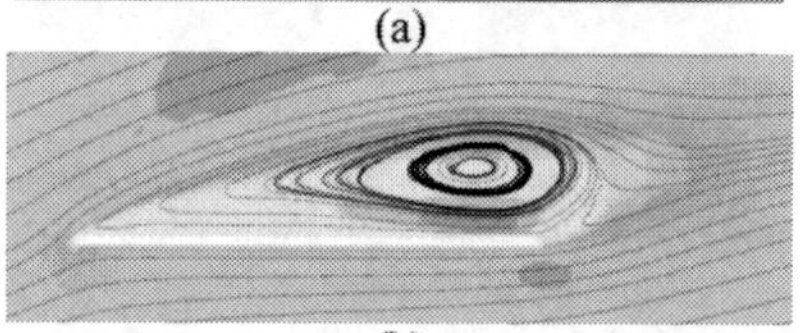

(b)

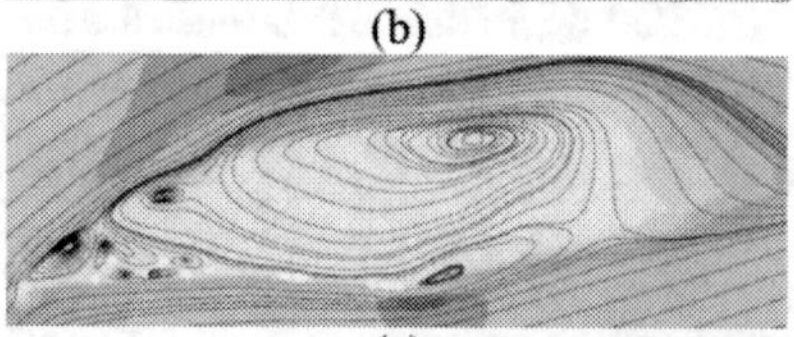

(c)

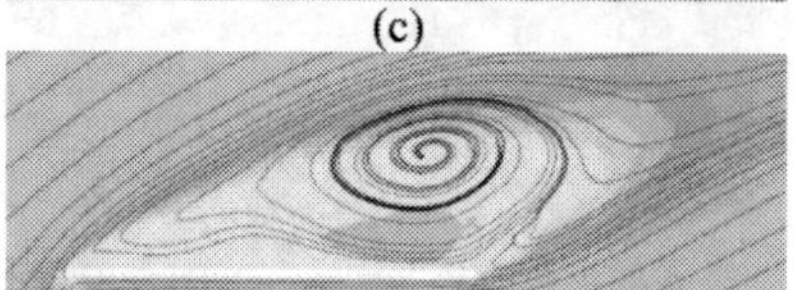

(d)

Figure 9. The streamline of airfoil with profile cross sections shown below at 0.2lrel, where lrel is the relative span length: (a) and (c) are dragonfly forewing at angles of attack 15° and 20° respectively; (b) and (d) are flat plate at angles of attack 15° and 20° respectively.

A leading edge vortex(LEV) is formed on first longitudinal vein of dragonfly forewing as shown in Figure10(a), A LEV is also formed on the tips of flat plate as shown in Figure10(b), Figure 10 show that the LEVs of dragonfly forewing is much higher than the LEVs of flat plate at same condition, so LEVs should be a source of high lift forces in dragonflies.

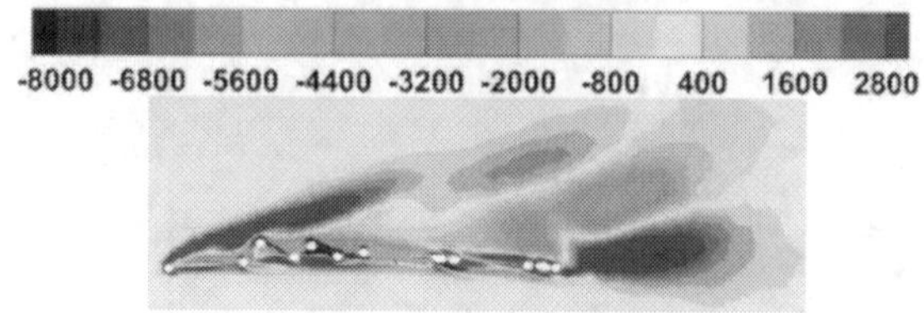

(a)

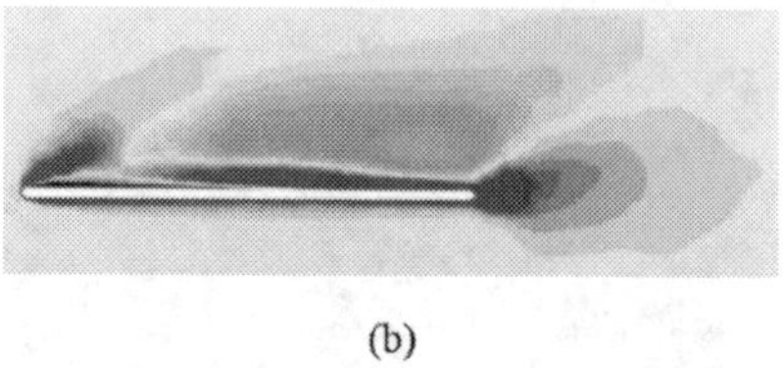

(b)

Figure 10. Vortex of airfoil angles of attack 15° with profile cross sections shown below at 0.2lrel, where lrel is the relative span length: (a) dragonfly forewing (b) flat plate.

4 Conclusion

Dragonflies have developed perfect wings through millions of years of adaptation,despite highly corrugated dragonfly forewing, it is found that the aerodynamic performance of the dragonfly wing is slightly better than the flat plate over the entire range of parameters tested, especially the effect of the corrugateions on the flow is more evident at large angle of attack.

This result can be a potential inspiration to the airfoil design of micro-aerial vehicles(MAVs), in this respect, nature provides a perfect design of wings that are suitable for flapping MAVs.

References

1. A. L. Thomas, G. K. Taylor, R. B. Srygley, J Exp Biol. J. **207**, 4299 (2004)
2. A. Vargas, R. Mittal, H. Dong, Bioinspir Biomim. J.**3**, 13-0 (2008)
3. R. Xiujuan, Scientia Sinica. J. **56**,884-897 (2013)
4. S. Sudo, K. Tsuyuki, T. Ikohagi, JSME Int J.J. **42**,721–729(1999)
5. H. Hertel, *Structure, form, movement*.M.(1966)
6. B. K. Antonia, P. Ute, N. Werner, Comput Biol Med. J. **28**,423-437(1998)
7. A. B. Kesel, J Exp Biol. J.**203**,31253-3135(2000)
8. C. J. C.Rees, Nat.J. 200–203 (1975)
9. X. G. Meng, M. Sun, Phy Flu.J. **25**, 200 (2013)
10. M. Okamoto, K. Yasuda,A. Azuma, J Exp Biol.J. **199**,281-294(1996)
11. J. G. Needham,*A genealogic study of dragonfly wing vention*.M. 703-706 (1993)
12. H. H. Ren,X. S. Wang,Y. L. Chen, Chin Phys B.J. **21**, 294-303 (2012)

Research of Boundary Layer Transition characteristics Induced by Three-Dimensional Discrete Roughness

Feng Li[1], Chao Gao[1], Zijie Zhao[1], Xudong Ren[1]

[1]Northwestern Polytechnical University, Xi'an 710072, China

Abstract: Roughness strip is a necessary technology for wind tunnel experiment. In order to improve the accuracy and reliability of transition simulation, a new fixed transition technology based on the three-dimensional discrete roughness elements has been established. The configuration parameters of roughness elements are calculated theoretically and the formula and manufacturing processes of roughness elements are developed. Using two-dimensional airfoil and three-dimensional combination models, the transition and additional resistance characteristics of discrete roughness elements are studied. Finally, the scale effect of roughness elements is analyzed and the influence laws of height, diameter, and spacing on transition characteristics have been obtained through numerical calculation. The results of this study indicate that this new discrete roughness is better in transition and additional resistance performance than conventional grit roughness. The results obtained in this paper has created a more reliable and accurate fixed transition technology for wind tunnel experiment and provided some reference for cross-flow transition mechanism.

1 Introduction

The boundary layer fixed transition technology has an important significance on the accuracy of wind tunnel experiment results. However, in high dynamic experimental environment, roughness strip may be damaged or grit is blown off, which leads to the failure of the experiment consequentially. The thickness of adhesive and emery grit is not easy to be held, so the height of roughness strip cannot be precisely controlled. The selection of dimensions and installation position of roughness strip lacks scientific reference standards in different flow conditions.

Discrete roughness strip is a three-dimensional boundary layer transition technology. the ability to resist air impact of discrete roughness is higher than emery strip. Secondly, the discrete roughness may exert a three-dimensional perturbation on the boundary layer. In addition, dimensions of roughness elements can be precisely controlled by special mould.

Huang Yong[1] proved the feasibility of this new fixed transition technology. Li Yueli[2] explored the influence of roughness elements on the standing wave of 3D boundary layer. Li Yiming[3] and Juillen[4] found that single roughness element can promote transition by accelerating the T-S wave. Radeztsky's research[5] indicated that cross flow boundary layer transition is more sensitive to three dimensional roughness elements. Muller[6] confirmed that the surface roughness is the most effective means of exciting cross flow vortex. Researches of Deyhle[7] and Reibert[8] also indicated that roughness elements aligned horizontally can stimulate the cross flow wave of corresponding wavelength. Collis[9] found that the increase of height of roughness elements can enhance the sensitivity of cross flow vortex. Robert's research[10] indicated that the effect of transition induction is more prominent when the Reynolds number enters the critical region. Nathaniel[11] discovered the influence of the distance between hemispherical roughness elements on the boundary layer flow. According to the Marxen's study[12], blunt and sharp roughness elements can induce transition and the surface curvature has little effect on the transition characteristics. Chan[13] found convective instability disturbed by discrete roughness elements is the main reason for the transition of the supersonic boundary layer. As shown by Plogmann [15], when the height of roughness elements is taken as 30% ~ 50% of boundary layer displacement thickness, transition effect is best.

2 Theoretical calculation of the height of roughness elements

The configuration parameters of roughness elements include height k, diameter d, spacing l, and k is a key factor to affect the boundary layer flow. According to the theory of boundary layer, height k should be greater than the critical height but less than boundary layer thickness. First of all, we need to calculate the height range of roughness elements. Height k is determined by Mach number Ma, Reynolds number Re and the installation position of roughness x. In this study, taking the NACA0012 airfoil and mid-wing combination as experimental models, the installation position is identified as 6% chord length from the leading edge of the models. Theoretical calculation method is as follows:

Checking the literature NACA TN4363[16], we know Reynolds number based on the critical height of roughness $R_{k,t}$=600. Reynolds number based on the position of roughness R_x can be obtained by formula (1).

$$R_x = \frac{\rho_\infty u_\infty x}{\mu_\infty} = \frac{u_\infty x}{v_\infty} \quad (1)$$

According to $\frac{R_{k,t}}{\sqrt{R_x}}$ and Ma, we can find out the dimensionless height of roughness element ηk by the literature NACA TN4363 and the height of roughness elements can be obtained by formula (2).

$$k = \frac{2x\eta_k}{\sqrt{R_x}} \quad (2)$$

3 The formula and manufacturing processes of roughness elements

Ingredients of roughness elements are constituted by putty and hardener, and the mixture ratio is 50:1. This formula has the characteristics of fast curing, easy molding and strong adhesion. The high precision manufacturing of roughness elements is realized through the mould method. The way to make it is to put the ingredients into the taper hole on the mould which is pasted on the surface of the model. After the putty becomes stiff, the plastic mould is removed off and a series of roughness elements are formed on the surface of the test model. The mould is made by three-layers drilling technology. The three layers are composed of protective layer, mould layer and spacing layer. The protective layer is used to avoid the burrs on the edge of the hole and the cutting depth is controlled by the spacing layer. The finished mould is shown in Figures.3. The height, diameter and spacing of roughness elements can be controlled precisely by mould. According to the results of theoretical calculation, the thickness of the mould layer is designed as 0.1mm which is the height of roughness elements. The discrete roughness elements installed on the surface of model are shown in Figure 1.

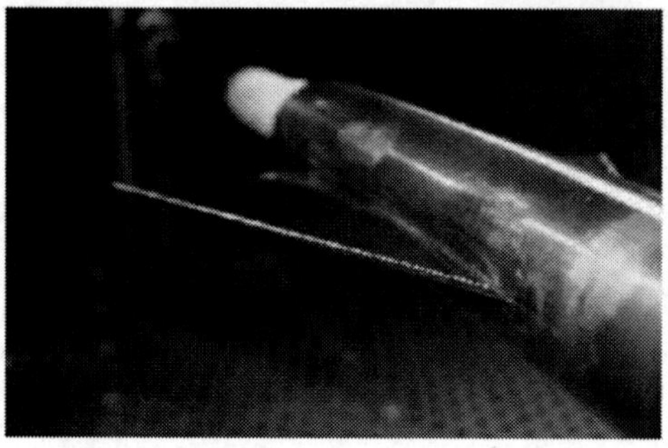

Figure 1. Discrete roughness elements installed on model

4 Experimental Results

The present study was carried out in NF-6 pressured continuous transonic wind tunnel at the Northwestern Polytechnical University, Xi'an, China. This tunnel has two test sections: one is 2D section with the size of 0.4m(width)×0.8m(height)×3m(length) and the other is 3D section with the size of 0.8m(width)×0.6m(height)×3m(length). The stagnation pressure and stagnation temperature of the tunnel air can be controlled from 1.0 to 5.5 bar and between 283 K and323 K. The Mach number range is $Ma = 0.3 \sim 1.2$. The sublimation technique was used to visualize the location of laminar to turbulent boundary-layer transition on the model surface. Photographs were taken in real time from a camera through an optical window mounted on the ceiling wall of the test section. The height, diameter and spacing of roughness elements are 0.1mm, 1.2mm and 2.5mm respectively.

4.1 Transition characteristics

Figure 2 shows the naphthalene-film sublimation experiment result of the NACA0012 airfoil model at Ma = 0.80, α = 0°, Re = 3.0 × 10^6. As the air flows, the elevated shear and temperature associated with high turbulence of the boundary layer at the location of transition cause the naphthalene to increase its rate of sublimation in the turbulent region with respect to that of the laminar region. Consequently, the naphthalene is removed in the turbulent area before the naphthalene in the laminar area is appreciably affected. Under the fixed transition, the white coating after the roughness strip immediately disappears leaving a dark band which spreads downstream towards the airfoil trailing edge, while under the free transition, a dark area across the airfoil span appears near the airfoil trailing edge. It is seen that the roughness elements have played a better role in promoting the boundary layer transition.

Figure 3 shows the naphthalene-film sublimation experiment result of the AGARD-B mode at Ma = 0.75, α = 0°, Re = 2.0 × 10^6. Under the fixed transition, the white coating after the roughness strip immediately disappears, while under the free transition, a dark area across the wing span appears near the airfoil trailing edge with acute-angled dark band. It verifies that the roughness elements have a same transition excitation effect for 3D model.

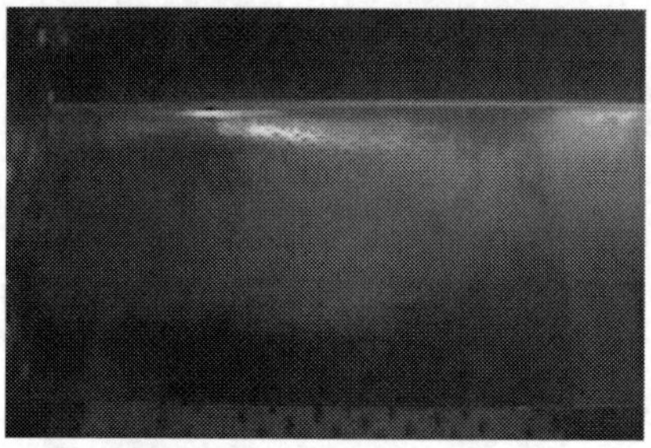

Figure 2. The naphthalene-film sublimation experiment result of NACA0012 airfoil mode, Ma = 0.80, α = 0ο, Re = 3.0 × 106

https://doi.org/10.1051/matecconf/201815103002

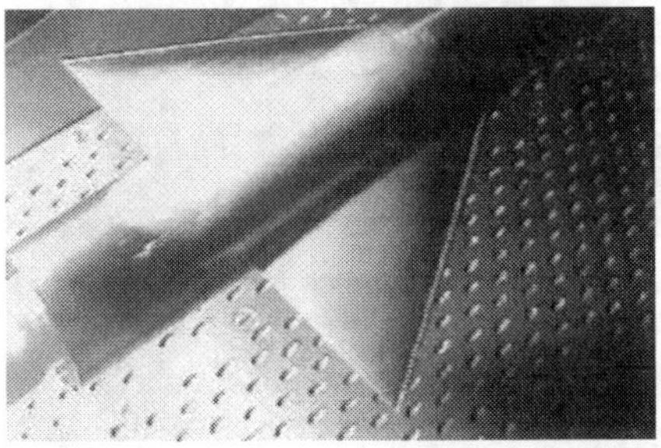

Figure 3. The naphthalene-film sublimation experiment result of AGARD-B model, Ma = 0.75, α = 0∘, Re = 2.0 × 106

4.2 Reliability verification in the high dynamic pressure condition

In order to verify the reliability of roughness elements under the high dynamic flow condition, verified experiment was performed through the using of total pressure regulating technology. The experiment model is a supercritical airfoil and the total pressure p0 is 2bar. Figures 4 and presents the surface pressure distribution the supercritical airfoil under high total pressures. It can be seen from the figures that the experimental curves are smooth and pressure distribution is reasonable under high total pressure flow condition. This illustrates that the property of roughness elements are stable and reliable and this transition technology can satisfy the requirement of pressured wind tunnel experiment.

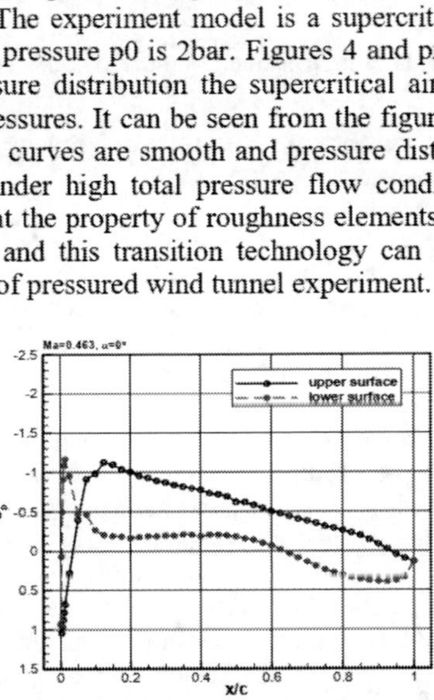

Figure 4. Pressure distributions of supercritical airfoil under high total pressures, p0=2bar

4.3 Additional resistance characteristics

Additional resistance is the key indicators of roughness strip. Figure 5 gives the resistance coefficients curves of 2D airfoil models under different transition patterns. The "new spot roughness strip" curve is the experimental result of discrete roughness elements and the "120# grit roughness strip" curve is the experimental result of 3mm wide 120# emery strip. The result shows a good agreement between the discrete elements and emery roughness strip and the additional resistance produced by the two transition patterns have the same magnitude. As shown in Figure 5, in the angle of attack from −4° to 9°, the resistance of free and fixed transition are basically identical, suggesting that in the range of small angle of attack, additional resistance produced by roughness strip can be ignored. But at an angle of attack greater than 10°, resistance of fixed transition patterns increases significantly and, consequently, the additional resistance need to be subtracted from the total resistance.

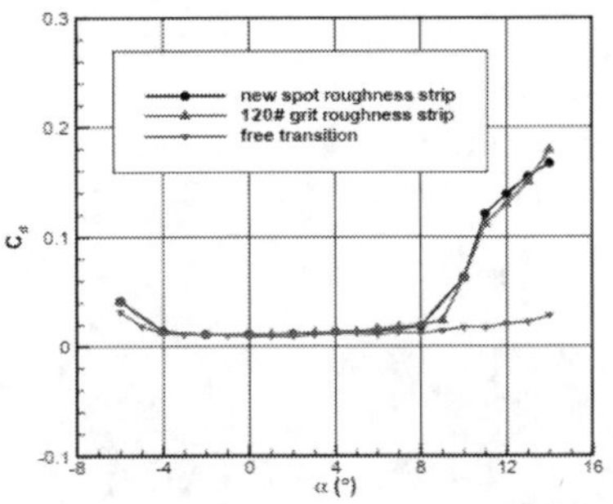

Figure 5.Resistance coefficients curves of supercritical airfoil models under different transition patterns

5 Scale effect of roughness elements

Roughness elements' height k diameter d, and the spacing l (the distance between the neighboring elements), is the key parameters affecting the transition characteristics. Taking 3D NACA0012 straight wing, with a chord length of 200mm and aspect ratio of 0.5, as the calculating model, the influence of configuration parameters of roughness elements on the transition characteristics was investigated through the numerical simulation method. By using the $\gamma-\overline{\mathrm{Re}}_{\theta}$ transition model, according to surface friction drag distribution, the transition position has been predicted successfully. Computational grids are composed of prism boundary layer grid and tetrahedral spatial grid, and the height of first boundary layer grid is 10-5m, as shown in Figure 6. Calculation parameters are as follows: Ma = 0.6, α = 2∘, Re = 2.74 × 106.

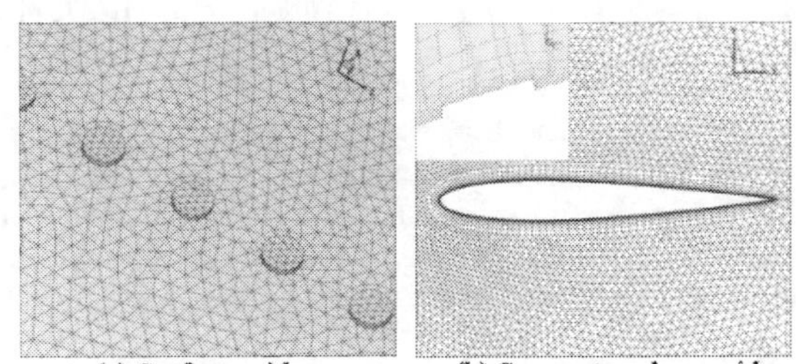

(a) Surface grid (b) Symmetry plane grid

Figure 6.Computational grids of discrete roughness elements fixed transition model

5.1 Influence of height k on the transition characteristics

The roughness elements height k is chosen to be 0.08mm、0.1mm、0.12mm、0.16mm、0.2mm, and diameter d of 1mm and spacing l of 2.5mm remain constant. Roughness elements are mounted on the upper surface of the wing from the leading edge of 6% chord and there is no roughness element on the lower surface. Figure 7 shows the calculation results of k = 0.1 mm. It can be seen that the position of the skin friction drag increasing suddenly on the upper surface is more close to the leading edge, which indicates that the transition position is moved forward under the action of roughness elements. Calculation results of other four cases also show a similar tendency. Figure 8 presents the transition position variation with roughness elements height. With the increase of height, transition position is moved to the leading edge of the wing. Near the k = 0.1 mm, transition position is very sensitive to the height. But when k is greater than 0.15 mm, the transition position are not easy to be changed by adjusting the roughness elements height.

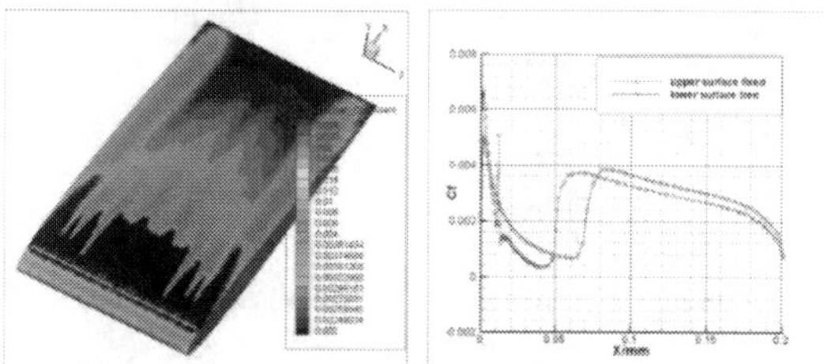

(a) Skin friction contour (b) Skin friction coefficients

Figure 7. Calculation results of k = 0.1 mm, d=1mm, l=2.5mm

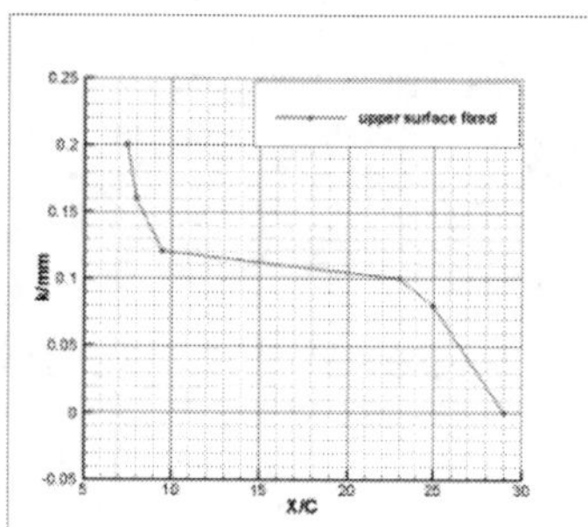

Figure 8. Transition position vs roughness elements height

5.2 Influence of diameter d on the transition characteristics

The roughness elements diameter d is chosen to be 1mm、1.2mm、1.5mm、2mm, and height k of 0.1mm and spacing l of 2.5mm remain constant. Roughness elements are mounted on the upper surface of the wing from the leading edge of 6% chord and there is no roughness element on the lower surface. Figure 9 shows the calculation results of d = 1.5mm. It can be seen that the position of the skin friction drag increasing suddenly on the upper surface is more close to the leading edge, which indicates that the transition position is moved forward under the action of roughness elements. Calculation results of other threer cases also show a similar tendency. Figure 10 presents the transition position variation with roughness elements diameter. With the increase of diameter, transition position is moved to the leading edge of the wing. Furthermore, a linear correlation has been found between the transition position and roughness elements diameter.

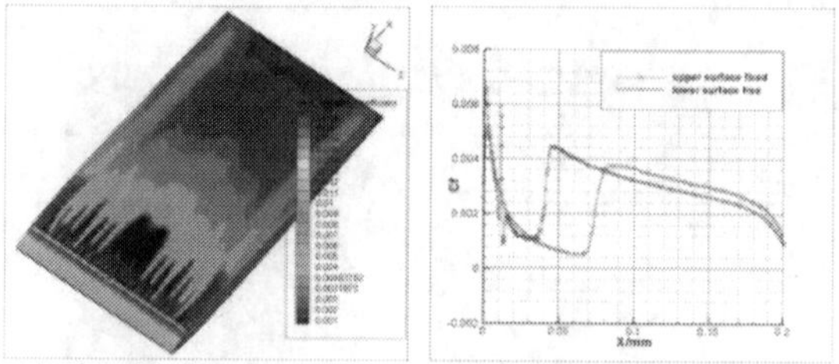

(a) Skin friction contour (b) Skin friction coefficients

Figure 9. Calculation results of k = 0.1 mm, d=1.5mm, l=2.5mm

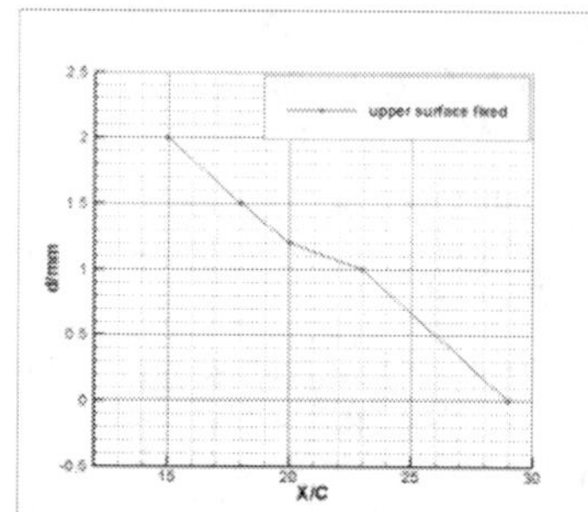

Figure 10. Transition position vs roughness elements diameter

5.3 Influence of spacing l on the transition characteristics

The roughness elements spacing l is chosen to be 2mm、2.5mm, 3.5mm, 5mm, and height k of 0.1mm and diameter d of 1mm remain constant. Roughness elements are mounted on the upper surface of the wing from the leading edge of 6% chord and there is no roughness element on the lower surface. Figure 11 shows the calculation results of l=3.5mm. It can be seen that the position of the skin friction drag increasing suddenly on the upper surface is more close to the leading edge, which indicates that the transition position is moved forward under the action of roughness elements. However, when the spacing increases to 5mm, the transition position has moved in opposite directions. Figure 12 presents the transition position variation with roughness elements spacing. With the increase of spacing, transition position is moved to the trailing edge of the wing. When l is greater than 5 mm, roughness elements will not be able to induce the transition. So the roughness elements spacing should be carefully selected.

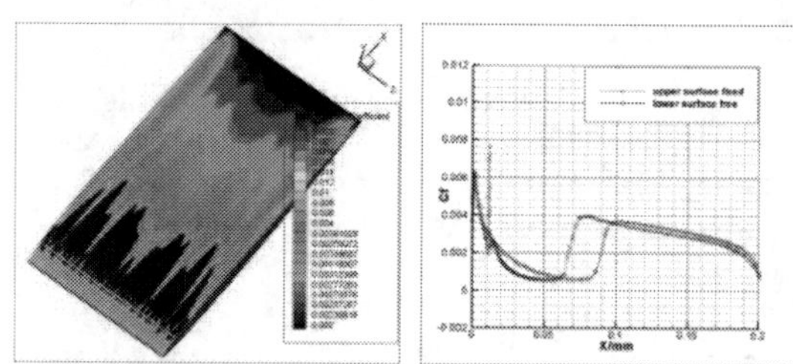

(a) Skin friction contour (b) Skin friction coefficients

Figure11. Calculation results of k = 0.1mm, d=1.5mm, l=3.5mm

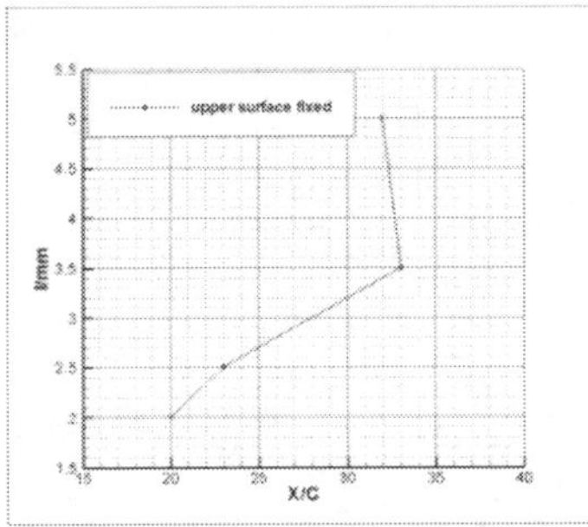

Figure 12. Transition position vs roughness elements spacing

5. Conclusion

In this paper we have reported the study of the boundary layer transition technology based on the three-dimensional discrete roughness elements. The fast curing, easy molding and strong adhesion formula and manufacturing processes of roughness elements have been developed The configuration parameters of roughness elements were calculated and the height range was determined. The main conclusion can be drawn as follows:

1) The roughness elements can play a better role in the boundary layer transition for 2D and 3D wind tunnel experiment model and will not damage the inherent aerodynamic characteristics of models.

2) The property of roughness elements is stable and reliable in the high dynamic pressure condition and this transition technology can satisfy the requirement of pressured wind tunnel experiment.

3) In the range of small angle of attack, the additional resistance of roughness elements is very small and can be ignored.

4) With the increase of height k and diameter d of roughness elements, transition position is moved to the leading edge of the wing, but spacing l presents a contrary tendency. When l is greater than 5 mm, roughness elements will not be able to induce transition.

Acknowledgments

The present work is supported by the National Natural Science Foundation of China, 11402209, the Aviation Science Foundation of China, 2015ZD53053, Shaanxi Natural Science Foundation, 2016JQ1023, Equipment Research Funds of China, 61422010401, and the Fundamental Research Funds for the Central Universities, 3102016ZY001.

References

1. Y. Huang, F. Qian, K Yu, and et.al. Experimental Investigation on Boundary-layer Artificial Transition Based on Transonic Trip Disk. Journal of Experiments in Fluid Mechanics (In Chinese), Vol. **20**, No. 3, pp. 59～62. (2006).
2. Y. Li, D. Li, Y. Yang, and et.al. Study on the Swept Wing Crossflow Stationary Wave and Its Harmonics. Journal of Experiments in Fluid Mechanics (In Chinese), Vol. **24**, No. 3, pp. 25～28.(2006),
3. Y. Li, D. Yan. An Experimental Investigation on the Effect of Single Two-Dimensional Roughness Elements on Boundary-Layer Transition. Journal of Peking University (In Chinese), Vol.**41**, No.1, pp.71～75(2005)
4. J. C. Juillen, D. Arnal. Experimental and Theoretical Study of Transition Phenomena on an Infinite Swept Wing[R].Rapp.Final No.51/5018, **5** , CERT/ONERA, Toulouse(1990)
5. R. H. Radeztsky Jr, M. S. Reibert, W. S. Saric. Development of Stationary Crossflow Vortices on a Swept Wing [R]. AIAA 2373(1994)
6. B. Muller, H. Bippes. Experimental Study of Instability Modes in a Three-dimensional Boundary Layer. In: Proc AGARD Symp On Fluid Dynamics of Three-Dimensional Turbulent Sheer Flows and Transition[R]. Cesme, Turkey,AGARD C-P 438(1988).
7. H. Deyhle, H. Bippes. Disturbance Growth in an Unstable Three-dimensional Boundary-layer and Its Dependence on Environmental Conditions[J]. Journal of Fluid Mechanics , 316: 73-113(1996)
8. M. S. Reibert. Nonlinear Stability Saturation, and Transi-tion in Crossflow-dominated Boundary layer[D]. USA: Arizona State University(1996)
9. S. S. Collis, S. K. Lele. Receptivity to Surface Roughness Near a Swept Leading Edge. Journal of Fluid Mechanics,, 380: 141-168(1999)
10. S. Robert, Downs III, B. Edward, White, A.Nicholas. Denissen. Transient Growth and Transition Induced by Random Distributed Roughness[J].AIAA Journal, 46(2):451~462(2008)
11. D.Nathaniel ,Varano, L. Roger, Simpson. Structure of Turbulent Boundary Layers and Surface Pressure Fluctuations with Sparse Roughness[R].AIAA 700, (2010)
12. Olaf Marxen, Gianluca Iaccarino. Numerical simulation of the effect of a roughness element on high-speed boundary-layer instability[R]. AIAA 4400(2008)
13. Chau Lyan Chang, Meelan M. Choudhari. Hypersonic Viscous Flow over Large Roughness Elements[R]. AIAA 0173 (2009).
14. B. Plogmann, W. Wuurz, E. Kruamer. Interaction of a Laminar Boundary Layer with a Cylindrical Roughness Element near an Airfoil Leading Edge[R]. AIA-3077(2012)
15. A. Braslow, E.Knox, Simplified Method for Determination of Critical Height of Distributed Roughness Particles for Boundary-Layer Transition at Mach Numbers from 0 to 5. NACA TN 4363, 82～83(1958)

Synchronous Surface Pressure and Velocity Measurements of standard model in hypersonic flow

Sun Zhijun [1], GU Yunsong[1], CHENG Keming[1], Di peng[2], Yingzheng Liu[2]

[1]*College of Aerospace Engineering, Nanjing University of Aeronautics and Astronautics(NUAA), Nanjing, China*
[2] *Gas Turbine Research Institute/School of Mechanical Engineering, Shanghai Jiao Tong University, Shanghai, China*

Abstract. Experiments in the Hypersonic Wind tunnel of NUAA(NHW) present synchronous measurements of bow shockwave and surface pressure of a standard blunt rotary model (AGARD HB-2), which was carried out in order to measure the Mach-5-flow above a blunt body by PIV (Particle Image Velocimetry) as well as unsteady pressure around the rotary body. Titanium dioxide (Al_2O_3) Nano particles were seeded into the flow by a tailor-made container. With meticulous care designed optical path, the laser was guided into the vacuum experimental section. The transient pressure was obtained around model by using fast-responding pressure-sensitive paint (PSP)sprayed on the model. All the experimental facilities were controlled by Series Pulse Generator to ensure that the data was time related. The PIV measurements of velocities in front of the detached bow shock agreed very well with the calculated value, with less than 3% difference compared to Pitot-pressure recordings. The velocity gradient contour described in accord with the detached bow shock that showed on schlieren. The PSP results presented good agreement with the reference data from previous studies. Our work involving studies of synchronous shock-wave and pressure measurements proved to be encouraging.

1 Introduction

The development of optical measurement technology shows the potential advantages in the research of flow field measurement. Since the advent of Laser Doppler Velocimetry (LDV), Optical measurement techniques have been applied and developed for a long time in the aerodynamic experiments. The Particle Image Velocimetry (Particle Image Velocimetry, PIV) highly desirable for capturing the flow structures due to their low intrusiveness and high spatial resolution, gradually becoming predominant technology in flow field measurements [1-4]

By using the method of laser speckle to capture the image of the particle displacement at a certain time interval, the velocity of particle motion was obtained by the cross-correlation method. The particles are distributed uniformly in the flow field, along with the fluid, so that the velocity of the particles can indicate the velocity of the current fluid. There for, the following properties of particles and the capture of the particle image become the key factors in the accuracy of PIV measurements.

In recent years, with the advent of high-energy lasers and high resolution digital image recorders, the PIV technology made supersonic flow field measurement possible. But for hypersonic flow, the high speed, high gradient and compressibility bring a lot of difficulties to the application of PIV in hypersonic flow field measurement.

The hypersonic flow field has a large velocity gradient, especially around the shock wave, which has a strong deceleration effect on the shock wave surface. Due to the effect of inertia, the particles in flow field cannot remain completely follow, the precision of velocity measurement has a great influence [5,6].Flow compressibility introduces two difficulties, namely the large spatial variation of the particle density and the inhomogeneous refractive index throughout the flow field, results in a distortion of laser propagation.[7]

As a global pressure testing technology, PSP (Pressure-Sensitive Paint) has been widely used in wind tunnel test.[8] However, PSP measurement in hypersonic flows is usually challenged by the paint response time. The fast-responding PSP typically have frequency response of a few hundred Hz which is sufficient for steady measurement [9-11].

With the continuous improvement of wind tunnel measurement and control technology, the problem in flow was becoming more and more complicated. For flow field research, a variety of testing methods were often required. Synchronous measurement technology can be used to simultaneously obtain test content (e.g. the aerodynamic force, surface pressure) in a single test.[12,13] The test results obtained by this method are relevant in time and make up for the defects of independent measurement technology in time synchronization, which is helpful for the study of the unsteady aerodynamic problems. On the other hand, the synchronous measurement can reduce the number of tests and improve the economy of single wind tunnel test.

The current study focused on overcoming the challenges in simultaneous shock-wave and pressure measurements in transient hypersonic flows by using optical technology. The combined PIV/PSP measurement was applied in synchronous test on the AGARD hb-2 standard model at Ma = 5.

2 Experiment setup

2.1 Hypersonic facility

The experiment was conducted in a conventional hypersonic blow-down facility (NHW[14]) at Nanjing University of Aeronautics and Astronautics. As shown in Fig. 1, NHW is an intermittent blow-down wind tunnel with vacuum suction to produce hypersonic flow. This tunnel mainly consists of a high-pressure gas source, a heater, a stabilizing section (with nozzle), a vacuum test section, a diffuser, a vacuum system and an electrical control system. A support mechanism with variable angle of attack is set in the test section. Four sets of axisymmetric supersonic nozzles with nominal Mach number of 5, 6, 7 and 8 are available.

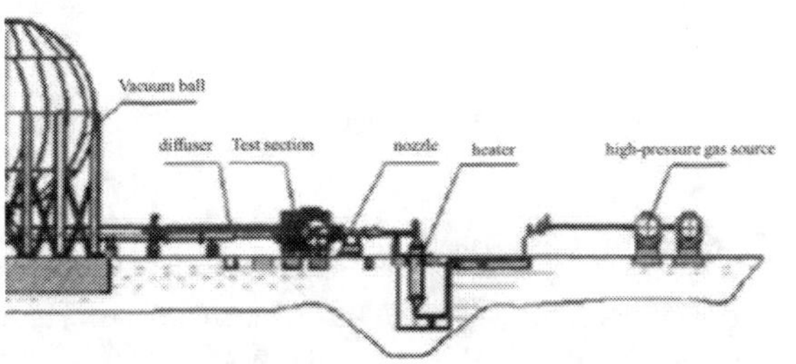

Figure 1. Sketch of Nan-hang Hypersonic Wind tunnel (NHW)

The tunnel has a 2m×2m×2.5m rectangular plenum chamber. The front and back ends of the plenum chamber are connected to the nozzle and the vacuum respectively. A pair of observation windows (Φ400 mm) is located on both side walls of the test section, which has been used to monitor the shock wave structures around the model. On top of the experimental section has a 0.6 × 0.6 m2 optical observation window with germanium glass, used in optical measuring system. A free jet - type test section (Φ500 mm) was installed inside the plenum chamber. The model support mechanism can accurately provide the Angle of attack required during the test.

The typical operating parameters of this wind tunnel at Ma = 5 are listed in Table 1. The wind tunnel is equipped with a 300-mm-diameter color Schlieren system for flow visualization, which was used to provide the flow structures around the nose of the model in the current study.

Table 1. Operating parameters of hypersonic wind tunnel

Mach number	5
Atmosphere pressure	101325Pa
Total pressure/static pressure	530
Dynamic pressure coefficient	0.03177
Total pressure	~500kPa
Total temperature	~500K
Static pressure in test section	~950Pa
Dynamic pressure in test section	~16000Pa

2.2 AGARD HB-2 test model

An AGARD HB-2(Hypervelocity ballistic model 2) standard model was selected for the current experimental study, which is a widely used aerodynamic force& moment measurement model in hypersonic experiments, and abundant wind tunnel test data are available from previous research for validation [15].

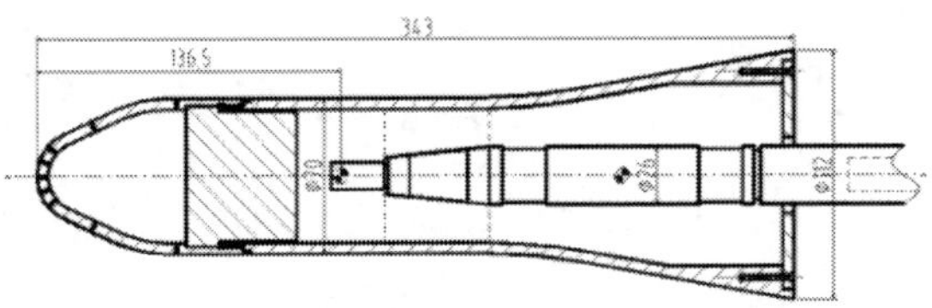

Figure 2. Schematic of AGARD HB-2 model

The test model got a shape similar to a reentry-body, at 0 Angle of attack, the flow field has a symmetrical structure. To prevent the surface from having intense laser reflection, the blackening treatment was obtained on the surface of test model, which reduces the reflection and improves the particle image signal to noise ratio (figure 3).

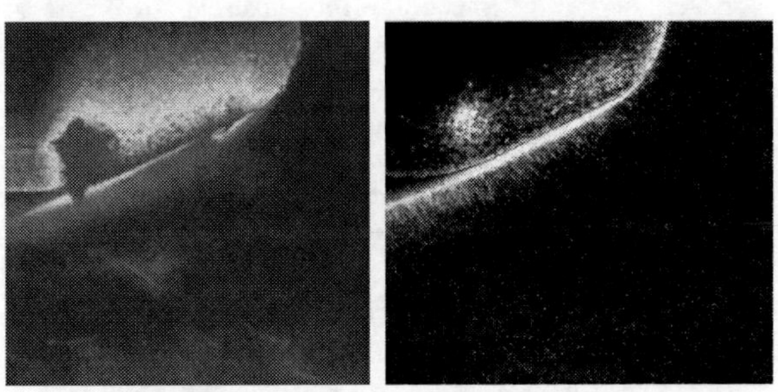

Figure 3. Particle image in the test area (left: baseline, right: model surface blackening treatment)

https://doi.org/10.1051/matecconf/201815103003

2.3 Pressure sensitive measurement techniques

The surface pressure was measured by PSP, which consists of a polymer–ceramic (PC) binder and four (pentafluorophenyl) porphyrin platinum (PtTFPP) as pressure sensor. The PC binder is composed of high concentration ceramic particles with a small amount of polymer to physically hold the ceramic particles to the surface [16]. The binder formed porous surface, which can interact immediately with oxygen, resulting in a high-frequency response.

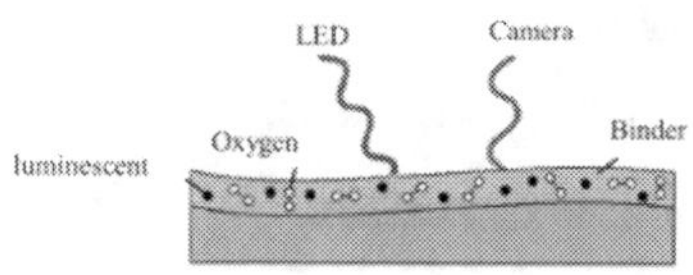

Figure 4. Technical principle of Fast-responding PSP[8]

Several pressure reference taps are seted on the surface of the model to calibrate the PSP measurement results(Fig.5).The pressure reference taps uses Kulite sensor for dynamic measurement, and the stagnation pressure reference tap is set at the head of the model to measure the change of head stagnation pressure during the whole test process.

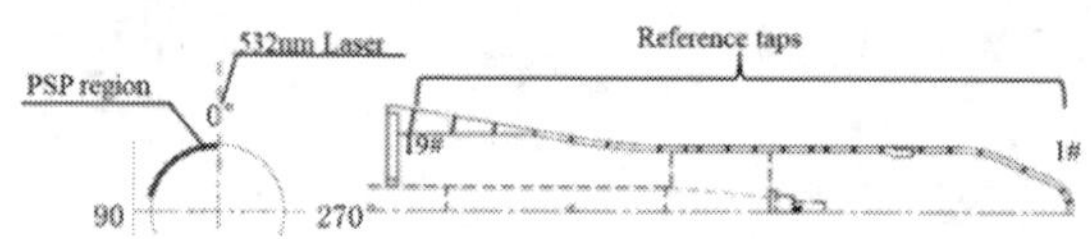

Figure 5. Paints and reference taps on the AGARD HB-2 model(left: PSP and PIV laser position, right: reference taps position on the model)

The experimental setup for the wind tunnel experiments is shown in Fig. 6. The HB-2 model coated with PSP was placed in the test section on the support system. In order to achieve the synchronization measurement of PIV and PSP, the experimental optical path was carefully designed. In order to meet the complex test conditions, the fluidized bed form of particle generator and the injection orifice upstream of the nozzle are designed. The solid nanoparticles are injected into the mainstream by the particle generator, which flows through the expansion of the nozzle.

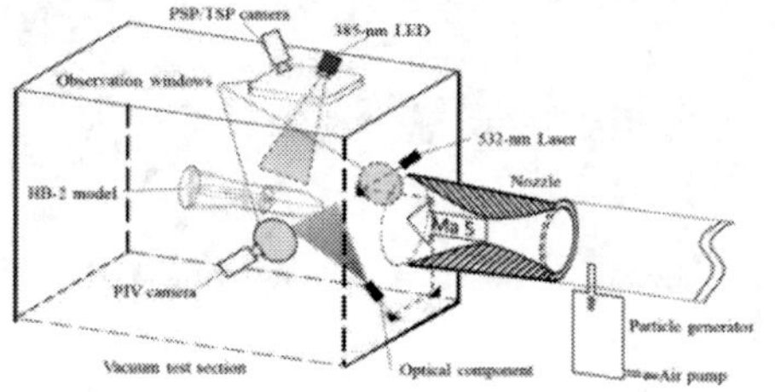

Figure 6. Schematic of the particle seeding device and optical path in the NHW wind tunnel.

The laser beam passes through the observation window, and the mirror group reaches the bottom of the model. A set of lenses turns the beam of light into a piece of light, shining on the symmetrical surface of the model. The PIV camera with a 200mm lens and a 532±5nm filter is set in another observation window for shooting, which provide 2048 by 2048 pixels images in a 75 × 75mm2 area. Continuous illumination was provided by the 385-nm LED through a glass window on the top of the test section. The luminescent signal was recorded continuously by a 12-bit high-speed camera at 500 Hz through a 570-nm long wave pass filter.

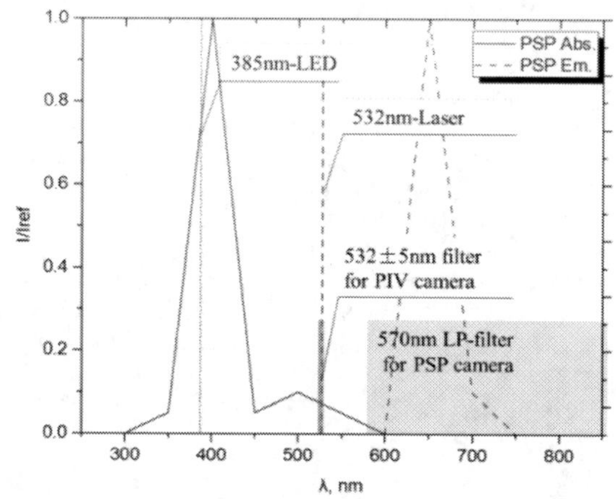

Figure 7. Absorption and emission of PSP (filter for PSP&PIV camera are shown with filled blocks)

2.4 Data acquisition

The PIV and PSP data were recorded simultaneously for a period of about 20s during each wind tunnel running. The external trigger signal for data collection is given by the synchronizer. When the wind tunnel starts working, the initial trigger signal is provided. The synchronizer sends a collection signal simultaneously to each collector after receiving the initial trigger signal (Fig 8). Single pulse triggering for particles generator, 2 Hz square signal for PIV camera, 500 Hz square signal for PSP camera. All the control signals are started in the same time and recorded by the National Instruments data acquisition (NI DAQ) system.

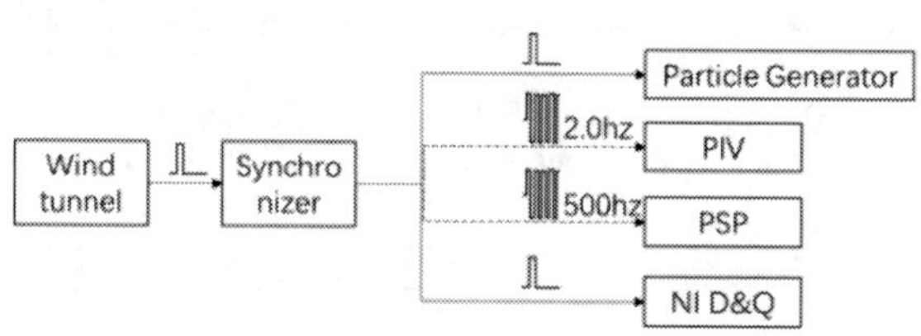

Figure 8. Schematic of system control links

3 Results and discussion

3.1 Flow structures

In the case of Ma = 5, the flow velocity needs to be calculated by the isentropic relation of the compressible flow. The formula is as follows:

$$V_\infty = a \cdot Ma = \sqrt{\gamma RT} \cdot Ma = Ma \cdot \sqrt{\frac{\gamma RT_0}{1+\frac{\gamma-1}{2}Ma^2}}$$

During the experiment, T_0(Total temperature) was about 500K, and γ=1.4, R=287.05. So the flow velocity can theoretically be calculated as V_∞= 915.0 m/s.

Figure 9 is the result of the velocity distribution of the head section after the test flow field is established. The measured velocity in front of the head section is about 900.3m/s, which is consistent with the theoretical value of the calculation.

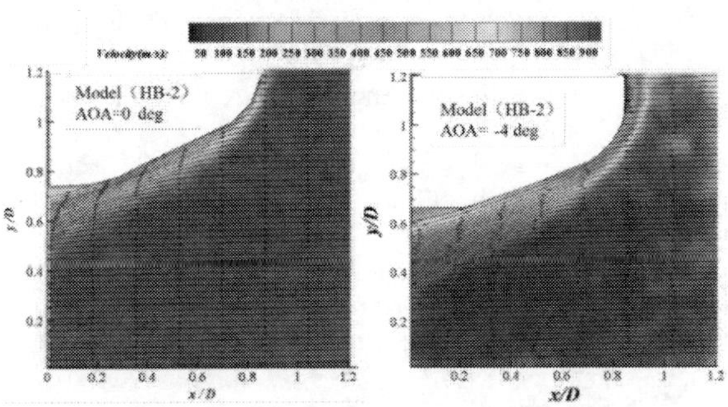

Figure 9. PIV result of velocity field around the dead area of AGARD HB-2 model(The average of 20 images in the steady stage of flow field)

The velocity gradient of the head area is shown in figure 10. The head section of the model is enveloped by an arched slows down area, which extends and gradually moves away from the model in downstream direction. The gradient is larger in the middle area near the symmetrical surface, and gradually decrease in downstream direction. According to the distribution of velocity gradient, it is easy to see that there is a bow shock wave that surrounds the head of the model.

The color Schlieren images recorded at Ma = 5 are shown in Fig. 11 for α = -4°. The spatial position of the bow shock in schlieren images were well matched with PIV measurements.

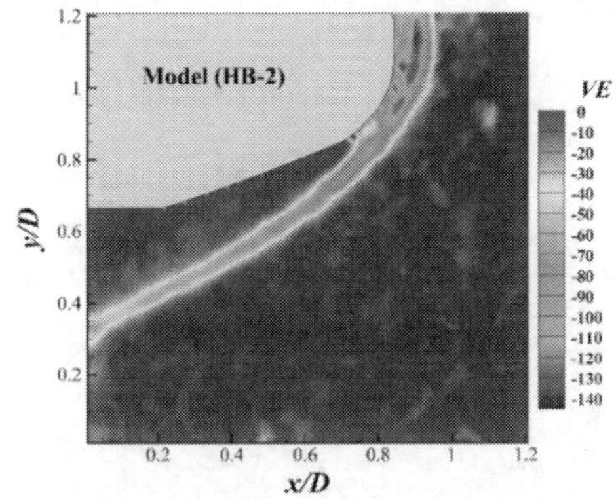

Figure 10. Result of velocity gradient（AOA=-4）

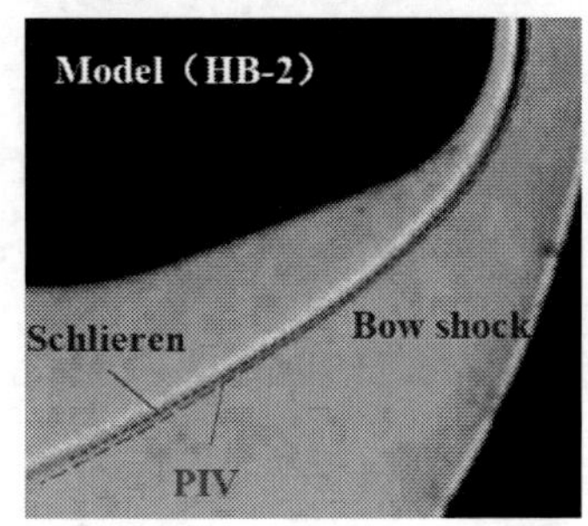

Figure 11. PIV result of bow shock wave（color schlieren taken at AOA=-4）

3.2 Surface Pressure

The surface pressure field of the cylindrical section measured by PSP is presented in Fig. 12. Due to the high surface shear stress and particles in hypersonic flow, the PC binder coating on the cone section surface was destroyed. Note that PSP data were not available on most of destroyed area.

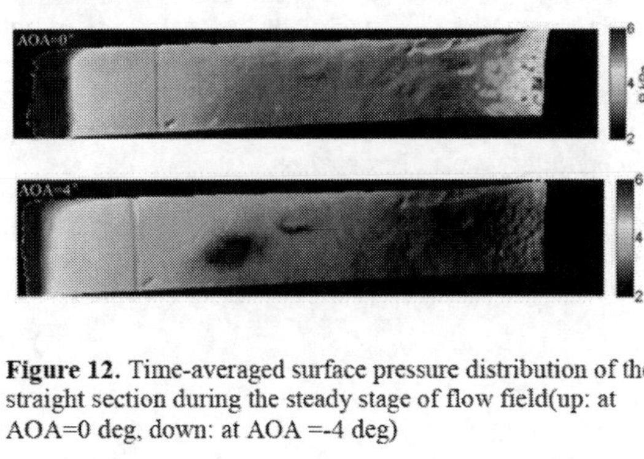

Figure 12. Time-averaged surface pressure distribution of the straight section during the steady stage of flow field(up: at AOA=0 deg, down: at AOA =-4 deg)

The time-averaged pressure field on the generatrix of the model measured by PSP has been normalized by the stagnation pressure (P0 from reference tap-1#) and is presented in Fig.5. The hollow triangle symbol in the diagram represents the pressure data of the reference taps(in Fig.13), measured by Kulite absolute pressure transducer.

The streamwise pressure profile is compared well with data obtained by pressure taps at the same Mach number in DAL(Douglas Aero-physics Laboratory[15]), as shown in Fig. 14. The pressure falls suddenly at the junction of cylindrical and cone sections due to an obvious change in the model profile. Then it decreases gradually and reaches a minimum between x/L = 0.5 and 0.6, finally climbs up at the start of the tail section. Noted that the measurement error between PSP and reference tap is about 0.2% (400 Pa), this error may have been caused by the following factors:(1) sensor paint inhomogeneity of PSP which results in accidental error, (2) system error due to air flow heating effect, (3) the uneven distribute of air density caused by shock waves in the flow filed resulted in optical measurement error.

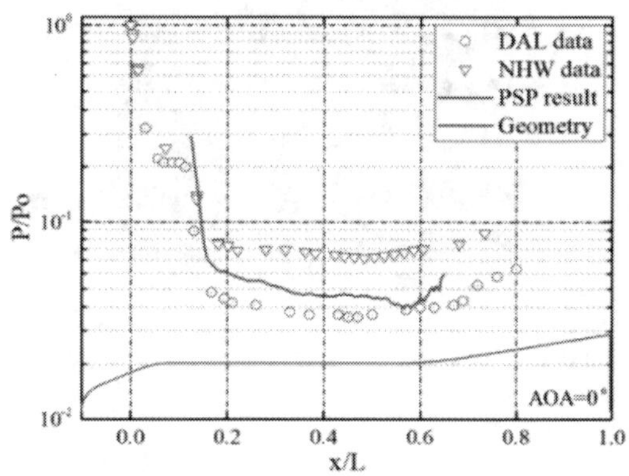

Figure 13. Results of surface pressure profile compared with previous data from DAL

4 Conclusions

Simultaneous PIV and PSP measurements in hypersonic flows are quite challenging mostly due to complex flow conditions. In the current work, the spatial position of the bow shock is accurate captured with PIV measurements. The measurements of the PSP are also consistent with previous studies, the evolutions of pressure during the test period were clearly visualized.

The combined PIV/PSP measurement technique has shown great potential for synchronous shock-wave and surface pressure measurements in hypersonic flows.

Acknowledgement

The author would like to thank all members of Innovation Laboratory of Flight Measurement and Flow Control of NUAA for the interesting and fruitful discussions during the course for this cooperation.

Funding

The authors disclosed receipt of the following support for the research, authorship, and publication of this article : The research work was supported by the National Science Foundation of China (No.11672134).

References

1. Westerweel J, Elsinga G E, Adrian R J. Annual Review of Fluid Mechanics, 2013, **45**(45):409.
2. Adrian R J.Experiments in Fluids ,2005 , **39**(2):159-169
3. Stanislas M, Kompenhans J, Westerweel J: *Particle Image Velocimetry* (Springer Berlin Heidelberg, 2007).
4. Fulvio Scarano . Topics in Applied Physics , 2008, 445—463.
5. Melling A. Measurement Science & Technology, 1999, 8(12):1406—1416.
6. Schrijer F F J,Scarano.7th International Symposium on Particle Image Velocimetry. 2007
7. Elsinga G E, Oudheusden B W V, Scarano F. Experiments in Fluids, 2005, **39**(2):246-256.
8. Liu T, Sullivan J P:*Pressure- and Temperature-Sensitive Paints*(Springer Berlin Heidelberg, 2005).
9. Gregory J W, Sakaue H, Liu T, et al. Annual Review of Fluid Mechanics, 2014, **46**(1):303-330.
10. Xiang X, Yuan M, Yu J, et al. Aiaa Aerodynamic Measurement Technology and Ground Testing Conference. 2013.
11. Okabe T, Miyazaki T, Sakaue H. Aerodynamic Measurement Technology, Ground Testing, and Flight Testing Conference. 2013:2201-2204 Vol.2.
12. Klein C: *Recent Developments of PSP and TSP at DLR for Application to Transonic Flows in Industrial Wind Tunnels*(Nein, 2008).
13. Egami Y, Fey U, Klein C, et al. Measurement Science & Technology, 2012, **23**(11):286-286.
14. XU X, CHENG K.M, WANG. Journal of Experiments in Fluid Mechanics (2009)
15. Gray, J. D. Summary report on aerodynamic characteristics of standard models HB-1 and HB-2. Aedc-tdr-64-137, July 1964.
16. Peng, Di, et al. Experiments in Fluids.**57**.12(2016).

Measurement of Mutual Interference Sound of Columnar Objects in Air Flow

Shohji Hamada[1], Yoshifumi Yokoi[2]

[1] Graduate student of Mechanical Engineering, National Defense Academy, Hashirimizu 1-10-20, Yokosuka-shi, Kanagawa 239-8686, Japan
[2] Department of Mechanical Engineering, National Defense Academy, Hashirimizu 1-10-20, Yokosuka-shi, Kanagawa, 239-8686, Japan

Abstract. When a columnar object is put in a flow, Karman vortex is formed, and aerodynamic noise is generated. It is known that when multiple columnar objects are put in a flow, the flow state becomes complex. This can be known by flow visualization experiment. On other hands, there are few researches on flow sound in the case of mutual interference, that it is not as far as the authors know. Measurement of flow sound is performed using a microphone. Therefore, it is necessary to confirm the sound of the interference flow field reaches the microphone outside the flow field without changing. In this research, experiments were measured to confirm flow sound transmits to a microphone placed outside the flow field without changing. Based on the results, the aerodynamic noise measurement from the columnar object was performed using a low noise wind tunnel experimental apparatus. As a result, it was obtained that some findings on the sound of flow in mutual interference flow field

1 Introduction

When a columnar object is put in a flow, a Karman vortex is formed. This phenomenon is a well-known phenomenon in fluid dynamics. Aerodynamic noise is generated by forming the Karman vortex. This matter is also well known phenomenon. There are several flow visualization experiments that when multiple columnar objects are placed in the flow, the flow behind the multiple columnar objects becomes complex [1-4]. In addition, vortex shedding characteristics which generated from multiple columnar objects are also investigated by measurement experiment using a hot wire anemometer [5-7]. It is due to the mutual interference of the flow based on the columnar arrangement state that the state of the flow becomes complex. By contrast, there are few researches on "sound of flow" in the case of mutual interference compared with the research investigated about "state of flow", that it is not as far as the authors know. In order to investigate the flow sound, the microphone measurement is performed. It is necessary to confirm that the sound of the field of interference flow transmits to the microphone placed outside the flow field without changing sound quality. In this research, experiments were measured to confirm "sound of flow" transmits to a microphone placed outside the flow field without changing sound quality. Based on the preliminary results, the aerodynamic noise measurement from the columnar object was performed using a low noise wind tunnel experimental apparatus. The main experimental contents are noise measurement in a single columnar object and noise measurement in multiple columnar objects arrangement which is a mutual interference flow field. As a result, some findings about "sound of flow" in mutual interference flow field were obtained.

2 Experimental apparatus and method

2.1 Experimental apparatus

The experiment apparatus consists of the low noise wind tunnel, the sound source device, the sound measuring device and two kinds of columnar objects (circular and square cylinders). The low noise wind tunnel has an air outlet and an air inlet in the anechoic chamber. The shape of air outlet is regular octagon. (one side of length is 200 mm; opposite side of length is 475 mm) The wind tunnel can cause air flow of up to 40 m/s and disorder strength at 20 m/s is 1 %. In this experiment, the air flow range is from 0 m/s to 30 m/s. The sound source device consists of the frequency generator, the integrated amplifier and the frequency counter. The sound measuring device consists of the condenser microphone (1/2 inch), the precision sound level meter and the FFT analyzer. The microphone is attached a wind screen (φ70, made of urethane foam). Two kinds of columnar objects are two circular cylinders and two square cylinders. The circular cylinder diameter and the side of the square cylinder are 20 mm, respectively.

2.2 Experimental method

2.2.1 Preliminary experiment

Four kinds of preliminary experiments were performed to confirm that the sound quality does not change by the airflow. The air flow was passed between the microphone and the sound source, and the sound was measured. Air velocity range was from 0 m/s to 30 m/s. Sound pressure level of the sound source range was from 40 dB to 80 dB. The frequency of sound source was 440 Hz. The microphone and the sound source were set apart from each opposite by 1000 mm from the center of wind tunnel outlet. The sound was measured by changing the frequency from the sound source. The placement of the experimental apparatus was the same as in the previous experiment. Air velocity range was from 0 m/s to 30 m/s. Sound pressure level of the sound source was 50 dB. Frequency range was from 130 Hz to 1960 Hz. These frequencies were set for each musical scale of 4 octaves. A cylinder was placed in the airflow and the sound passed through the turbulent airflow was measured. Air velocity range was from 0 m/s to 30 m/s. The sound pressure level of the sound source was 60 dB. The frequency of the sound source was 440 Hz. The cylinder was placed in the airflow. The microphone and the sound source were installed at the position of 1000 mm on both sides from the position of the installed cylinder wake flow of 500 mm. The sound source was installed in the airflow and the sound was measured. Air velocity range was from 0 m/s to 30 m/s. The sound pressure level of sound source range was changed from 50 dB to 70 dB. The frequency was 440 Hz. The sound source was installed in the airflow, and the microphone was installed at 1000 mm from the sound source.

2.2.2 Two circular cylinders case

Two parallel arrangenet circular cylinders were installed in the airflow, and the aerodynamic sound generated by the distance ratio was measured using a microphone installed outside the airflow. A schematic view is shown in Fig 1. Two cylinders were fixed at 500 mm of the outlet. Fig. 1 (b) showed the installation state of two cylinders. Two cylinders can be changed distance between cylinders by attatchment equiped on the outlet.The diameter of the two cylinders was same (D = 20 mm). The center of two cylinders was set the (500, 0). It had been confirmed that two cylinders did not resonate duaring experiment. The microphone was set the (500, -1000). The air velocity was set 20 m/s. Two cylinders were set the distance ratio indicating the characteristic flow (G/D = 0.25, 0.75, 1.75) [5-7]. G/D is called distance ratio, after here. G is the distance between the surface of cylinders. D is diameter. The characteristic flow in the previous research is that when the distance ratio is large, the vortices generated from the two cylinders are the same as the vortexes generated from independent each single cylinders. When the distance ratio decreases, each vortex generated from the two cylinders releases a different vortex. When the distance ratio is small, two cylinders emitt a vortex that the two cylinders are generated from like a single cylinder having twice the diameter.

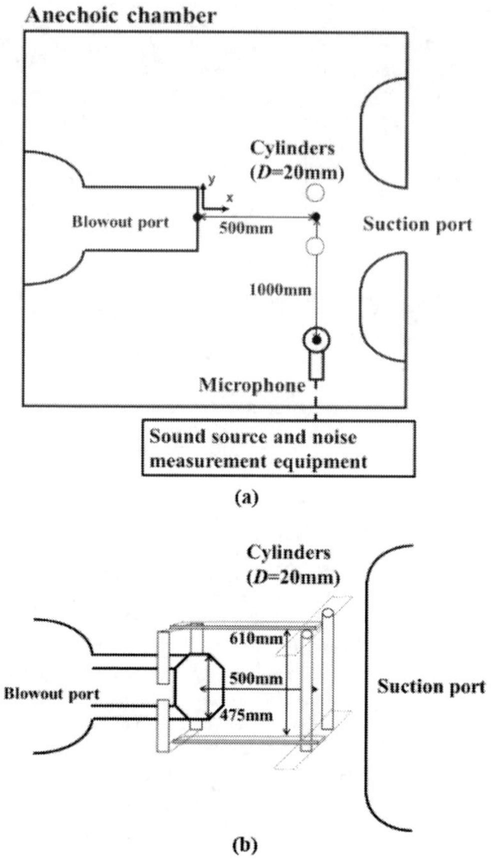

Figure 1. Experimental apparatus placement in anechoic chamber;(a) Top view, (b) Side view

2.2.3 Two square cylinders case

The experiment of using two square cylinders was performed for investigating influence of changing separation point. Because a separation point of circular cylinder is not constant in the airflow. On the other hand, a separation point of square cylinder is constant. Two parallel arrangement square cylinders were installed in the airflow, and the aerodynamic sound generated by the distance ratio was measured using a microphone installed outside the airflow. Two square cylinders were fixed in the rear stream of the outlet as the case of circular cylinders. The diameter of the two cylinders was same (D = 20 mm). The center of two cylinders was set the (500, 0). The microphone was set the (500, -1000). The air velocity was set 25 m/s. Two cylinders were set the distance ratio indicating the characteristic flow (G/D = 0.6, 1.7, 4.9). Previous research showed the characteristic flow [3]. When the distance ratio is large, the vortices generated from the two cylinders are the same as the vortexes generated from independent each single cylinders. When the distance ratio decreases, each vortex generated from the two cylinders releases alternating vortexes and twin vortex. When the distance ratio is small, two cylinders emitt alternating vortexes and one side of the vortex.

3 Experimental results and discussion

3.1 Preliminary experiment results

From the four preliminary experiments, the following was obtained. The sound does not change its quality even across the airflow. This phenomenon is not dependent on frequency. The same result can be obtained even if the airflow is turbulent. The sound generated in the airflow is transmitted outside the airflow without changing its quality.

3.2 Two circular cylinders case

Fig. 2 shows the relationship between vortex shedding frequency and sound pressure level obtained from a microphone measurement experiment. Airflow velocity was 20 m/s. The Strouhal number was about 0.18.($St = f\,d / U$, f: vortex shedding frequency, d: cylinder diameter, U: mainstream velocity) This is the same result as the case of using hot wire anemometer experiment. Fig. 3 shows the relationship obtained from two parallel arrangement circular cylinders experiment between vortex shedding frequency and sound pressure level at each distance ratio. In the case of previous study using hot wire anemometer, two frequency peaks can be seen at the distance ratio 0.75. But in each other distance ratios, there was only one frequency peak. Fig. 4 showed the relationship between the distance ratio and the Strouhal number. At a point of distance ratio 1.75, the present result shows the same Strouhal number with the case of a single cylinder. Other distance ratios were different from those of the previous research. It considered this difference was caused the influence of the background noise, the change of the sound, and the change of the separation point. However, they did not cause the difference. In the Fig. 3, the sound pressure level of the background noise is not large so it could not affect the result. From preliminary experiment results, the sound generated inside the airflow transmits to the outside the airflow with not changed its quality.

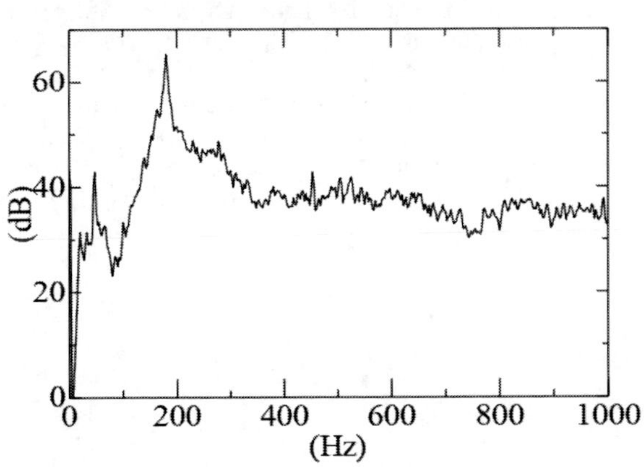

Figure 2. Single circular cylinder: Relationship between vortex shedding frequency and sound pressure level

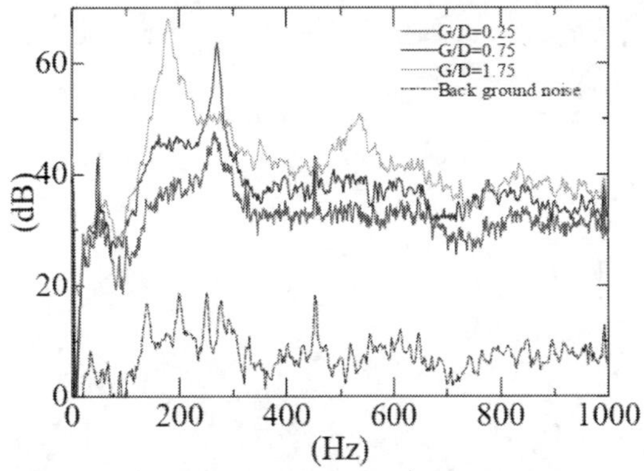

Figure 3. Two parallel circular cylinders: Relationship between vortex shedding frequency and sound pressure level

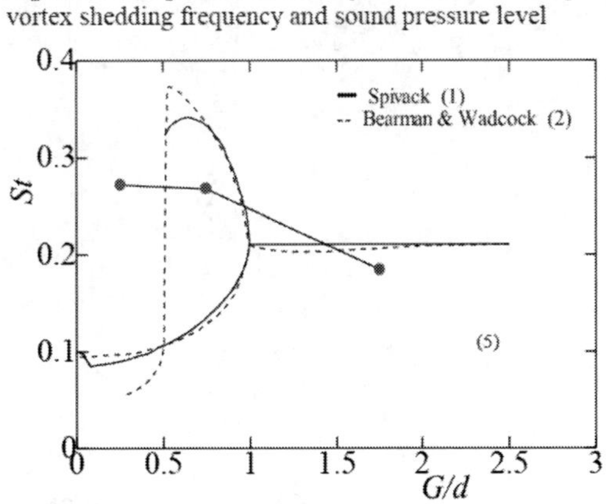

Figure 4. Two parallel circular cylinders: Relationship between distance ratio and Strouhal number

3.3 Two square cylinders case

The generated sound from single square cylinder was measured. Airflow velocity was 25 m/s. Fig. 5 shows the relationship between vortex shedding frequency and sound pressure level obtained from a single cylinder experiment. The Strouhal number was about 0.124. This is the same result as the result of using hot wire anemometer result. Fig. 6 shows the relationship obtained from this experiment between vortex shedding frequency and sound pressure level at each distance ratio. At distance ratio 1.7 and distance ratio 4.7, one frequency peak can be seen. At distance ratio 0.6, the frequency peak cannot be seen. Fig. 7 shows the relationship between the distance ratio and the Strouhal number. At a distance ratio 4.9, the result shows the same Strouhal number with the case of a single square cylinder. At distance ratio 1.7, multiple frequency peaks cannot be seen. At distance ratio 0.6, there is no frequency peak. This is also different from previous research. Experimental results in the case of a square cylinder and the case of a circular cylinder are same results. That is, the measurement result by the hot wire anemometer is different from the measurement result by the microphone.This result shows that the vortex shedding frequency is different from the frequency of sound of mutual interference; this shows it cannot be trusted to know the condition of objects only by the result of the sound. For example, even if two objects are present at any distance ratio, the frequency obtained by sound is one frequency:thus it cannot be not known the condition of objects.

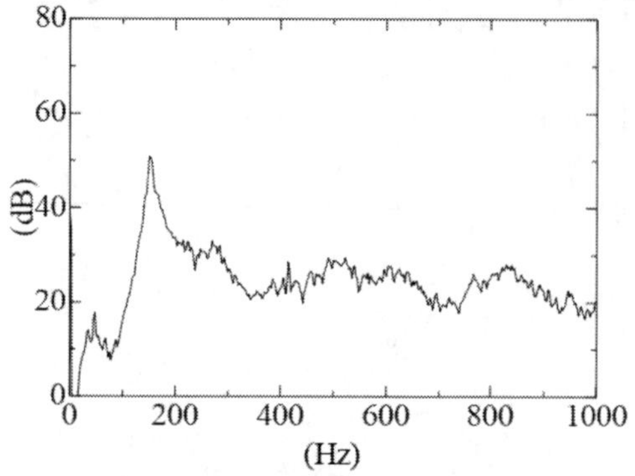

Figure 5. Single square cylinder: Relationship between vortex shedding frequency and sound pressure level

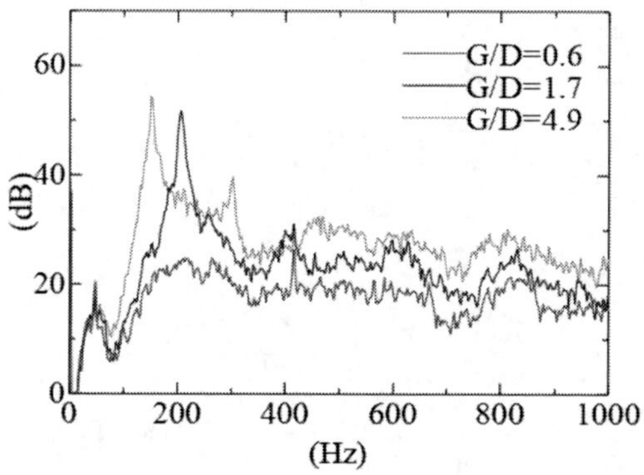

Figure 6. Two parallel square cylinders: Relationship between vortex shedding frequency and sound pressure level

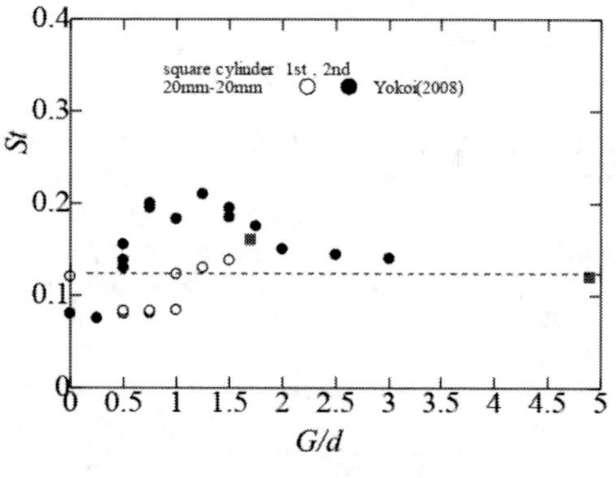

Figure 7. Two parallel square cylinders: Relationship between distance ratio and Strouhal number

4 Conclusions

The mutual interference flow from the two columnar objects installed in the airflow was measured by the microphone. The following conclusions were obtained.

(1) The quality of the sound generated in the airflow can be known outside the airflow without changing the quality.
(2) In the case of a single columnar object, it was confirmed that the measurement result of the hot wire anemometer and the measurement result of microphone became same result.
(3) As a result of measuring the sound of mutual interference flow and analyzing the frequency of the sound, it was shown that it is different from the result of the hot wire anemometer.
(4) In the case of square cylinder, also the same result as the conclusion (3) was obtained.

References

1. Ohumi, K., Imaichi, K. and Tada, E., On Vortex Streets Behind Circular Cylinders in Side-by-Side Arrangement, Journal of the Visualization Society of Japan, Vol. **10**, Suppl. No. 1 (1990), pp. 35-40.
2. Ohumi, K., Visualization Experiment and Numerical Calculation on Two Circular Cylinders in Staggered Arrangement, Journal of the Visualization Society of Japan, Vol. **15**, Suppl. No. 1 (1995), pp. 217-220.
3. Y Yokoi, K Hirao, The interation Vortex Flow Around a pair of Parallel Arranged Bluff Cylinders. (In the case of stationary setting), Proceedings of the Japan Society of Mechanical Engineers (Part B) 74 vol. **741**(2008-5)
4. Moriya, M. at al., Fluctuating Fluid Forces of Two Circular Cylinders in Parallel Arrangement at Close Spacing, Transaction of the Japan Society of Mechanical Engineers, Series B, Vol. **68**, No. 676 (2002), pp. 3310-3315.
5. Spivak, H. M., Vortex Frequency and Flow Pattern in the Wake of Two Parrallel Cylinders at Varied Spacings Normal to an Air Stream, Journal Aeronatutical Sciences, Vol. **13** (1946), pp. 289-297.
6. Bearman, P. W. and Wadcock, A. J., The Interaction Between a Pair of Circular Cylinders Normal to a Stream, Journal of Fluid Mechanics, Vol. **61**, Part 3 (1973), pp. 499-511.
7. Kobayashi, T., Characteristics of Fluid-Dynamic Forces Acting on Circular or Square Cylinders in Close Proximity, Transaction of the Japan Society of Mechanical Engineers, Vol. **42**, No. 357 (1976), pp. 1452-1461

The Gradational Route Planning for Aircraft Stealth Penetration Based on Genetic Algorithm and Sparse A-Star Algorithm

LI Maoquan[1], ZHANG Yunfei[1] and LI Shihao[1]

[1]School of Aeronautic Science and Engineering, Beihang University, Beijing 100191, China

Abstract. It is established for a gradational route planning algorithm which includes two layers. The first layer makes use of genetic algorithm to obtain the global optimal path by its global optimal characteristics. The second layer makes use of A* algorithm to obtain the local optimal path by its dynamic characteristic. When flying along the global optimal path, locating the new threat and confirming its performance, the aircraft can plan the local optimal path timely by A* algorithm. It is constructed for the cost function with two goals of the range and the average detection probability, which is used as the goal function for optimal path planning. Two paths that obtained from two optimal methods are merged to construct the optimal route comprehensively considering the threats and range. The simulation result shows that the cost of new optimal route is lower than the original optimal path obtained only by the genetic algorithm.It revealed that our algorithm could obtain an optimal path when a new radar threas occured.

1 Introduction

Aircraft path planning is to get an optimal or satisfactory flight path to ensure the success of missions, synthetically considering the aircraft arrival time, fuel consumption, threats and other factors[1][2]. In modern warfare, with the increasing performance of air defense systems, aircraft safety penetration is the key to achieving precision strike. So the low altitude penetration path planning is becoming more and more important.

In the artificial intelligence algorithms[3], the genetic algorithm has strong robustness, characterized by flexible and simple implementation, without being limited by the search space. It is a kind of parallel global search algorithm, but it has a long planning time and is not suitable for real-time path planning. A* algorithm due to the introduction of the heuristic function, is able to avoid the blind search[4], which greatly improves the efficiency of the search.

In this paper, a gradational route planning algorithm is established, including the first layer using genetic algorithm to obtain the global optimal path by its global optimal characteristics based on the known threats data, and the second layer using A* algorithm to obtain the local optimal path by its dynamic characteristic facing the new threat. Combining the global route optimization and dynamic route optimization, this algorithm can play advantages and avoid shortcomings.

2 Radar model and Radar network integrated detection model

In this paper the so-called "critical scattering cross-section" method is used to calculate the detection capability of single-step radar[5].

$$\sigma_{cr(\hat{p},P_{fa})}^{(c+n)} = \sigma_{cr(\hat{p},P_{fa})}^{(n)} + \frac{\sigma_c}{F^4}\left(\frac{S}{N}\right)_{\min(\hat{p},P_{fa})} \quad (1)$$

Considering the radar cross section of target, the signal to noise ratio(SNR) is.

$$\frac{S}{N} = \frac{\sigma}{\sigma_{cr(\hat{p}_{fa})}^{(c+n)}} \quad (2)$$

According to the relationship of detection probability with SNR for the pulse integration number of a given radar, the detection probability of the radar to the target under certain false alarm probability can be calculated.

The information fusion center of the radar network integrates the information detected by the single radar to obtain the comprehensive detection information[6][7]. Data fusion often uses the Rank K fusion rule, that is, assuming that there are n radars, and if there is at least K radars to determine the target, the radar network determines the target[8]. When K=1, data fusion is called "OR rule". For the distributed radar network, assuming the detection statistics are independent and the detection probability is $P_{d,k}$, then the detection probability of at least one radar found in the radar network is expressed as:

$$P_D = 1 - \prod_{k=1}^{n}(1 - P_d(k)) \quad (3)$$

3 Genetic algorithm

3.1 Gene code

In order to discretize the planning area based on simulation step size and referring to the above constraints, the change of course is not more than 45°. Divide the possible courses within each grid into five values[9], as shown in Figure 1.

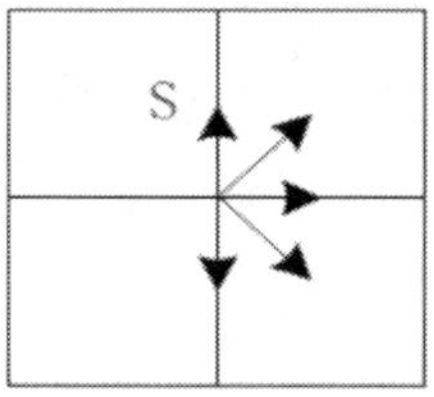

Figure 1. Node's directions

3.2 Selection operator

This paper combines the optimum maintaining tactics with the remainder stochastic sampling[4]. The specific process is as follows:

1) Calculate the expected number N_i in the next generation for each individual in the population.

$$N_i = M \cdot F_i / \sum_{i=1}^{M} F_i \qquad (4)$$

F_i is the fitness value of the current individual;M is the population size.

2) The integer part of N_i is the number of survivors of the individual in the next generation, so that N individuals in the next generation can be determined.

$$N = \sum_{i=1}^{M} floor(N_i) \qquad (5)$$

3.3 Crossover operator

This paper uses the method of reconstructing a small track near the intersection to ensure that the matching tracks can be crossed successfully, as shown in Figure 2.

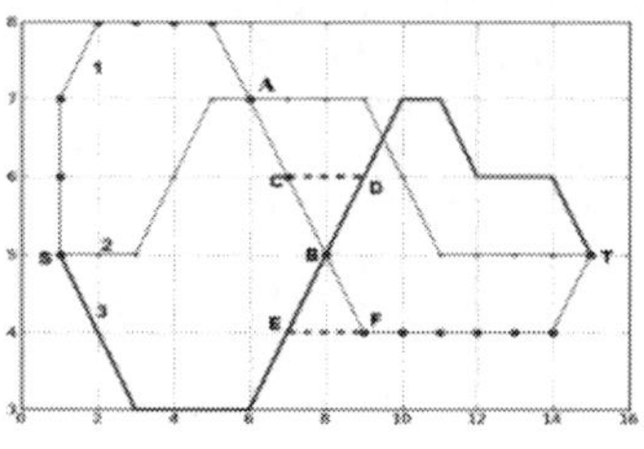

Figure 2. Crossover operator

In the above figure, the two paths 1 and 3, due to not meet the corner of the constraints at the B node which is invalid generally for the intersection, will give up the cross point. The following measures are adopted to complete the two tracks cross. C, D, E, and F are selected near the two track B nodes, and then the track segments (BE segment, DC segment) are truncated.

3.4 Mutation operator

In this paper, the mutation operator adopts the method of combining reconstruction operator and disturbance operator. In the mutation, select an operator randomly to operate. The specific operation is as follows:

Reconstruction operator: first select a node as the starting node of the mutation on the track, such as the A node in Figure.3. Then find a track segment (ABT segment) which is different from the original track from the beginning point of the mutation and the target node and satisfies the track constraints, and replace of the original track segment (ADT segment).

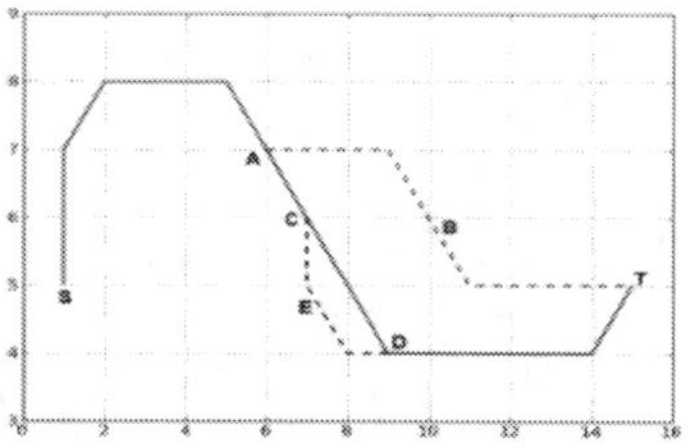

Figure 3. Mutation operator

Perturbation operator: Firstly, a node is randomly selected as the starting node of the disturbance operation, such as the C node in Figure 3. Then reconstruct a segment (CED segment) to the end node of the mutation (D). If the segment still satisfies the track constraints, the mutation is successful.

3.5 Cost function of genetic algorithm

Assuming that the radar network of n-radars has a comprehensive detection probability $P_{C,j}$ of the target at point j, the avoidance capability of a route can be evaluated by the average comprehensive detection probability of the route of m-point.

$$\overline{P_D} = \frac{1}{m} \sum_{j=1}^{m} P_{D,j} \qquad (6)$$

The length and the detection probability are commonly not on the same order of magnitude (0–1). Therefore, it needs to be normalized.

$$L_k = \frac{\sum_{j=1}^{n} L_{k,j} - L_{\min}}{L_{\max} - L_{\min}} \qquad (7)$$

Where:

$L_{\min}$ is the shortest range, that is, the straight distance from the starting point to the end point; $L_{\max}$ is the maximum range from the start to the end.

The cost function of the genetic algorithm is finally rewritten as follows:

$$J = w_1 L_k + w_2 \overline{P_D} \qquad (8)$$

4 A-Star algorithm

4.1 Sparse A* algorithm

The sparse A * algorithm introduces the constraints of the path planning into the A * algorithm[9], which can eliminate many sub-nodes that do not satisfy the constraint, and reduce the computational complexity. Since this article specifies that the maximum turning angle within a grid cell should be less than 45°, there are actually at most three sub-nodes that satisfy the constraints, as shown in Figure 4.

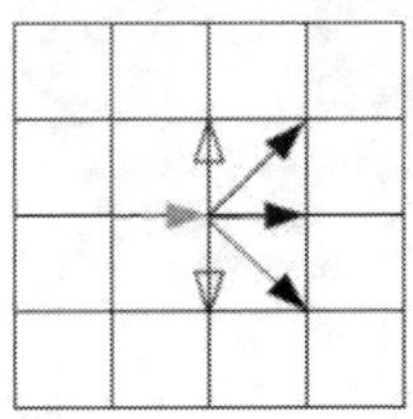

Figure 4. node's directions of sparse A* algorithm

4.2 Cost function of sparse A* algorithm

The evaluation function[10] of the A * algorithm is

$$f(n) = g(n) + h(n) \tag{9}$$

where $f(n)$ is the total cost from the start node to the current point, $g(n)$ is the actual cost from the start node to the current node, $h(n)$ as the heuristic function, represents the cost from the current node to the target point.

5 Gradational path planning

The above algorithms have their own advantages and disadvantages. The advantage of genetic algorithm is that the path generated is global optimum, but the disadvantage is that the process may not be convergence and the computation time increases with the increase of precision. So it is not suitable as the real-time planning algorithm. A* algorithm has the advantage of fast computation, and the computation time won't be affected by the calculation precision. But the result may be the local optimum. The gradational path planning is an optimization algorithm that combines the advantages of the two algorithms.

The gradational path planning requires the global optimal path is obtained by genetic algorithm based on the known threats and terrain data before flight, and the path is used as the standard path. During the flight, through the threat detection device to detect new threats. When the new threat occurs, the aircraft conduct real-time path planning by A* algorithm, and update the standard path by the new path to avoid new threat.

In Figure 5, the path SMBNE is the optimal path for the known threats by genetic algorithm, and S and E are the start point and end point of the path. A new threat C is detected when the aircraft is flying to point M. In order to avoid the new threat, the aircraft will use the A* algorithm to generate a new path, and then return to the original optimization path after bypassing the new threat to avoid other threats on the battlefield. The idea is that draw a circle with the center C and the radius CM and the intersection M and N of the optimization path are the start point and the end point of the A* algorithm. In this way, a track (such as MDN) is found between M and N to make the cost value lower than the original track MBN. If the distance of CM is larger than the distance between C and the end point E, then E is taken as the end of the A* algorithm.

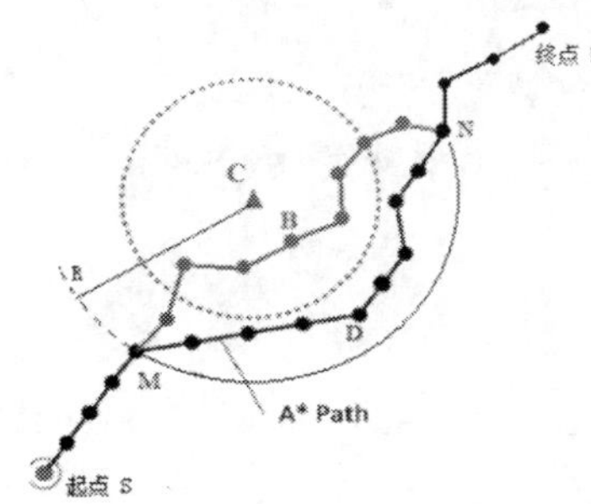

Figure 5. diagram of gradational path

6 Simulation results and analysis

6.1 Battlefield setting

Assuming the size of the planning area is $400km \times 400km$, and the number of grids is 100×100, so the size of each grid is $4km \times 4km$.

The ground threat system uses two types of radars as shown in Table.1. And the sites of radars are shown in Table.2.

The aircraft target is spherical target and its RCS=-5dBm^2. The start point is (0, 260, 3), and the end point is (400, 130, 3). The range factor $w_2 = 0.6$.

Table 1. Radar performance

Performance	AN/ SPS-48	TRMS
Center frequency (GHz)	3	5.6
Polarization mode	HH	HH
Maximum detection range (km)	230	100
Radar characteristic constants ($10^{21}m^2$)	8.203	0.409

Table 2. Coordinates of radar

No	Type	X/km	Y/km	Z/km
A	TRMS	140	100	0.5
B	TRMS	160	270	0.5
C	AN/SPS-48	280	380	0.5
D	TRMS	300	190	0.5

6.2 Simulation results of genetic algorithm

Figure 6 contains the cloud diagram of comprehensive detection probability and the optimized path computed by genetic algorithm in the case of known the distributions of radars A, B and C. This path is used as the original path in the dynamic planning.

Figure 7 is the cloud diagram after the new threat D is detected, which shows a portion of the original path entered the detection area of the radar D.

In contrast, the comprehensive detection probability of the original path after the addition of radar D is significantly improved as shown in Table 3.

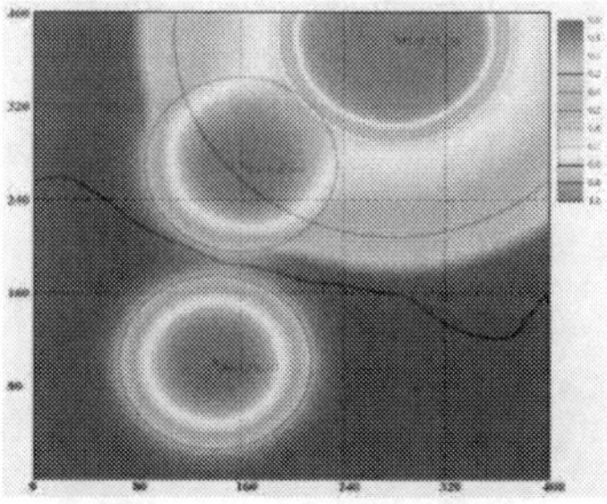

Figure 6. Standard path generated by genetic algorithm

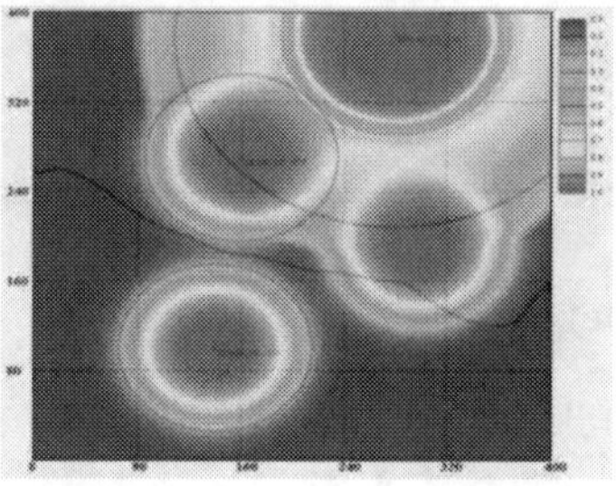

Figure 7. New detection probability distribution

Table 3. The comprehensive detection probability

Radars	Comprehensive detection probability
A,B,C	0.103
A,B,C,D	0.716

For the aforementioned radars distribution and the original path, it is assumed that the aircraft finds the new radar D and calculates its threat area before arriving at point M. The dynamic planning simulation is carried out on the aircraft at three positions M, N and Q respectively. (Figure 8.)

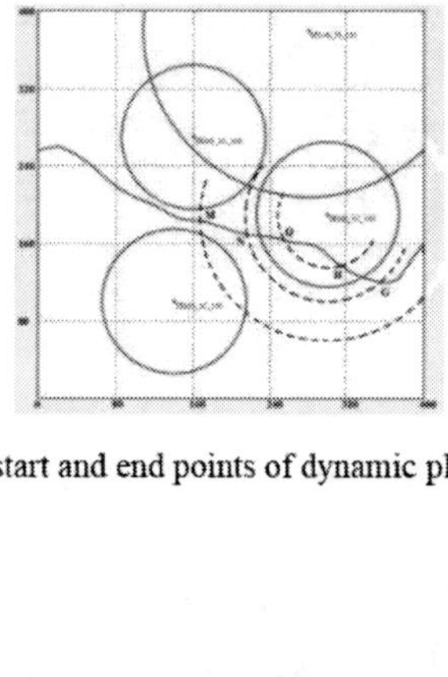

Figure 8. The start and end points of dynamic planning

With the point M as the start point for dynamic planning, the circle with the radius DM has no second intersection, then select the end of the path as the dynamic path planning end point. The length of the final path is 462km, the comprehensive detection probability is 0.143 and the cost value is 0.215. (Figure 9.)

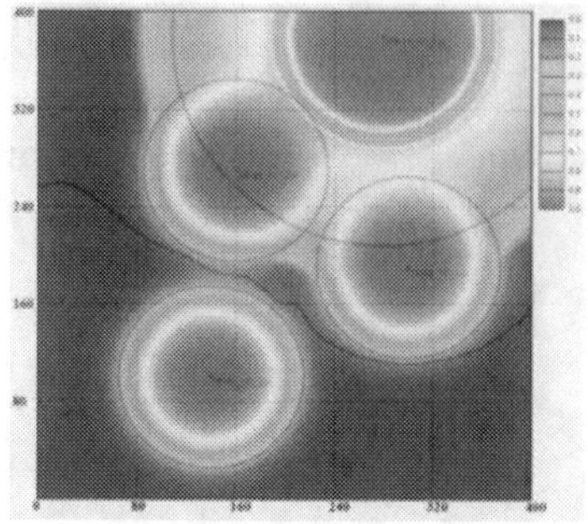

Figure 9. Gradational optimal path with point M

When point N is the start point for dynamic path planning, the point G is the end point. The length of the final path is 474km, the comprehensive detection probability is 0.242 and the cost value is 0.293. (Figure 10.)

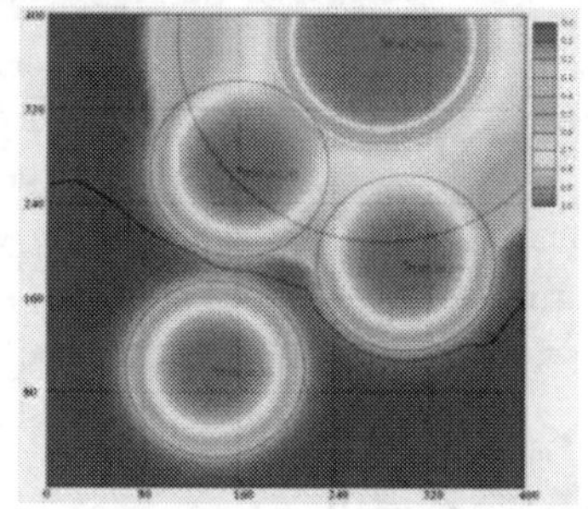

Figure 10. Gradational optimal path with point N

When point Q is the start point for dynamic path planning, the point H is the end point. The length of the final path is 472km, the comprehensive detection probability is 0.593 and the cost value is 0.429. (Figure 11.)

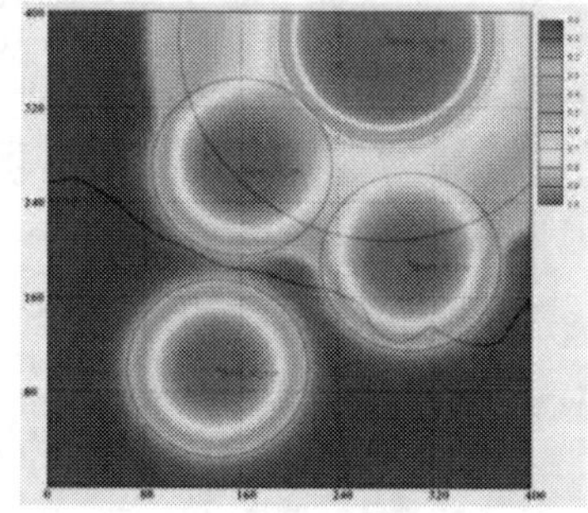

Figure 11. Gradational optimal path with point Q

In the three cases, the parameters of the final path are shown in Table.4. It can be seen that after the dynamic

https://doi.org/10.1051/matecconf/201815104001

path planning, the comprehensive detection probability is significantly reduced, but the range has increased. Compared with the original path, the cost value of the two paths with M and N as the start points is obviously reduced. For the case of point Q as start point, the dynamic planning is too late and the aircraft is too close to the new radar, although the dynamic planning makes the comprehensive detection probability is reduced, but the range increases, resulting in the cost is almost equal to the original path's.

Table 4. The parameters of different paths

Start point	Range (km)	Comprehensive detection probability	Cost
Original path	455.5	0.716	0.425
M	461.7	0.143	0.215
N	473. 6	0.242	0.293
Q	472.2	0.593	0.429

The simulation results show that the aircraft should observe the battlefield situation at any time and detect the new threat as soon as possible. After determining the enemy's new threat, the dynamic route planning should be carried out as soon as possible and integrated with the original optimization track.

7 Conclusion

A gradational algorithm is established that includes two layers of optimization algorithm. the first layer is the global path optimization completed by genetic algorithm using its global optimal performance. the second layer is a dynamic path optimization completed by A* algorithm using its dynamic characteristics. In flight, the aircraft flies along the global optimal path planed by the genetic algorithm. When a new threats was detected and its performance was determined, the A* algorithm was used to give a local optimal path. Two paths was linked to combining the gradational path planning algorithm .. The range of the new optimal path would be longer than the original path, but the comprehensive detection probability is much lower than the original path. Considering the range and comprehensive detection probability, the cost is lower than the original path if taking the dynamic path planning before detected by the new threat.

References

1. HE Pei. *Research on the Method of Low-altitude Penetration Track Planning*. Beihang University, Beijing, China, 2003.
2. ZHENG Changwen, YAN Ping, DING Mingyue. *Route Planning For Air Vehicles*. National Defense Industry Press, Beijing, 2008.
3. WANG Jun, ZHOU Shudao, ZHU Guotao. Research of Common Route Planning Algorithms for Unmanned Air Vehicle. Fire Control & Command Control, Vol. 37, No. 8 (2012), p. 5-9.
4. DU Ping, YANG Chun. Introduction of Air Vehicle Path Planning Algorithms. FLIGHT DYNAMICS, Vol. 23, No. 2 (2005), p. 10-14.
5. ZHANG Kao, ZHANG Yunfei, MA Dongli. *Calculation and Analysis of Radar Stealth Performance of Aircraft*. National Defense Industry Press, Beijing, 1997.
6. LIN Cuoyun, DONG Jiali. *Method and Theory of Multi-objects Optimization*. JiLin Education Press, Changchun, China, 1992.
7. LI Chunhua, ZHENG Changwen, ZHOU Chengping. Fast search algorithm for 3D-route planning. Journal of Astronautics, Vol. 23, No. 3 (2002), p. 13-17.
8. FAN Hongda, MA Xiangling, YE Wen. *Low Altitude Penetration Route Planning Technology for Aircraft*. National Defense Industry Press, Beijing, 2008.
9. LI Xia, XIE Jun, et al. Path Panning for UAV Based on Improved Heuristic A* Algorithm. The Ninth International Conference on Electronic Measurement & Instruments(3) (2009), p.488-493
10. HE Wang, QING Li, and NONG Cheng. Real-time Path Planning for Low Altitude Flight Based on A* Algorithm and TF/TA Algorithm*. 8th IEEE International Conference on Automation Science and Engineering(2012), p. 837-842

MATEC Web of Conferences 151, 04002 (2018) https://doi.org/10.1051/matecconf/201815104002
ACMAE 2017

Research and Application of Virtual Simulation Technology in the Aerospace Bearing Design and Manufacture

Liu Jiangshan[1] and Chen Ming[2]

[1]*Tongji University, School of Mechanical Engineering, 201804 Shanghai, China*
[2]*Tongji University, Sino-German College of Applied Sciences, 201804 Shanghai, China*

Abstract. Bearings are widely used in aerospace and other fields, its performance directly affects the production efficiency and safety. Nowadays, virtual simulation technology has become an indispensable part of intelligent manufacturing field. As a virtual simulation technology, FEA has been widely used in bearing design. China needs to import many aerospace bearings every year in aerospace area, Chinese national defense and other high precision technology is limited because the blockade of advanced bearing technology. We can use dynamics modeling and virtual simulation technology to achieve the predictive design, and strive to achieve foreign level. In this paper, the author proposed a method of bearing design based on virtual simulation technology. The factors of bearing which affect the dynamic characteristics are considered, the process of design bearing based on virtual simulation is also considered. According to the different design parameters, the simulation results are used to verify the rationality, these can reduce the cost and improve the reliability. The virtual simulation technology is applied to design the 7016C angular contact ball bearing which used in aerospace area, and supported decision-making in structure design and data analyze. Finally, The feasibility of this method is verified by experiments..

1 Introduction

With the development of computer technology and digital technology, product design engineers in the field of intelligent manufacturing can use virtual simulation technology and product design synchronization to carry out the design of human-computer interaction [1], and use the obtained simulation results to guide the design of the product. As an import technology of virtual simulation, CAE (Computer Aided Engineering) technology is a rapid development of computing technology, it is a core technology to achieve major engineering and industrial product design analysis, simulation and optimization. At present, the global industrial revolution, the manufacturing transformation and upgrading accelerate the CAE technology exploration and dependence, more and more enterprises take the simulation as an important part of manufacturing innovation [2]. The FEA (finite element analysis) process contains a large amount of initial information that needs to be processed, and also generates a certain amount of analysis result information, which facilitates the understanding and application to the analyst, and let them to observe the simulation results quickly and accurately [3]。

Rolling bearings are one of the most widely used parts in aerospace area and other modern productions. They are widely used in various rotating parts of aerospace products [4-5]. They have advantages of little starting torque, little friction and easy replacement. we need to learn from foreign technology and achieve our own design bearings. As we all know, it is difficult to accurately grasp the bearing design parameters in the design processes. The dynamic performance is very complex in the rotation processes between the parts of bearing, so it involves many disciplines and technical fields [6].

Bearing virtual simulation technology is a combination of mathematics, computer and signal processing combined disciplines [7]. Through the virtual simulation technology, you can get the dynamic parameters in the process of bearing rotation,and also can evaluate the performance of the bearing, so as to provide a reference for the design and optimization bearing. Bearing virtual simulation technology began in the late 1950s, Jones A B first applied the computer technology to analyze ball movement and friction characteristics of the ball bearing and proposed quasi-static analysis method of the rolling bearing [8]. With the development of bearing theory and computer technology, the simulation technology is really used in the rolling bearing analysis field after the middle 1970s [9]. In the process of virtual design, it emphasizes designing, analyzing, debugging and manufacturing, the digital virtual simulation in the virtual environment formed by the computer before the product is put into operation and completing the problem in the process of designing.

2 Analyze the structure and select the design parameters

The 7016C angular contact ball bearings contain four parts: outer ring, inner ring, roller and cage. It is widely used in aerospace area and the structure diagram is shown in Fig. 1.

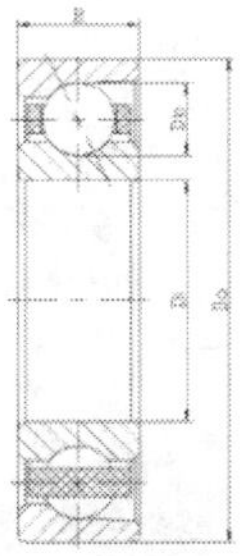

Figure. 1. The structure model of 7016C bearing

In order to verify the rationality of bearing design, 7016C (self-designed) means bearing designed by self, 7016C (FAG) means bearing imported from FAG company. The main structural parameters of the two kinds of bearings are shown in Table 1.

Table 1. The main structure parameters of designing

Name	*7016C (self-designed)*	*7016C (FAG)*
outer diameter	125mm	125mm
Inner diameter	80mm	80mm
width	22mm	22mm
roller diameter	13.494mm	14.288mm
Number of roller	20	19

3 Establish vibration equation and mathematical model of bearing

In order to study the dynamic characteristics of the bearing, Gupta considered the influence of the ball mass and the displacement of the Hertz load in the research process, a six degree of freedom model is used to simulate the steel ball movement process [10]. Walter analyzed the motion model of ball bearing under the action of lubricating oil [11].

The vibration of the ball bearing is mainly caused by non-linear force in the rotation process, mainly including three categories:

(i) The inherent vibration of bearing structure.

$$\begin{cases} f_x = \dfrac{\sqrt{K_x/M}}{2\pi} \\ f_y = \dfrac{\sqrt{K_y/M}}{2\pi} \end{cases} \tag{1}$$

The natural vibration frequency of the bearing can be calculated by:

Where M is the mass of the bearing system.

(ii) The vibration caused by unbalance force.

The expression of the vibration frequency produced by the unbalanced force is as follows and its value is equal to the frequency of shaft rotation:

$$f = \frac{\omega}{2\pi} \tag{2}$$

(iii) The vibration caused by the rolling element through the bearing area

When the outer ring fixed, the inner ring rotated, the bottom roller rolled from the load area to the non-load area, it will cause variable stiffness vibration.

When the number of compression rollers pass through the bearing area is odd and there is a roller on the load line, it is called "odd pressure" state. When the number of compression rollers is even and symmetrical on both sides of the load line, it is called "even pressure". The pressure distribution is non-linear and its expression is as follows:

$$f_{vc} = \frac{\omega_{vc}}{2\pi} = \frac{\omega_b N_b}{2\pi} \tag{3}$$

In order to simulate the bearing rotation process, this paper bound the axial direction (Z-axis direction), loaded the radial direction (Y-axis direction) when analyzed, the working system model is shown in Fig. 2.

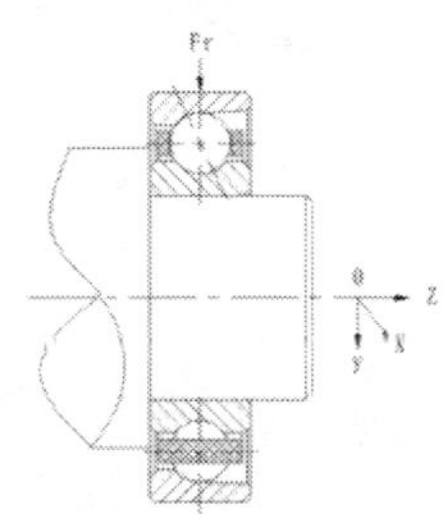

Figure. 2. The working system model of 7016C bearing

The vibration equation is established according to Fig. 2 showed as follows:

$$\begin{cases} M\ddot{x} + C\dot{x} + K_n \sum_{j=1}^{N_b} (x\cos\theta_j + y\sin\theta_j - \\ \gamma)^{3/2}\cos\theta_j = F_r + Me_m\omega^2\cos\omega t \\ M\ddot{y} + C\dot{y} + K_n \sum_{j=1}^{N_b} (x\cos\theta_j + y\sin\theta_j - \\ \gamma)^{3/2}\cos\theta_j = Me_m\omega^2\sin\omega t \end{cases} \tag{4}$$

In the formula above: M is the quality of bearing system; C is the damping of bearing system, K is the imbalance of the eccentricity of the bearing system. it can be seen from the formula above. The vibration excitation force mainly from the Non-linear bearing force and unbalanced force.

4 Establish the dynamic characteristics and dynamics equation of bearing

In solving the dynamic simulation of rolling bearings, the central difference method is generally used, because this method does not need to balance the iteration and does not need to directly solve the tangent stiffness, so it speeds the solution and it also saves a lot of computing time.

It is assumed that the displacement of any node has two components in direction, it is the displacement component u along the radial direction and the displacement component w in the axial direction and regardless of the influence of the damping factor. At time t = 0, the displacement, velocity and acceleration are known and the time domain n is divided into Δ t. Then the system is solved by the finite element method likes this:

$$M\ddot{a}_t + C\dot{a}_t + Ka_t = Q_t \tag{5}$$

In the formula above: M, C, K and Q_t are the mass matrix, damping matrix, stiffness matrix and the load vector of the nodes. $\ddot{a}_t$, $\dot{a}_t$ and a_t are the acceleration vector, velocity vector and displacement vector of the system nodes.

The displacement do Taylor expansion at time t and take the highest quadratic polynomial to have its approximation:

$$a_{t+\Delta t} = a_t + \Delta t\,\dot{a}_t + \frac{\Delta t^2}{2}\ddot{a}_t \tag{6}$$

The relationship between the acceleration and the displacement is obtained by solving the above equation:

$$\ddot{a}_t = \frac{1}{\Delta t}(\dot{a}_{t+\Delta t} - \dot{a}_t) = \frac{1}{\Delta t^2}(a_{t+\Delta t} - 2a_t + a_{t-\Delta t}) \tag{7}$$

The relationship between speed and displacement is expressed as:

$$\dot{a}_t = \frac{1}{2\Delta t}(a_{t+\Delta t} - a_{t-\Delta t}) \tag{8}$$

Put the above two formulas into the formula (5) you get the following formula:

$$Q_t - (K - \frac{2}{\Delta t^2}M)a_t - (\frac{1}{\Delta t^2}M - \frac{1}{2\Delta t}c)a_{t-\Delta t} = (\frac{1}{\Delta t^2}M + \frac{1}{2\Delta t}c)a_{t+\Delta t} \tag{9}$$

The equation above is the classical formula of the central difference method, it can be seen from above, if we know $a_{t-\Delta t}$ and a_t, so $a_{t+\Delta t}$ can be solved, and then can be solved the speed and acceleration at time t.

5 Establish the 3-D model and finite element model of bearing

5.1 Establish the 3-D model of bearing

Using Pro/E to establish the 3-D model of 7016C angular contact ball bearing, as is shown in Fig. 3. Consider the complexity of the actual bearing model and the limitations of simulation, the chamfer and oil film and other factors have little influence to the dynamic characteristics, deal with the model as following:

(i) Ignore the chamfer of the bearing.

(ii) Ignore the radial and axial clearance of bearing.

(iii) Ignoring the effect of lubricating oil to bearing vibration.

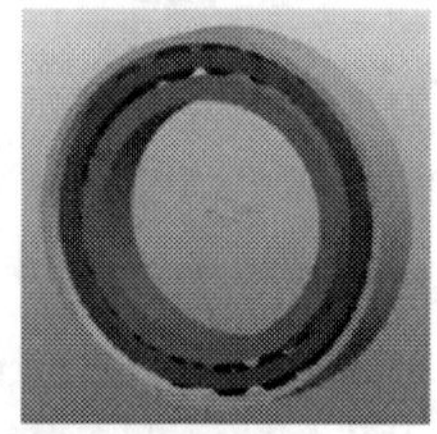

Figure. 3. The 3-D model of 7016C bearing

5.2 Establish the finite element model of bearing

Considering that the strain of the roller is the largest, the deformation of the inner and outer ring is small, the deformation of the cage can be neglected. Therefore, when select the material model, the cage is defined as rigid body, the roller, inner ring and outer ring are defined as elastoplastic. It will greatly reduce the computational time after defined as rigid body because the freedom degrees of all nodes in the rigid body are coupled to the mass center. No matter how many nodes are defined by the finite element model, the rigid body has only six freedom degrees, The motion is calculated from the centroid and passes the corresponding displacement value to the node.

Transform the 3-D model established by Pro/E into IGES format and import into finite element software. Select shell 163 unit and solid 164 unit. The material of inner and outer ring is GCr15 steel, the cage is phenolic laminated cloth, the density is ρ = 7800 × 10^{-6}kg/mm^3, the elastic modulus E is 208 GPa, the Poisson's ratio μ is 0.3 [12]. The mesh is divided by scanning and mapping. Establish a contact between the inner and outer ring and the cage. The contact type is automatic face-to-face contact and the static friction coefficient is 0.35, 0.35 and 0.2, the dynamic friction coefficient is 0.16, 0.16 and 0.1. The rigid surface of the outer ring is set to full constraint. The working bearing receives the pressure and speed. The pressure is reduced to uniform force applied to the lower half surfer of the inner ring and the speed applied to the rigid surface of inner ring.

The simulation model of bearing is shown in Fig. 4.

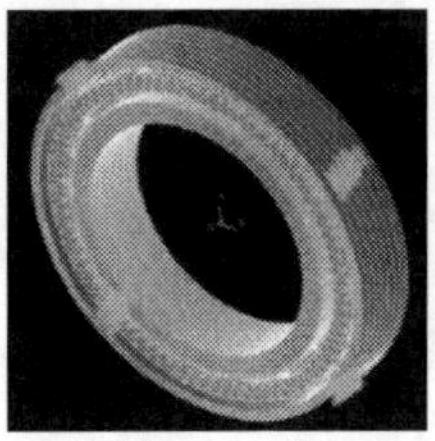

Figure. 4. The simulation model of 7016C bearing

6 Contrast analysis and simulation of the vibration characteristics of bearing

6.1 Analysis of the simulation results of bearing vibration of 7016C (self-designed)

Under the load of 10 KN and the speed of 1800 r/min, the analysis results of 7016C (self-designed) bearing is shown in Fig. 5 to Fig. 7. Three moments of the calculation results are analyzed (0.02 s, 0.05 s, 0.08 s) .

Figure.5. The equivalent stress plot of vibration bearing in 0.02 seconds

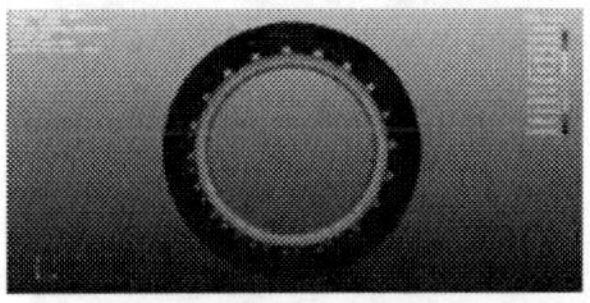

Figure.6. The equivalent stress plot of vibration bearing in 0.05 seconds

Figure.7. The equivalent stress plot of vibration bearing in 0.08 seconds

It can be seen from the above Fig.5 to Fig.7, the larger stress of 7016C (self-designed) bearing generally appear in the contact area of roller and the inner and outer ring. The maximum stress value and the location of the maximum stress will be changed with the bearing rotation. As the time increases, the stress on the bearing is gradually reduced, because the bearing will start to receive a greater impact. The stress value tends to be stable until the steady operation, the stress value slowly becomes smaller when the bearings gradually stop running.

It can be seen from Fig.8, when the 7016C (self-designed) bearing begins to rotate, it produces a strong vibration because the roller receives a sudden load. With the stable operation of the bearing, the vibration value tends to be stable. The maximum stress value exhibits a certain periodicity, which is related to the periodic entry of the bearing element into the load bearing area and the non-load area.

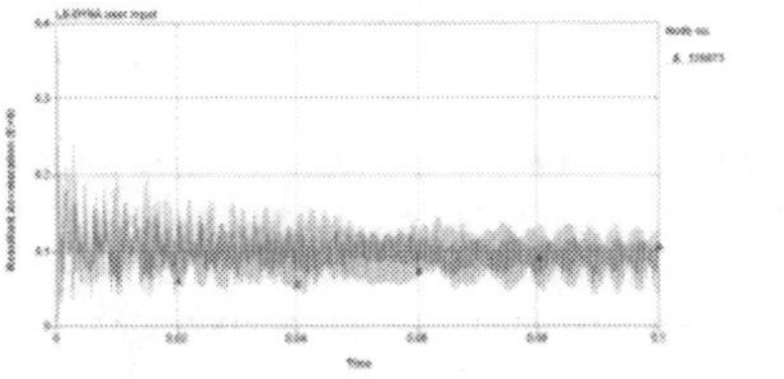

Figure.8. The vibration distribution of bearing over time

6.2 Analysis the simulation results of bearing vibration of 7016C (FAG)

Under the load of 30 KN and the speed of 3000 r/min, the analysis results of 7016C (FAG) bearing is shown in Fig. 9 to Fig. 11. Three moments of the calculation results are analyzed (0.02 s, 0.05 s, 0.08 s) .

Figure.9. The equivalent stress plot of vibration bearing in 0.02 seconds

Figure.10. The equivalent stress plot of vibration bearing in 0.05 seconds

Figure.11. The equivalent stress plot of vibration bearing in 0.08 seconds

It can be seen from the above Fig.9 to Fig.11, the stress is small when the 7016C (FAG) bearing begins to rotate. The stress gradually becomes larger with the rotation process. The maximum stress appears below a certain area in the contact surface of roller and the inner and outer ring , The value gradually decreases from the inside to the outside, the stress is mainly distributed in the bearing area. The stress value in the bearing area is significantly greater than the stress value in the non-bearing area. The area where the stress occurs is elliptical.

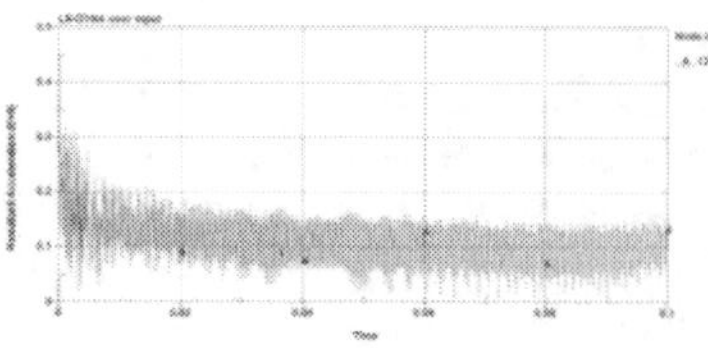

Figure.12. The vibration distribution of bearing over time

It can be seen from Fig.12, when the 7016C (FAG) bearing begins to rotate, it produces a strong vibration because of the roller received a sudden load, with the stable operation of the bearing, the vibration value tends to be stable.

6.3 Contrast analysis the simulation results of bearing vibration between 7016C (self-designed) and 7016C (FAG)

In order to verify the rationality of self-designed bearing, applied the load of 30 KN and the speed of 8000 r/min on the 7016C (self-designed) bearing and 7016C (FAG) bearing. After dynamic simulation, three moments of the calculation results are analyzed (0.02 s, 0.05 s, 0.08 s). The results are shown in Fig. 13 to Fig. 18.

Figure.13. The equivalent stress plot of vibration (self-designed) bearing in 0.02 seconds

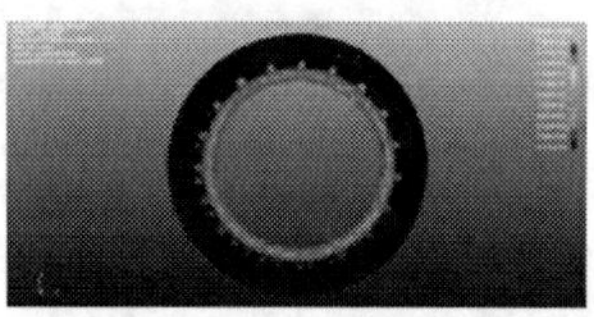

Figure.14. The equivalent stress plot of vibration (FAG) bearing in 0.02 seconds

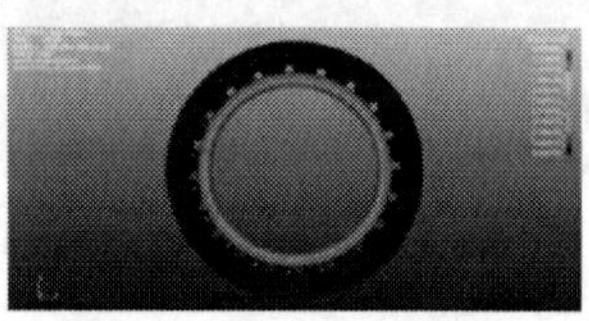

Figure.15. The equivalent stress plot of vibration (self-designed) bearing in 0.05 seconds

Figure.16. The equivalent stress plot of vibration (FAG) bearing in 0.05 seconds

Figure.17. The equivalent stress plot of vibration (self-designed) bearing in 0.08 seconds

Figure.18. The equivalent stress plot of vibration (FAG) bearing in 0.08 seconds

It can be seen from Fig. 13 to Fig. 18, the stress of the 7016C (self-designed) bearing is bigger in the beginning, but as the operation progresses, the stress becomes smaller and tends to be smooth. The stress of the 7016C (FAG) bearing is smaller in the beginning, but the stress increases with the rotation process. The stress value is three times higher than 7016C (self-designed) bearings.

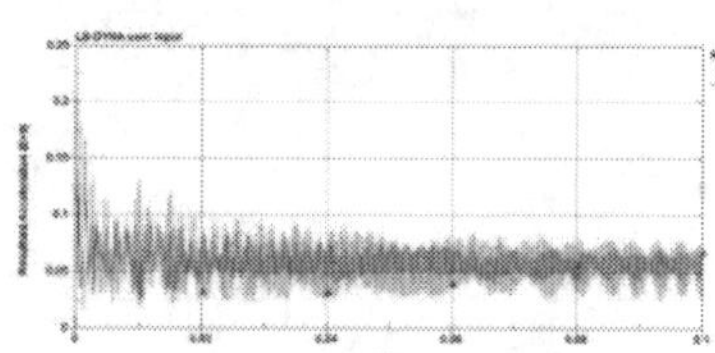

Figure.19. 30 KN,8000 r/min,the vibration distribution of 7016C (self-designed) bearing over time

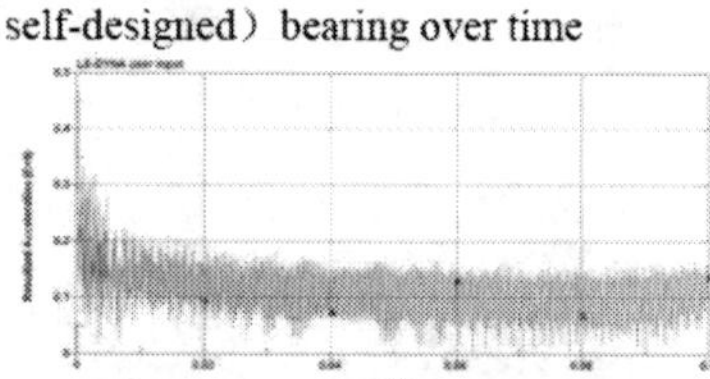

Figure.20. 30 KN,8000 r/min,the vibration distribution of 7016C (FAG) bearing over time

It can be seen from Fig. 19 to Fig. 20, when the bearing starts to rotate, the vibration is increased due to the effect of a sudden applied load. As the bearing is stable, the vibration value tends to be stable, and the vibration value of 7016C (FAG) bearing is greater than the vibration value of 7016C (self-designed) bearing .

7 Experiment analysis

In order to verify the feasibility of the method in designing bearings, select the GCr15 steel material, the process into bearing after a certain heat treatment process and the processing parameters shown in Table 2.

Table.2. The table of processing parameters

Name	Processing parameters
Datum flatness（um）	0. 42
Cutting speed V（m/min）	130
Feed speed f（mm/r）	0.03
Cutting amount a_p (mm)	0.025

The performance evaluation indicators of the bearings are mainly include surface roughness, channel roundness, hardness, surface burns, residual stress and etc on. The results are shown in Table 3. All the indicators meet the requirements.

Table.3. The results of experimental

Name	Test value
Roughness（um）	<0.1
Roundness（um）	0.8
hardness（HV）	774.5
Burn degree	Nothing
Residual stress situation	All are compressive stress, controllable

8 Conclusions

The paper is based on the computer virtual simulation technology and uses the 7016C angular contact ball bearing as the research object. Firstly, the 3-D model is established by using Pro/E. Secondly, introduce the 3-D model into the finite element software ANSYS to obtain the finite element model. Finally, uses the simulation analysis software LS-DYNA to analyze the dynamic simulation of 7016C (self-designed) bearing and 7016C (FAG) bearing, get the stress and dynamic characteristics of the two kinds of bearing which changed with the pressure and the speed. Prove the structural parameters are reasonable through the experiment of 7016C (self-designed) bearing, it shows that the method is feasible. The method can provide the basis theoretical and analysis foundation for the virtual simulation design and fault diagnosis of the bearings.

References

1. Zhang Liyun, Liu Yi, Liu Jia. Maintenance time prediction using virtual reality[J]. Journal of Computer-Aided Design & Computer Graphics, 2016, 8(8): (in Chinese):1383-1391.
2. Yi Ran. Intelligent Simulation for Intelligent Manufacturing[J]. Aeronautical Manufacturing Technology, 2015, (18): 64-65.
3. Xie Chun, Ma Lan, Peng Yinghong. Visualization of the FEM analysis data[J]. Journal of Computer-Aided Design & Computer Graphics, 2000, 2(2): 81-84.
4. Li Guochao, Peng Wei, Li Yongcai, et al. Simulation and dynamic Analysis of outer ring fault on rolling bearings using explicit finite element method[J]. China Mechanical Engineering, 2012(12): 23(23): 2825-2829.
5. Zhang Jianjun, Wang Zhongsheng, Lu Yuhua. Nonlinear dynamic modeling of localized defects in rolling element bearing[J]. Journal of vibration and shock, 2010, 11(29): 30-34.
6. Cui Bo. Study on dynamic simulation method for rolling bearing based on finite element[D]. Taiyuan: Taiyuan University of Technology, 2010.
7. Zhu Chen, Gao Lixin, Wei Zhiheng. Summary of bearing simulation[J]. Plant Maintenance Engineering, 2016, (6): 41-42.
8. Jones A B. Ball motion and sliding friction in ball bearings[J]. Trans ASME, Journal of Basic Engineering, 1959, 8: 1-4.
9. Huang Qiangui, Deng Sier, Teng Hongfei. Status and Development of Rolling Bearing Simulation System[J]. Bearing, 2002, (4)):34-38.
10. Gupta P K. Dynamics of roiling element bearings Part II: Ball bearing analysis[J]. ASME Journal of Lubrication 1979, 101:305-311.
11. Wailers C T. Dynamics of ball bearing[J]. ASME Journal of Lubrication Technology. 1971, 93(1): 1-10.
12. Liu Jiangshan, Song Li, Lu Chao. Analysis on dynamic simulation of deep groove ball bearing based on LS -DYNA[J]. Bearing, 2015(2): 35-38.

Structure design of an innovative adaptive variable camber wing

An-Min Zhao[1,3], Zou Hui[1], De-Shan Liu[1,3] and Bing-Fei Liu[2]

[1]*School of Aeronautical Science and Engineering, Beihang University, Beijing, 100191, China*
[2]*Airport college, Civil Aviation University of China, Tianjin, 300300, China*
[3]*Beijing Dashan brothers Unmanned Aerial Vehicle Technology Company Limited, Beijing, 100191, China*

Abstract. In this paper, an innovative double rib sheet structure is proposed, which can replace the traditional rigid hinge joint with the surface contact. On the one hand, the variable camber wing structural design not only can improve the capacity to sustain more load but also will not increase the overall weight of the wing. On the other hand, it is a simple mechanical structure design to achieve the total wing camber change. Then the numerical simulation results show that the maximum stress at the connect of the wing rib is 88.2MPa, and the double ribs sheet engineering design meet the structural strength requirements. In addition, to make a fair comparison, the parameters of variable camber are fully referenced to the Talon Unmanned Aerial Vehicle (UAV). The results reveal that the total variable camber wing can further enhance aircraft flight efficiency by 29.4%. The design of the whole variable camber wing structure proposed in this paper has high engineering value and feasibility.

1 Introduction

For low speed aircraft, the adaptive variable camber wing can effectively improve the lift and drag coefficients than conventional wing, delay the flow separation of the wing surface, and significantly improve the performance of the aircraft. Therefore, a lot of the adaptive variable wing research has been carried out at home and abroad.

In the 1980s, NASA and Boeing Launched the mission adaptive wing technology research, the abolition of the traditional aircraft control surface, the use of flexible composite material skin and the digital flight control system [1]. In 2003, in order to satisfy the needs of long flight time, the United States FlexSys Inc company designed a piezoelectric ceramic structure that used as a driver to alter the camber of wing [2]. In 2006, Kota et al proposed a flexible skin covering in the trailing edge of the wing and achieving the deformation of the entire wing by the smooth bending of the flexible trailing edge [3]. In 2009, Beihang University used shaped memory alloys as drive the trailing edge of the wing under temperature induction. The average lift of the wing was greater than 20% in the wind tunnel experimental [4]. In 2011, the University of Naples Federico in Italy adopting shape memory alloy (SMA) actuators designed with variable camber wing model [5]. At the same time, Abdullah, E. J adopted shape memory alloy (SMA) as the actuator in the adaptive airfoil control system. Results from the numerical simulation of the proposed controller are presented which demonstrate the potential of implementing SMA actuators in an adaptive airfoil control system for Unmanned Aerial Vehicle (UAV)[6].

In 2013, Liu weidong et al proposed a novel 0-Poisson's ratio cosine honeycomb support structure of flexible skin, which can make the power consumption and driving force is lower [7]. In 2016, Li, Hongda design an innovative adaptive variable camber compliant wing, with skin that can change thickness and tailing edge morphing mechanism that is designed based on a new kind of artificial muscles. The results show that this method is capable of designing airfoils for this special wing in a quick and effective way [8].

From the design of the adaptive variable camber adaptive wing in domestic and abroad, it can be seen that in the realization of the adaptive deformation of the wing, from the traditional rigid structure to the compliant structure, from the conventional material to new functional material has certain research. But generally speaking, smart material drive cost is high, efficiency is low, it is difficult to realize engineering demand in present. While the traditional complex mechanical drive usually pays a greater price on the structural weight.

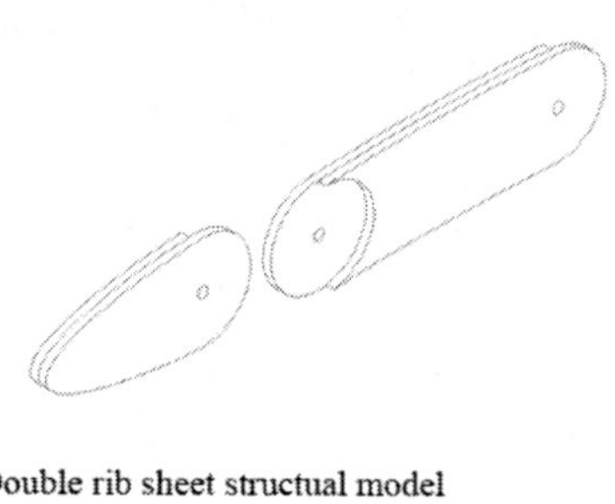

Figure 1. Double rib sheet structual model

In this paper, a double rib sheet structure was designed in Fig. 1, which can achieve the overall wing variable camber. On the one hand, it is designed with a simple mechanical structure will not increase the weight of the wing weight, and the other side, the load transfer between the sections of the wing structures is converted from the point contact to surface contact, which leads to strong structural bearing capacity.

2 Structural design of whole variable camber wing

The Fig. 1 shows that the wing rib model are double rib sheet structure, one of which is a semicircular groove and the other layer is a circular ear piece which is in close contact. In this paper, the design of the compete wing rib is connected by four double ribs sheet, as shown in Fig. 2.

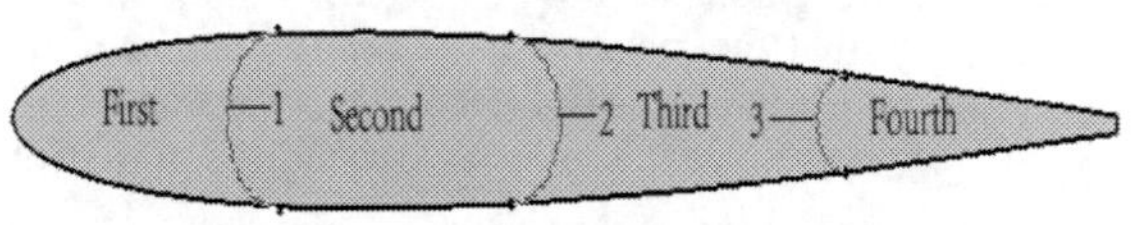

Figure 2. Four section wing rib structure models

The entire wing structure consists of four different wing structural sections in Fig. 3, which include the wing leading edge section, wing trailing edge section, wing mid-section and wing mid-aft section, which are connected to each other by a double wing rib sheet structure. The relative rotation between the ribs of the wing to achieve the overall variable camber wing.

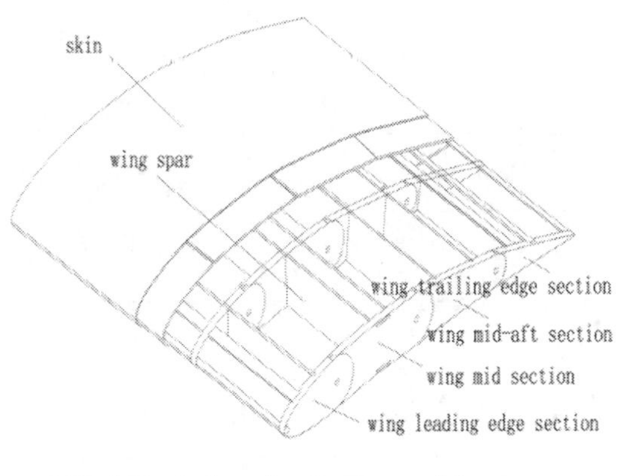

Figure 3. Variable camber wing structure model

The designed adaptive variable camber wing structure is same as that of the conventional wing, without any weight gain except the driving device.

3 Engineering finite element analysis

Whether the overall design of the strength of the variable camber wing structure can be qualified for large transport aircraft and civil aircraft. According to the technical parameters of the Lockheed C-130 "Hercules" transport aircraft (see Fig. 4) in Table 1 [9-12], the structural strength design of the whole variable ca4mber is carried out.

Figure 4. Lockheed C-130 Hercules transport plane

The analysis of the transmission of the aerodynamic load of the variable camber wing structure is simplified. The maximum bearing structure can be determined as the second wing rib structure section and the third wing rib section. Therefore, the area where the double ribs sheet is connected is considered to be a dangerous cross section. Based on the data Table 1 and the simplified results of the load, the aerodynamic force acting on the third segment can be calculated to be 339.4 Pa.

It is assumed that the safety factor is 2, the strength residual coefficient is 1.14, the overload is 3g, and the design thickness of the fin is 15mm. Fig. 5 shows a finite model of the second section of the ribs completely fixed and the third section of the ribs in the loading direction with translational degree of freedom.

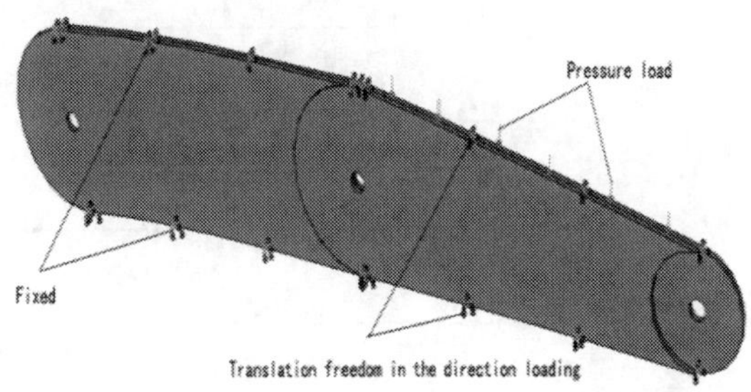

Figure 5. Boundary conditions and loads of finite element model

In this paper, the dynamic analysis method is adopted to complete the structural finite element analysis under the action of the surface pressure load. Fig. 6 shows that the maximum stress on the double rib sheet connection is 88.2 MPa, which is much smaller than the yield stress of the material, as shown in Table 2. Therefore, the double rib sheet with a single thickness of 15mm are safe to meet the actual carrying capacity.

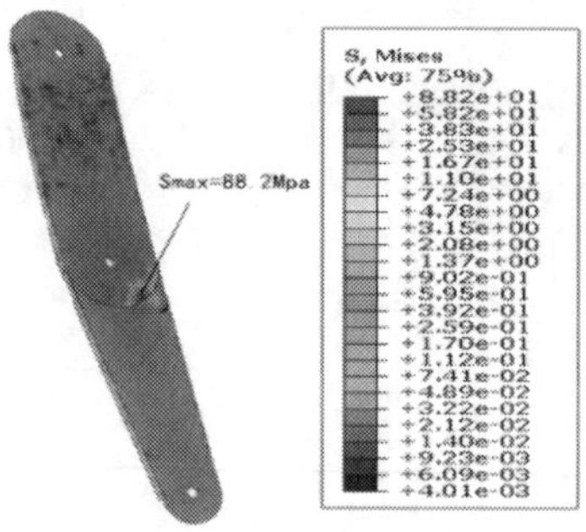

Figure 6. Stress contour of double rib sheet mode

4 Flight experiment

A flight test that Talon UAV and adaptive wing aircraft of the same design parameters were performed respectively is as follows. First, take a 5000mAh 4S model battery, which has a full voltage of 16.4V to the variable camber wing aircraft power. In the adaptive flight state of the wing, respectively, each flight of 12m/s, 14m/s, 16m/s for 5 minutes, the voltage drop of the battery was 1.2V. Then, we adopted the same battery in the case of full voltage for Talon UAV to compete the same flight mission, measured the voltage drop of battery was 1.7V. The specific flight data is shown in Table 3.

As shown in Table 3, the variable camber wing in adaptive flight state is 0.5V lower than that of the Talon UAV. That is, the variable camber adaptive flight than the same parameters Talon UAV flight efficiency increase by 29.4%.

5 Conclusions

In this paper, an innovation double rib sheet structure is studied that allows the overall structure of the wing to change according to the relative rotation the double rib sheet joint. This comparison study indicates that this adaptive variable camber wing has better aerodynamic performance than the tradition wing. Main conclusions for the innovation double rib sheet research of this paper are as follows:

1. An innovative double rib sheet structure is designed, which achieves the whole variable camber of wing with a simple mechanical structure, and does not increase the overall structural weight of the wing.
2. The load transfer mode of the new designed of variable camber wing structure is surface contact, which makes the structure more loading-bearing.
3. Taking the Lockheed C-130 Hercules transport aircraft as the engineering prototype, its aircraft variable camber wing designed with double rib sheet structure. Based on the numerical simulation results show that the maximum stress at the connect of the wing rib is 88.2MPa, and the double ribs sheet design satisfies the structural strength requirements.
4. This paper proposes a new design method which can realize the overall variable camber of wing. Compared with the design scheme of traditional fixed wing Talon UAV, the total variable camber wing can further enhance aircraft flight efficiency by 29.4%.

Table 1. The technical parameters of Lockheed C-130 Hercules transport plane.

Span(m)	Wing area(m^2)	Maximum take-off weight (kg)	Maximum flight speed (km/h)	Wing load (kg/m^2)
40.4	162	70300	592	434

Table 2. Material property of double rib sheet.

ρ (kg/m^3)	E (N/m^2)	σ (N/m^2)	υ
40.4	162	70300	592

Table 3. Comparison of flight test of adaptive camber aircraft and Talon aircraft.

Test	Aircraft type	Cruise setting	Voltage drop	Dissipative energy
Test Ⅰ	Adaptive variable camber wing	5min 12m/s 5min 14m/s 5min 16m/s	1.2V	21600J
Test Ⅱ	Talon	5min 12m/s 5min 14m/s 5min 16m/s	1.7V	30600J

Acknowledgement

The authors gratefully acknowledge the financial support of National Natural Science Foundation of China (No. 11502284) and it is also supported by "the Fundamental Research Funds for the Central Universities (3122016C006)" of China.

Reference

1. POWERS S, WEBB L, FRIEND E, LOKOS W, editors. Flight test results from a supercritical mission adaptive wing with smooth variable camber. 28th National Heat Transfer Conference; 1992.
2. Zhu H, Liu WD, Zhao CS. Morphing Aircraft and Its Morph-driving Techniques[J]. Machine Building & Automation. 2010.
3. Kota S, Osborn R, Ervin G, Maric D, Flick P, Paul D. Mission Adaptive Compliant Wing – Design , Fabrication and Flight Test Mission Adaptive Compliant Wing[J]. 2006.
4. Li Jun, Qin YanHua, et, al Development of a morphing wing with adaptive capability[J]. Acta Aerodynamica Sinica. 2009,27(5):505-8.
5. Barbarino S, Pecora R, Lecce L, Concilio A, Ameduri S, De Rosa L. Airfoil Structural Morphing Based on SMA Actuator Series: Numerical and Experimental Studies[J]. J Intell Mater Syst Struct. 2011,22(10):987-1004.
6. Abdullah, E.J., C. Bil, and S. Watkins, Adaptive airfoil control system using shape memory alloy actuator for unmanned aerial vehicle. Incorporating Sustainable Practice in Mechanics of Structures and Materials, ed. S. Fragomeni, et al. 2011, Boca Raton: Crc Press-Taylor & Francis Group. 141-146.
7. LiuWeidong, ZhuHua, ZhouShengqiang, BaiYalei, WangYuan, ZhaoChunsheng. In-planecorrugatedcosinehoneycombfor1Dmorphingskinanditsapplicationonvariablecamberwing[J]. Chinese Journal of Aeronautic. 2013,26(4):935-42.
8. Li HD, Ang HS. Preliminary airfoil design of an innovative adaptive variable camber compliant wing[J]. J Vibroeng. 2016,18(3):1861-73.
9. Spicer CW, Holdren MW, Cowen KA, Joseph DW, Satola J, Goodwin B, et al. Rapid measurement of emissions from military aircraft turbine engines by downstream extractive sampling of aircraft on the ground: Results for C-130 and F-15 aircraft[J]. Atmos Environ. 2009,43(16):2612-22.
10. Haywood J, Francis P, Osborne S, et al. Radiative properties and direct radiative effect of Saharan dust measured by the C-130 aircraft during SHADE: 1. Solar spectrum. 2003; 108: 8577.
11. Haywood JM, Osborne SR, Francis PN, et al. The mean physical and optical properties of regional haze dominated by biomass burning aerosol measured from the C-130 aircraft during SAFARI 2000. Journal of Geophysical Research Atmospheres. 2003; 108: 225-31.
12. Highwood EJ, Haywood JM, Silverstone MD, Newman SM, Taylor JP. Radiative properties and direct effect of Saharan dust measured by the C-130 aircraft during Saharan Dust Experiment (SHADE): 2. Terrestrial spectrum. Journal of Geophysical Research Atmospheres. 2003; 108: SAH 5-1.

High Accuracy Positioning using Jet Thrusters for Quadcopter

ChenHuan Pi [1], JinKwan Lin [2] and Stone Cheng [3]

[1,2,3] National Chiao Tung University, Department of Mechanical Engineering, Taiwan

Abstract. A quadcopter is equipped with four additional jet thrusters on its horizontal plane and vertical to each other in order to improve the maneuverability and positioning accuracy of quadcopter. A dynamic model of the quadcopter with jet thrusters is derived and two controllers are implemented in simulation, one is a dual loop state feedback controller for pose control and another is an auxiliary jet thruster controller for accurate positioning. Step response simulations showed that the jet thruster can control the quadcopter with less overshoot compared to the conventional one. Over 10s loiter simulation with disturbance, the quadcopter with jet thruster decrease 85% of RMS error of horizontal disturbance compared to a conventional quadcopter with only a dual loop state feedback controller. The jet thruster controller shows the possibility for further accurate in the field of quadcopter positioning.

1 Introduction

A quadcopter (Quadrotor) is a Micro Aerial Vehicle (MAV) driven by four rotors. The control technique and applications of a quadcopter are popular research topics in recent years due to their low price, small size and easy to modification. Their maneuverability and ease to use make them fit to provide stable flight, navigate into places that ground robots or human cannot reach and perform tasks.

Some applications such as entertainment and search and rescue do not need extremely accurate positioning. However, for interaction with the environment at close range, such as in canopy sampling [1] or factory assemble line delivery, high accuracy positioning is required.

Traditionally, horizontal translation of quadcopter relies on vector component of force changed by tilting the quadcopter a certain angle. Increasing the tilt angle increases lift loss and wind cross-section area. To ameliorate this situation, Bannwarth et al. [2] designed a wind disturbance rejection nonlinear controller on quadcopter. Their experimental results showed that the controller had a position error below 1 meter. Chau and William proposed a continuous robust controller of attitude trajectory tracking for a quadcopter in the presence of uncertainty and unknown disturbance [3]. For these controllers, minimizing the error in the position of the quadcopter under unpredictable circumstances is essential.

In a flight vehicle, many different control methods can be used. The jet system is one of the actuator among these methods and commonly used in flight vehicle attitude control system several years. Jet engines can be classified as air-breathing engines such as turbojet engines and non-air-breathing engines such as rocket engines.

The cold gas thruster is the simplest case in the rocket engines. Unlike other engines used fuel burning, there is no combustion chamber in this system. The exhaust gas was driven by the pressure difference between the tank and outside, not by a chemical reaction. These unbalance pressure push and accelerate the gas flow to drive the system movement. Although this engine cannot supply as much power as fuel burning engines, using for attitude control is enough. [4, 5] show the different usage of fuel burning engine from cold gas thruster, which drive quadcopter by gears and transmissions.

This system can be realized by using components such as pressurized gas tank, regulator valve, solenoid and nozzle to create thrust force needed to perform translational or rotational maneuver of vehicle. Many researches provide the feasibility and effectiveness of this system. John G. Furumo [6] demonstrated his system design and cold gas propulsion system that generate 1-D rotation movement of test platform. Assad Anis [7] compared performance values for several cold gas propellants and hardware designs. In [8], a flying vehicle is developed using nitrogen as propellants which is applied in exploration to other celestial bodies.

The rest of this paper presents a control strategy of the combination of quadcopter system and cold gas thruster system. Four cold gas thrusters mounted perpendicular to each other on the quadcopter generate the horizontal force needed to achieve the control goal. However, increasing control inputs in original quadcopter system makes it more complex and more challenge in controller design. Section 2 describes the physical configuration and mathematical system model. Section 3 discusses the controller design. Section 4 shows the simulation results. Conclusion and future works are presented in Section 5.

2 Dynamic model

2.1 Jet Thruster Quadcopter

Unlike the traditional quadcopter in Fig. 1, the jet thruster quadcopter adds four nozzles along the quadcopter arms which can supply additional force on its horizontal plane. These thrusters can be used to improve horizontal maneuverability for more accurate positioning and disturbance rejection.

The model dynamic equations would become Eq. 1,

$$\begin{bmatrix} m\ddot{x} \\ m\ddot{y} \\ m\ddot{z} \end{bmatrix} = R_{IB} \begin{bmatrix} -T_1 + T_3 \\ -T_2 + T_4 \\ \sum F_i \end{bmatrix} + \begin{bmatrix} 0 \\ 0 \\ g \end{bmatrix} \quad (1)$$

where four nozzles are placed as Fig. 2, which generate constant force while it turned on. Since the thruster force vector passes through the center of mass in the quadcopter, it does not affect the rotation motion of the quadcopter and the dynamic equation in rotation is still as same as the conventinal quadcopter.

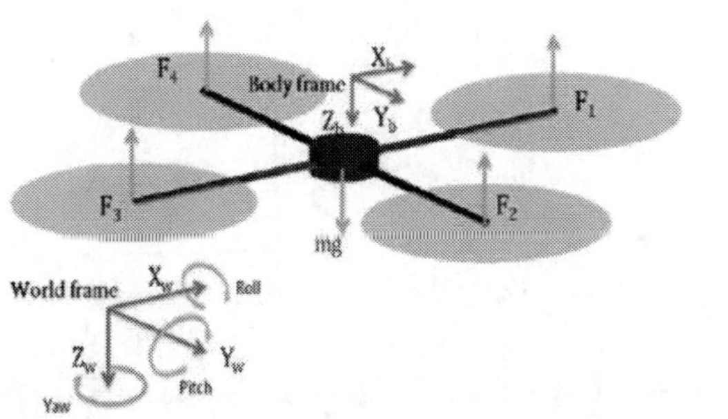

Figure 1. Structural sketch of the coordinate system and forces acting on the coventional quadcopter.

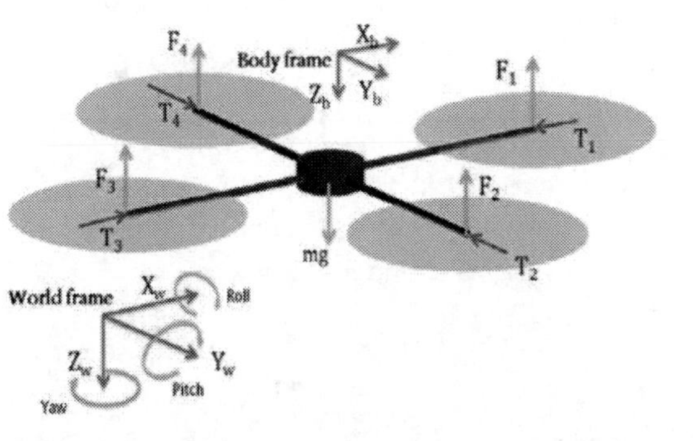

Figure 2. Structural sketch of the coordinate system and forces acting on the quadcopter with additional four jet thrusters.

3 Controller Design

This section presents the dual loop state-feedback controller and jet thruster controller used in the proposed quadcopter. Since the quadcopter is known to be an under-actuated system, it cannot control 6 degree-of-freedom at the same time. Due to the restrict condition, the controller of the horizontal positioning is separated from the z-axis translation, roll, pitch and yaw 4 degrees of freedom.

In inner loop state feedback controller, the states are chosen as Eq. 2

$$\begin{aligned} x_i &= \begin{bmatrix} z & \dot{z} & \theta & \dot{\theta} & \phi & \dot{\phi} & \psi & \dot{\psi} \end{bmatrix}^T \\ \dot{x}_i &= A_i x_i + B_i u_i + \begin{bmatrix} 0 & 1 & 0 & 0 & 0 & 0 & 0 & 0 \end{bmatrix}^T g \end{aligned} \quad (2)$$

where

$$A_i = \begin{bmatrix} 0&1&0&0&0&0&0&0 \\ 0&0&0&0&0&0&0&0 \\ 0&0&0&1&0&0&0&0 \\ 0&0&0&0&0&0&0&0 \\ 0&0&0&0&0&1&0&0 \\ 0&0&0&0&0&0&0&0 \\ 0&0&0&0&0&0&0&1 \\ 0&0&0&0&0&0&0&0 \end{bmatrix},$$

$$B_i = \begin{bmatrix} 0 & \frac{-\cos\theta\cos\phi}{m} & 0 & 0 & 0 & -\frac{l}{I_{yy}} & 0 & -\frac{K_f}{K_m}\frac{1}{I_{zz}} \\ 0 & \frac{-\cos\theta\cos\phi}{m} & 0 & -\frac{l}{I_{xx}} & 0 & 0 & 0 & \frac{K_f}{K_m}\frac{1}{I_{zz}} \\ 0 & \frac{-\cos\theta\cos\phi}{m} & 0 & 0 & 0 & \frac{l}{I_{yy}} & 0 & -\frac{K_f}{K_m}\frac{1}{I_{zz}} \\ 0 & \frac{-\cos\theta\cos\phi}{m} & 0 & \frac{l}{I_{xx}} & 0 & 0 & 0 & \frac{K_f}{K_m}\frac{1}{I_{zz}} \end{bmatrix}^T$$

$$u_i = \begin{bmatrix} F_1 & F_2 & F_3 & F_4 \end{bmatrix}^T$$

and B_i is function of θ, φ and ϕ The inner loop controller uses conventional state-feedback, which u_i is design to be Eq. 3.

$$u_i = -k_i x_i - B_i^{\dagger} \begin{bmatrix} 0 & 1 & 0 & 0 & 0 & 0 & 0 & 0 \end{bmatrix}^T g \quad (3)$$

where $B_i^{\dagger}$ is Moore–Penrose pseudo inverse of B_i. The stability of the system is ensured by using a linear–quadratic regulator to find a k_i that minimizes the cost function by solving the algebraic Riccati equation.

The proposed design for the inner loop controller enables control of rolling and pitching, which means the position of quadcopter can be changed by tilting it and have component force acting on its side. To control the position, an outer loop is designed by using another state-feedback controller, which the state is chosen as Eq. 4.

$$\begin{aligned} x_o &= \begin{bmatrix} x & \dot{x} & y & \dot{y} \end{bmatrix}^T \\ \dot{x}_o &= A_o x_o + B_o u_o \end{aligned} \quad (4)$$

Through the system dynamic equations, the relationships between rolling and pitching angle θ and φ to x and y axis acceleration in inertia frame are

$$\begin{bmatrix} \dot{x} \\ \ddot{x} \\ \dot{y} \\ \ddot{y} \end{bmatrix} = \frac{\sum F}{m} \begin{bmatrix} 0 \\ \sin\psi \sin\phi + \cos\psi \sin\theta \cos\phi \\ 0 \\ \sin\psi \sin\theta \cos\phi - \cos\psi \sin\phi \end{bmatrix} \quad (5)$$

Since θ and φ would be in small angle when the quadcopter is hovering, the equation can be simplified by linearization of the dynamic equations, which can be written as

$$\begin{bmatrix} \dot{x} \\ \ddot{x} \\ \dot{y} \\ \ddot{y} \end{bmatrix} = \frac{\sum F}{m} \begin{bmatrix} 0 & 0 \\ \cos\psi & \sin\psi \\ 0 & 0 \\ \sin\psi & -\cos\psi \end{bmatrix} \begin{bmatrix} \theta \\ \phi \end{bmatrix} \quad (6)$$

and the dynamic equations Eq. 4 can be written in state space representation, where

$$A_o = \begin{bmatrix} 0 & 1 & 0 & 0 \\ 0 & 0 & 0 & 0 \\ 0 & 0 & 0 & 1 \\ 0 & 0 & 0 & 0 \end{bmatrix}, \quad B_o = \frac{\sum F}{m} \begin{bmatrix} 0 & 0 \\ \cos\psi & \sin\psi \\ 0 & 0 \\ \sin\psi & -\cos\psi \end{bmatrix}$$

$$u_o = \begin{bmatrix} \theta & \phi \end{bmatrix}^T = -k_o x_o$$

and by using same method to determine inner loop controller design, a proper k_o can be found.

In this structure, the jet thrusters are considered an auxiliary force that increases positioning accuracy. A variable structure controller design method is used to control jet thrusters. This method is chosen because the jet thruster only has two states, turned on or turned off. And a pair of thrusters can be set to provide force $+T$, 0, $-T$ applied on the quadcopter. The phase portrait analysis is used to design the controller and present in following simulation.

Eq. 7 is a system with mass m applied an external force F.

$$\begin{aligned} \ddot{x} &= \frac{F}{m} = u \\ \dot{x} &= ut + C_0 \\ x &= \frac{1}{2}ut^2 + C_0 t + C_1 \end{aligned} \quad (7)$$

where C_0 and C_1 are constants that depend on the initial system state.

Define the variables x_1 and x_2

$$\begin{aligned} x_1(t) &= x(t) \\ x_2(t) &= \dot{x}(t) \end{aligned}$$

Hence, the phase plane trajectory can be plotted with the relation between x_1 and x_2

$$x_1 = \frac{1}{2}\frac{m}{F}x_2^2 + \frac{1}{2}\frac{m}{F}C_0 - \frac{m}{F}C_0^2 + C_1 \quad (8)$$

In Fig. 3, the solid lines show the phase trajectories when the positive thruster is turned on, and the dotted lines show the phase trajectories when the negative thruster is turned on.

The next steps are designing the controller and choosing the switching surface. The main concept is when the system state reached the designed switching surface, the states would keep on it by changing u. It leads to the origin in any initial state [9]. In Fig. 4, the switching surface S (Eq. 9) is separated into two structures, region1 and region 2.

$$S(x_1, x_2) = \begin{cases} x_1 + \frac{1}{2}\frac{m}{F}x_2^2 = 0, & \text{when } x_1 < 0 \text{ and } x_2 > 0, \\ x_1 - \frac{1}{2}\frac{m}{F}x_2^2 = 0, & \text{when } x_1 > 0 \text{ and } x_2 < 0. \end{cases} \quad (9)$$

When the system state is in region 1, u is set to $-T$ which would lead the state following the dotted trajectory and reach the designed surface. However, u would be set as T and state follows the solid trajectory reaching the surface.

In Fig.5 shows the overall system block diagram of the controller. The pose controller is designed as a dual loop controller which inner loop controls z-axis, roll, pitch, yaw and outer loop controls x-axis and y-axis. Jet thruster controller is considered as an auxiliary component for higher accuracy positioning.

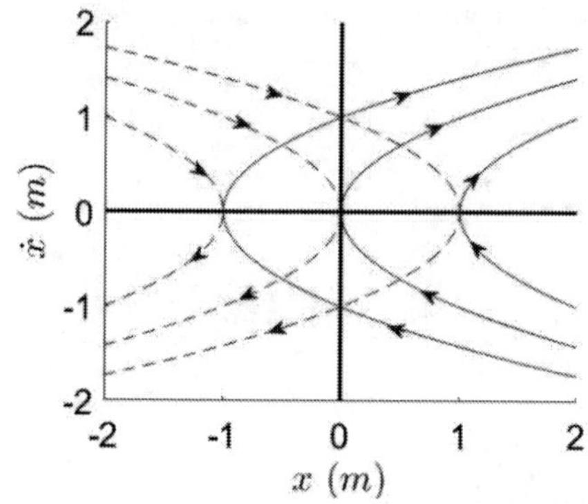

Figure 3. System state trajectory with an external force applied.

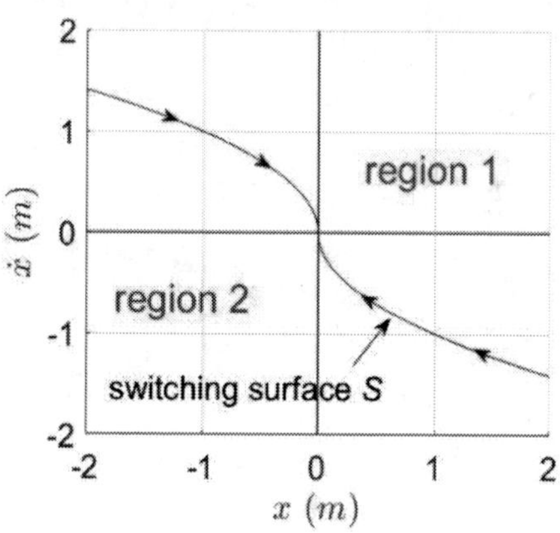

Figure 4. Switching surface S of the jet thruster controller.

$$u(x_1,x_2)=\begin{cases} T, & \text{if } (x_1,x_2) \text{ in rigion 1} \\ 0, & \text{if } (x_1,x_2) \text{ on switching surface} \\ -T, & \text{if } (x_1,x_2) \text{ in rigion 2} \end{cases} \quad (10)$$

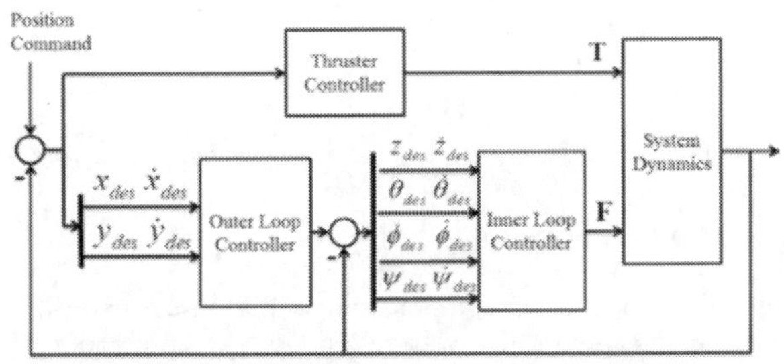

Figure 5. The overall system block diagram of the controller.

4 Simulation Results

This section compares simulation results between the proposed quadcopter and a conventional quadcopter that only uses a dual loop state feedback controller and the quadcopter with jet thruster controller is applied. The system equation of motion described in section 3 is implemented in simulations performed by Simulink in MATLAB.

Simulations of two conditions are set up and presented. One is step input response, which is commonly used to evaluate a system performance. Another is testing loiter with external random force applied on the quadcopter to simulate environmental disturbance in order to observe positioning performance and wind rejection.

$$u(x_1,x_2)=\begin{cases} u(x_1,x_2), & \text{if } x_1^2+x_2^2 \geq \sigma \\ 0, & \text{if } x_1^2+x_2^2 < \sigma \end{cases} \quad (11)$$

where σ is the threshold that control the deadzone region.

In order to reduce the cost of gas while state near the origin and cause oscillation of jet thruster output, dead zone is added which follows the Eq. 11.

4.1 Step Input Response Results

The initial position of the quadcopter is placed at $\{x,y,z\}=\{0,0,0\}$ and set the goal to $\{x,y,z\}=\{1,1,0\}$. Fig.6 shows step response of positioning. It can be observed that quadcopter with thruster have shorter convergence time compare to conventional one and have less overshoot. Fig.7 shows that the phase trajectory follows the designed switching surface to the setting goal directly without having spiral phase trajectory. Fig.8 shows the output of nozzles on x and y axis.

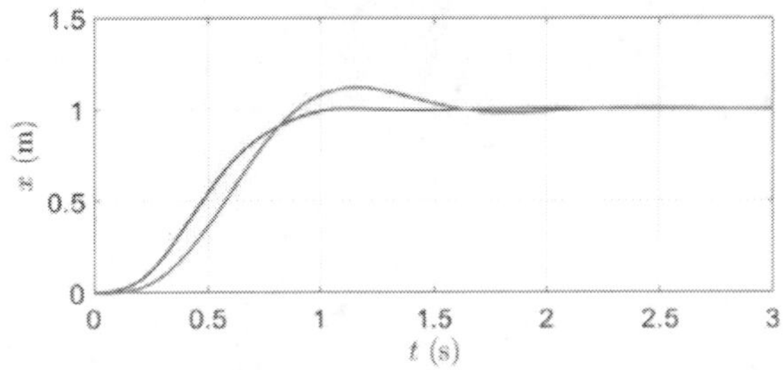

(a)

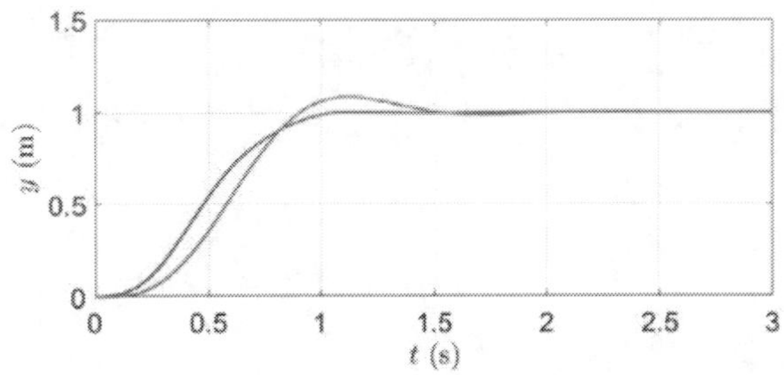

(b)

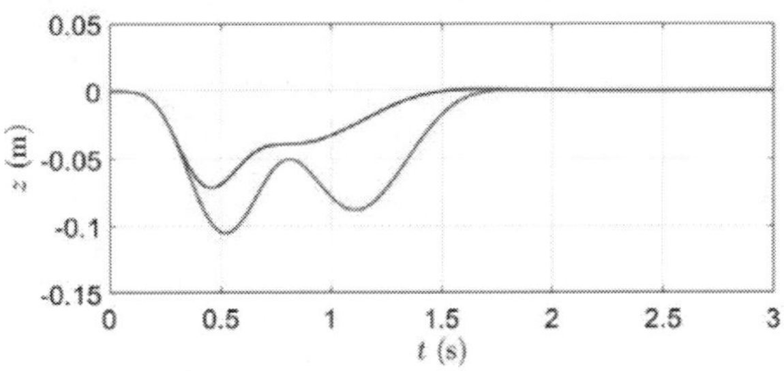

(c)

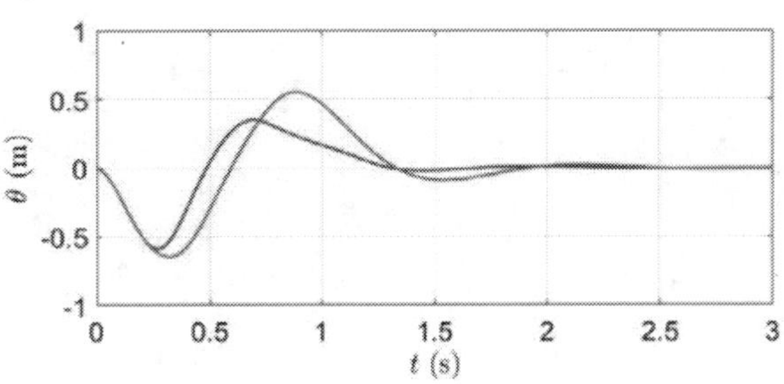

(d)

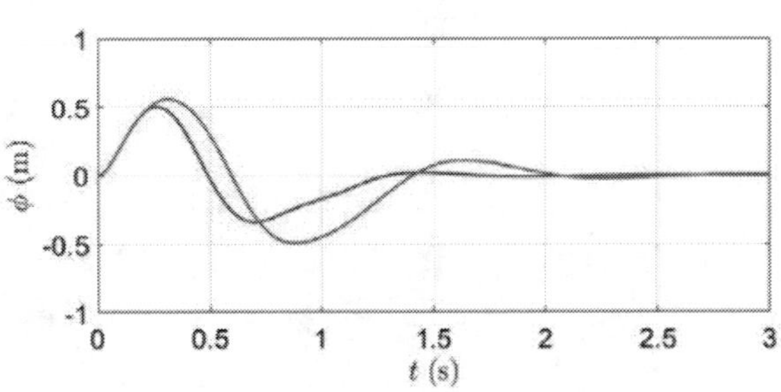

(e)

https://doi.org/10.1051/matecconf/201815104004

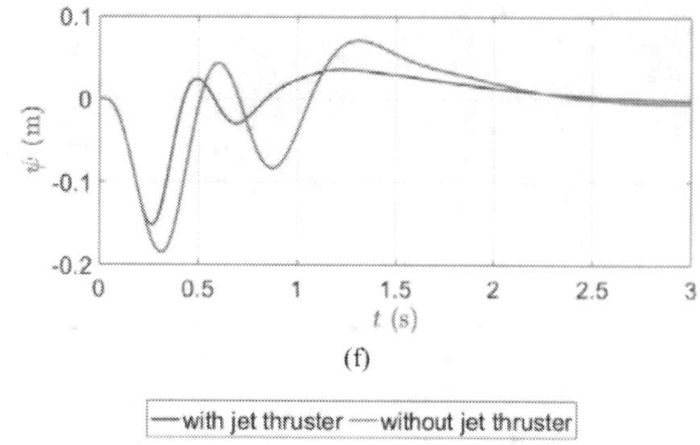

(f)

Figure 6. Simulation of system response to a step input.(a)(b)(c) translation motion, (d)(e)(f) rotation motion.

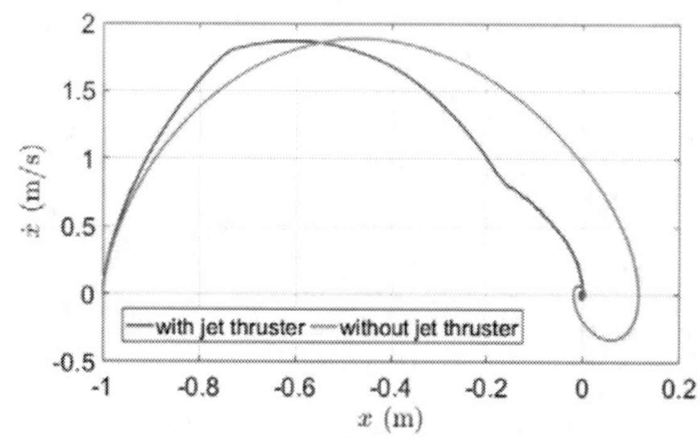

(a)

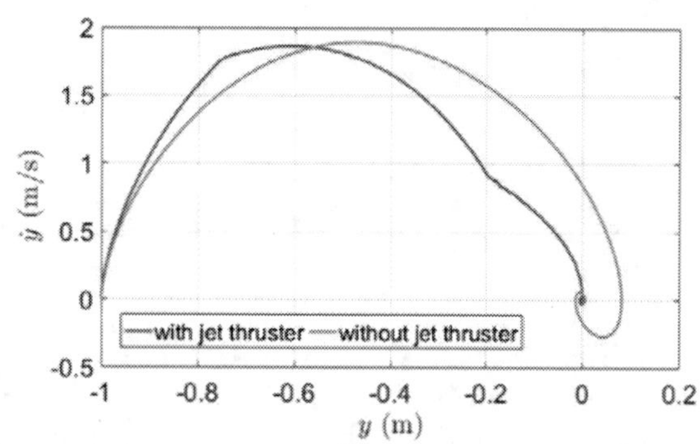

(b)

Figure 7. State error trajectory on (a) x-axis (b) y-axis.

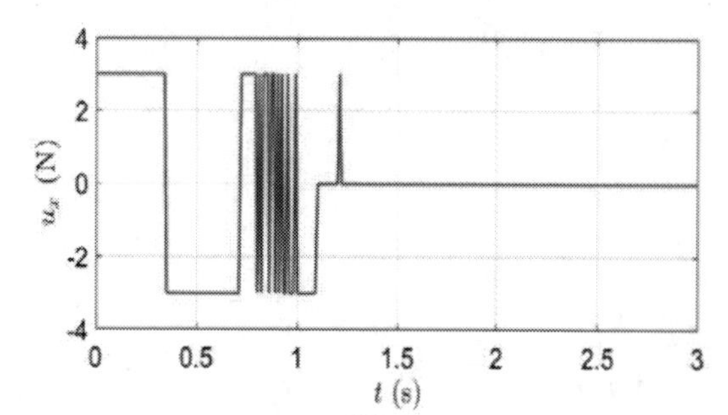

(a)

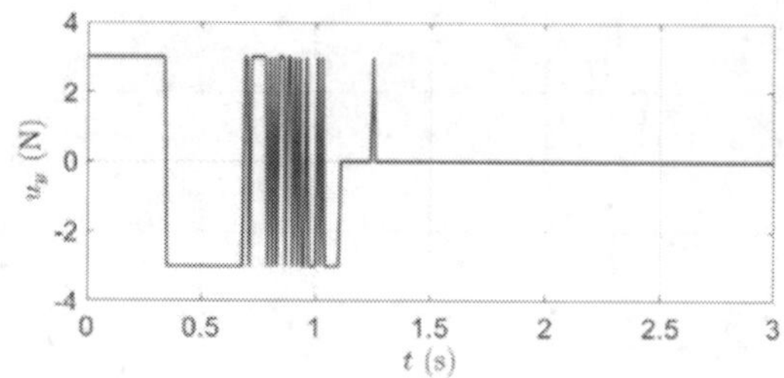

(b)

Figure 8. Control input of 3N jet thruster (a) x-axis (b) y-axis.

4.2 Loiter Results.

The high accuracy positioning performance of the jet thruster controller was tested by simulating loiter in a disturbed environment. In Fig. 9 shows that the quadcopter equipped with jet thrusters resisted external force over 10s. To simulate the wind disturbance, a random force within maxima 5N is applied on the x-axis and y-axis of quadcopter. A conventional quadcopter requires a large pitch angle when the external force is high in the horizontal direction. It cost longer time to create enough resist force than by adding a thruster on its side. Table.1 summarizes the RMS error of the quadcopter. The simulation results show that the quadcopter equipped with jet thruster can have better positioning than the conventional one which decreases about 85% of RMS error.

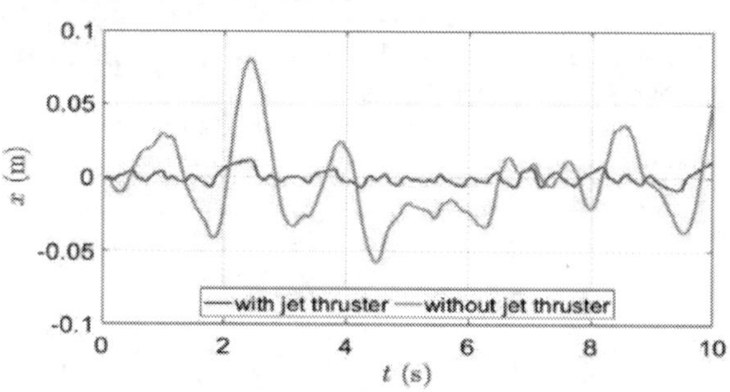

(a)

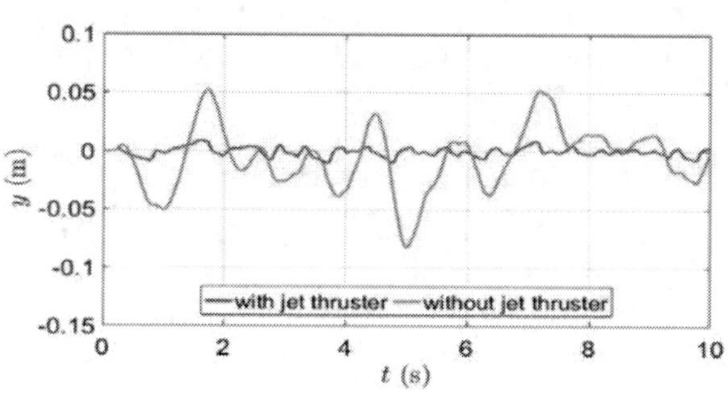

(b)

 https://doi.org/10.1051/matecconf/201815104004

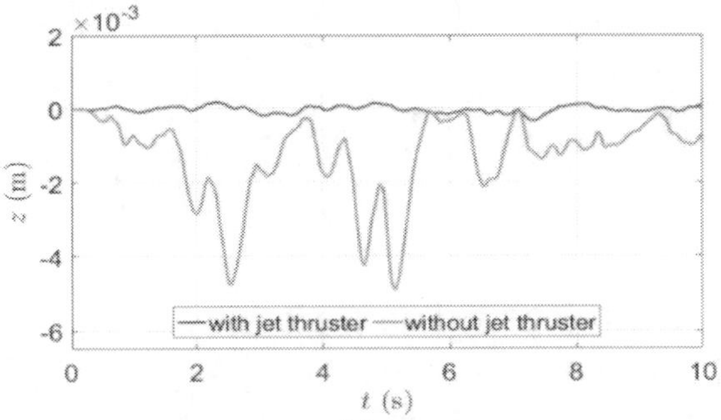

(c)

Figure 9. Simulation of positioning performance with disturbance over 10s. (a) x-axis (b) y-axis (c) z-axis.

Table 1. RMS error over 10s loiter simulation under external force shown in Fig. 9.

	e_x RMS (cm)	e_y RMS (cm)	e_z RMS (cm)
Traditional quadcopter	2.69	2.78	0.17
Quadcopter equipped with jet thruster	0.41	0.38	0.01

5 Conclusions

To investigate the control problem of an over actuated system, a quadcopter model with simple dual loop state feedback controller with the auxiliary jet thrusters is created and simulated. Based on the dynamic equations for a second order system, the jet thruster controller is designed to lead the phase trajectory of state error to the origin. In the simulation result it shows that it has better performance compared with the conventional system. The horizontal RMS error in loiter simulation can decrease about 85%. However, keeping the state on the designed trajectory requires excessive energy due to cancel out the force generated from the simple outer loop of the state feedback.

In future works, we will improve the dual loop controller which enter the switching surface smoother and keep tracking on it. Further studies are also needed to analyze how the switching frequency of the thruster affects fuel consumption. Further hardware tests are also needed.

References

1. James R. Kutia, Karl A. Stol and Weiliang Xu, *Canopy sampling Using an Aerial Manipulator: A Pre-liminary Study* (ICUAS, Denver, Colorado, USA, Jun. 9-12, 2015)
2. J.X.J. Bannwarth, Z. J. Chen, K.A.Stol, *Disturbance Accomodation Control for Wind, International Conference on Unmanned Aircraft System*(ICAUAS, Arlington, VA, USA, June 7-10, 2016)
3. Chau T. Ton and Willam Mackunis, *Robust Attitude Tracking Control of a Quadrotor Helicopter in the Presence of Uncertainty*, (51st IEEE Conference on Decision and Control, Maui, Hawaii, USA, Dec. 10-13, 2012)
4. Pang Tao, *DESIGN, PROTOTYPING AND AUTONOMOUS CONTROL OF GASOLINE-ENGINE VARIABLE-PITCH QUADCOPTER,* Electical and computer engineering national university of singapore (2016)
5. Chin E. Lin, Thanakorn Supsukbaworn, *Development of Dual Power Multirotor System*, International Journal of Aerospace Engineering Volume **2017**, ID 9821401, 19 pages (2017)
6. John G. Furumo, *COLD-GAS PROPULSION FOR SMALL SATELLITE ATTITUDE CONTROL, STATION KEEPING, AND DEORBIT*, Hawai'i Space Grant Consortium Undergraduate Fellowship Reports, HSGC Report No. 13-22. Hawai'i Space Grant Consortium, Honolulu. 67-78. (2013)
7. Assad Anis, *Design & Development of Cold Gas Propulsion System for Pakistan Remote Sensing Satellite (PRSS)*, (2nd ICAST, Islamabad, Pakistan, 29th – 30th Nov. 2008)
8. Sarah L. Nothnagel, Professor Jeffrey A. Hoffman and Brett J. Streetman, *Development of a Cold Gas Propulsion System for the TALARIS Hopper*, Science in Aeronautics and Astronautics, Massachusetts Institute of Technology. Retrieved from http://ssl.mit.edu/publications/theses/SM-2011-NothnagelSarah.pdf (2011)
9. Michal Athans: *Minimum-Fuel Feedback Control Systems: Second-Order Case*, IEEE Trans. Appl. Ind., vol. **82**, no. 65, p. 8–17, Mar. (1963).

Aerodynamic Analysis of Low Speed Wing Design using Taguchi L_9 Orthogonal Array

Kenneth Witcher[1], Ian McAndrew[2] and Elena Vishnevskaya[3]

[1]*Dean, Embry Riddle Aeronautical University, College of Aeronautics, Worldwide, Florida, USA*
[2]*Associate Dean, Embry Riddle Aeronautical University, College of Aeronautics, Worldwide, Chichester, UK*
[3]*Assistant Professor, Embry Riddle Aeronautical University, College of Arts and Sciences, Worldwide, Bitburg, Germany*

Abstract. The study of aerodynamics has been preoccupied with understanding flight at increasing speeds and ultimately supersonic. Today, this pursuit has advanced the science for both Hypersonic and Transonic flight to near Mach 1 supporting economical commercial flight operations. This research presents the data from a Taguchi array on low speed with twin wing designs to establish the design parameters for their use in low speed and high altitude. Also presented is how aerodynamic advantages can be achieved through understanding the interactions of parameters and their use. This is compared to operational effectiveness when applied to remotely piloted aircraft that are not constrained by direct requirements. The research concludes with suggestions for improved designs and further work that may enable higher altitudes with low speeds.

1 Background

There has been a collective effort since the Wright Flyer's first flight to fly faster, higher and further. This was achieved with advances and developments in design and improved propulsion systems through the 1960s. In a few decades, aviation progressed from the Wright Flyer to jet passenger aircraft, Comet, TU 104, B 707 and finally Concorde. Now, supersonic flight on a commercial basis is no longer possible, efficient transonic commercial speeds are the aims for manufacturers, and economical considerations drive designs and operations [1]. Most modern defence manufactures strive to achieve stealth and unmanned flight where the aircraft will no longer be limited by g-forces a pilot can withstand. It could be argued current aviation philosophies are at a cross-roads in direction. There are developments for space travel tourism and hypersonic flights to reach anywhere on the globe within 2 hours. Unmanned Commercial flight is also a real possibility [2].

With Unmanned Aerial Vehicles (UAV), the traditional development rules have changed and requirements need to align with current design capabilities. In this paper the theory, or lack of theory, relating to bi-planes aerodynamics and stability at low speeds is addressed and explores how legacy theories do not consider current design capabilities. Bi-planes were used as the ability to create lift was not as efficient as current designs and the materials available were limited. Having two wings offered extra lift for the weight of the aircraft and overcome the engine power concerns and limited propeller theory.

Early Bi-planes were still working with the infancy of aerodynamics and often limited by the propulsion systems weight and capability. Eventually, multi-engine concepts were recognised as way to overcome reliability concerns. A twin-wing can generate more lift in certain phases of flight; it also has disadvantages from the construction perspective. In Fig. 1, below, the construction is shown of the Bi-plane and the methods to make the structure rigid are clearly shown with the struts between the two wings [3]. These struts cause excessive drag and limit maximum speed. One major advantage that twin-wing has over the modern mono-wing it the low speed and manoeuvring stability in turns, where drag is not greatly influential. Bi-plane fighters were still in use at the start of WWII as combat tactics mainly focused on out-manoeuvring the enemy [4]. There were fighter successes against faster monoplanes; however, the operational ceiling became their aerial weakness for attack where the ability to turn in a smaller circle lost its advantage.

Figure 1. A classic twin-wing designed aircraft.

A standard feature of all twin-wing aircraft is that the top wing is off-set forward (negative stagger), see Fig. 1 above. This stagger is for several reasons: first to allow the pilot to see the ground when taxing and on approach to landing. Secondly, to allow the pilot to get into the cockpit; and finally for the pilot to see above and around for all stages of flight. The latter was critical in the early fighters as if spotted first you were vulnerable. Stagger was, however, a weakness. The extra drag from this configuration reduced the top speed and the stall speed was increased, which was a reason for many crashes by early inexperienced aviators. Given all aircrafts were designed and manufactured in this manner, then these disadvantages disappeared.

2 Evaluating Wing Designs

Early aerodynamics focused on obtaining more speed resulting in the actual behaviour at low speeds not being thoroughly documented. It was always the goal to fly faster and initial developments achieved this aim quickly. Further, aerodynamics has not been concerned with low speeds and certainly not twin-wing. This research is investigating if twin-wing designs can be a realistic solution for UAVs to generate sufficient lift at low speeds for high altitude flight. As the roles and possibilities of UAVs expand, this is an area where solutions are needed and current theory does not offer practical solutions.

Given modern materials for monoplanes can produce a twin-wing without the needs for struts, an experimental design approach is suggested to determine what are the advantages or disadvantages for wing stagger, height, and operations. There are four principal inputs, shown in table 1: the height of the top wing above the lower, the staggering of the top wing, speed of flight and altitude of flight. The wing selected for modelling is NACA6412, primarily as this can be validated against the CFD model (Micro CFD®) and it is a classic wing profile. As little or no theory is obtained for these parametric designs, a three level parameter was selected that requires a Taguchi L9 array, shown below in table 1. Using three level will enable the result to determine if any non-linear responses are established and what might be the response, i.e., a curve.

Table 1. Parameter settings and values.

	Low (1)	Medium (2)	High (3)	
A	0.7	1	1.3	Height as % of Chord
B	-0.5	0	0.5	Off-set as % of Chord
C	0.3	0.325	0.35	Speed Mach
D	3000	5000	8000	Altitude ft

Taguchi's L9 array is statistically indeterminate; it is of a high resolution that does not model for interactions between any of the factors at two level. This research is primarily concerned with the principal factors, and further research for interactions is planned.

3 Methodology

An experimental array is a design matrix to maximise the possibility of determining the way inputs (factors) have on output. In this array (L_9), no interactions are considered among the parameters as these are based on Bernoulli's principle where static and dynamic pressures are related to the speed of the airflow on the upper and lower surfaces of the aerofoil. If interactions are to be included, then a separate array is needed [5]. The CFD runs (rows) were completed in numerical order, the first experiment was a single wing NACA6412 to validate the result of a wind tunnel test with respect to Mach, Density, Pressure and Temperature. Table 2, below, shows the L_9 array and results. The maximum lift and drag are the two outputs shown in the final column of the table.

Table 2. L_9 array and output results.

A	B	C	D	Max lift	Max drag
1	1	1	1	18	15
1	2	2	2	19	15
1	3	3	3	21	14
2	1	2	3	19	15
2	2	3	1	20	16
2	3	1	2	22	14
3	1	3	2	19	17
3	2	1	3	21	16
3	3	2	1	23	15

Lift and drag in N/dm^2

An L9 array has nine experimental runs, not repeated, as this is a computer-generated model not susceptible to variation. CFD simulation cannot offer variables and thus the range, standard deviations, and ultimately signal to noise ratios are not able to be established. Each run generated a maximum lift and drag value for the design. As no operation of an aerofoil will generate a constant pressure underneath, as on the upper surface, the combined output of the pressures are averaged to give a defined lift value for that unique experiment. The drag is directly proportional to the calculations and directly stated as an output. Hence, a high confidence from the datum run comparing to a wind tunnel set up, as discussed earlier. These calculations of the effective lift and combined drag respectively, at each set up, are shown in the last two columns of table 2. The distribution of these (speed, density, pressure and temperature) were modelled to determine the complete influences. These are explained and discussed fully in the following section, as Mach, density, and pressure; all inter-related though Bernoulli's equation [6]. Classic aerodynamics does not model temperature in this equation, it does, however, influence lift and drag significantly in certain situations and has to be included as directly proportional to lift and drag. The tunnel length modelled was 12.5 m and boundary conditions for 2D modelling are shown in Fig. 7 below.

4 Results and Discussions

Experimental run two, where the wings are without stagger and closest together, is shown below in Fig. 2, pressure distribution is displayed in the figure below. Dark blue indicates low pressure and green, yellow and orange indicate higher pressure levels. Clearly, the higher pressures are needed under the aerofoil and lower above to generate lift.

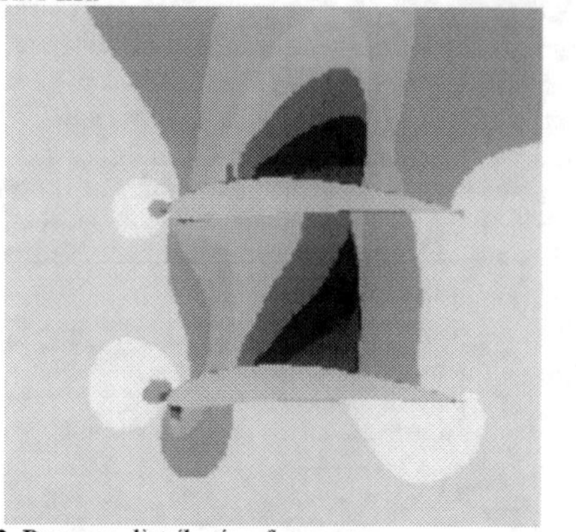

Figure 2. Pressure distribution for run two.

This clearly shows the upper pressures from the bottom wing are influencing the lower surface on the upper wing. In effect, the lift generated by the upper wing is considerably reduce from its theoretical maximum. The leading edge with its high pressure is in effect reducing the total lift as this design of aerofoil has most of the leading edge on the upper profile.

If the negative stagger is modelled under the same parameters, the results are shown below in Fig. 3. This time the upper wing will influence (reduce) the possible theoretical lift of the lower wing. The upper wing exhibits classic pressure profiles for lift [7]. Downwash pressure from this profile is altering the pressure on the lower and very little lift generated by the aerofoil [8]. Its combined pressure lift is considerably lower than what theory suggests and could, under certain conditions, act more as a spoiler (down force) than an aerofoil.

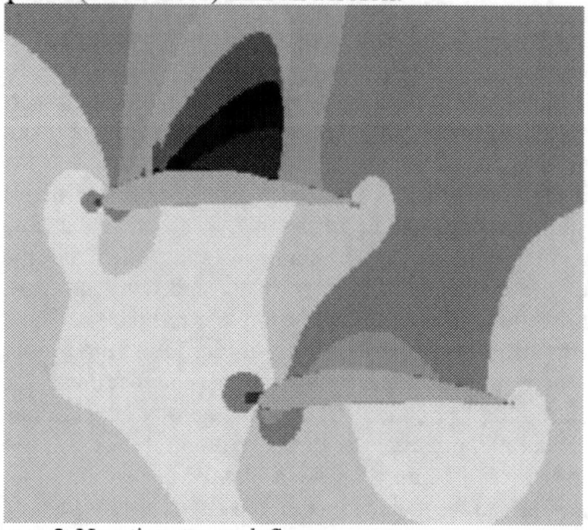

Figure 3. Negative stagger influence.

Stagger has a significant influence on the lift generated, and also the legacy drag from these parameters. It is the extent from the other inputs that need evaluation. The height is of interest as if infinite then there would be zero influence on the output lift and drag. That is not a design solution for this research question.

Each column needs evaluating on its own and in combination. If column A is used as an example, the average of the outputs at each of the three levels is needed. Thus at level 1 the average is (18 + 19 + 21/3) = 19.3. Fig.s 4 & 5 shows the summary of each column for lift and drag respectively. The summary of all for lift is shown below in Fig. 4.

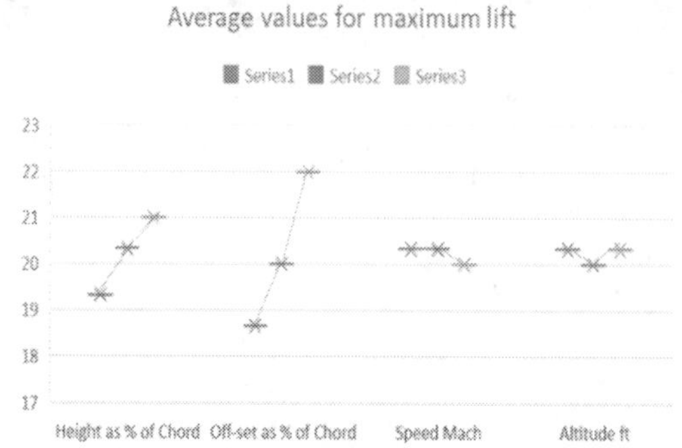

Figure 4. Summary of Lift for each column.

Fig. 4 above shows clearly that as the height separation between the wings becomes larger the lift generated also increases; it could be linear in response outside this range. It could be assumed this height increase is not an influence rather, at lower speeds, the wash down effect from the upper wing has no negative impact on the pressure distribution [9]. Likewise, off-set (stagger) is worst when negative, it improves if directly above but increases the most when a positive stagger A and B suggest high separation and positive stagger are the favourable settings.

Speed has a significantly lower influence with the lift being limited from the positive stagger by a small percentage. Altitude is not significant in these ranges and it might be shown statistically that variation is random and not spurious. Overall, the off-set (stagger) is critical and the separation less critical.

Fig. 5 shows the evaluations for the drag influences from this experiment. Each point is calculated as an average in the same format as for the lift, and average of the three values for each setting. It is presented in the same format as for lift.

Consider the setting with the highest drag and lowest lift, run 7. This is where the difference between average lift and drag is at its smallest level [10]. This is where the height is at the maximum, off-set is negative stagger, speed was set high and the altitude at medium. The negative stagger performance could have been predicted, the height is perhaps more enlightening, and the lesser influences are dependent on possible interaction. It is not possible to determine from this experiment and the outputs. Interaction analysis requires a separate experiment. It does suggest that the settings are sensitive to individual values and could easily change the positives to negatives, which is lower lift and increased drag. The opposite of the focus of this research. For example, if lift is sensitive to speed, any increase in headwind could result in loss of altitude.

 https://doi.org/10.1051/matecconf/201815104005

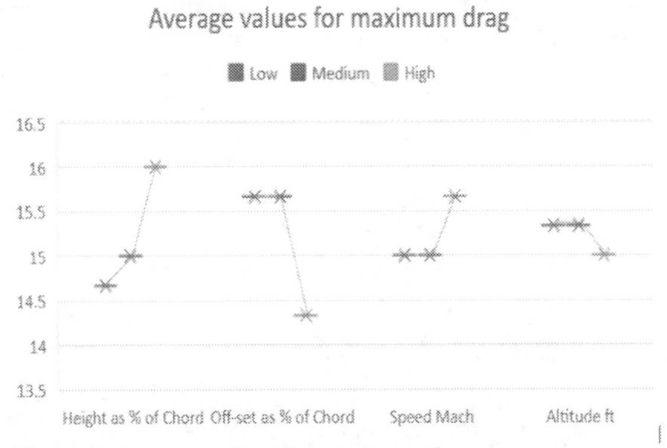

Figure 5. Average values for maximum drag.

Height separation does not appear to be linear. The height increase may produce more lift, but it is also producing higher drag, the instant advantage is tempered by this disadvantage. However, the stagger is of more interest [11]. Whilst the negative and zero stagger play no significance in reducing drag, the positive stagger has a disproportional influence on the reduction of drag that appears far more than the induced drag from height separation – this seems at first contradiction to theory.

Increasing the air speed increases drag just as increases in altitude naturally reduces drag as the air becomes thinner and lower friction [12]. These latter two are not significant in influence as the height and stagger, in the same way as is for lift; the sensitivity of each on the principal inputs needs to be known.

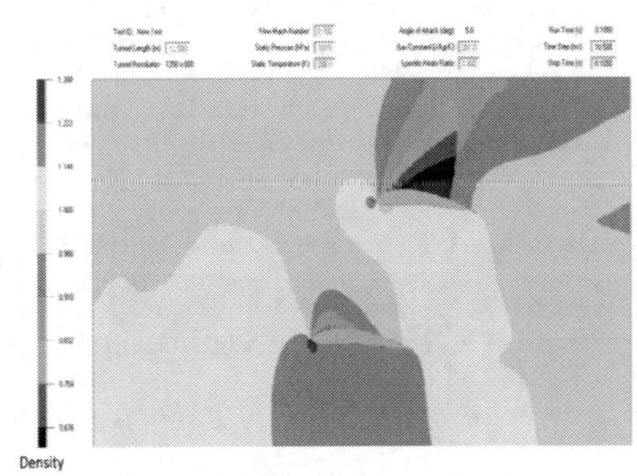

Figure 6. Optimum setting – Density.

When the heigth between the two aerofoils are maximum, see Fig. 6 above, with a positive off-set the influence of one on the other wing is drastically reduced. First, the lower one is almost totally unaffected from the upper, the density difference is greater and that remains constant at these speeds for the length of the underside. Secondly, the upper wing has a low density compounding the difference between the top and bottom. Finally, the difference in density between the two wings is relatively consistent and allows for the two wings to be accepted as not significantly influening each other. Removing this interference is key to lift at altitude where the air density is lower and pressure resulting will be closer between the surfaces and the wings.

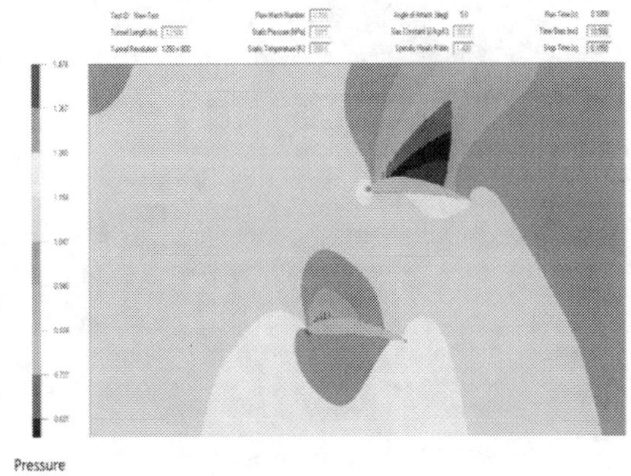

Figure 7. Optimum setting – Pressure.

The pressure distributions shown in Fig. 7above, that the density influences the pressure directly and follows that of Bernoulli's equation [13]. Most importantly is the consistency of pressure between the upper surface of the lower wing to the underneath of the upper wing. Other settings have been directly influenceing this channel between the wings to the detriment of generating pressure. It would appear that the height setting may be close to optimum; although that needs to be verified by further defining. The positve stagger being correct is reinforced further by observing the response values shown in Fig. 4.

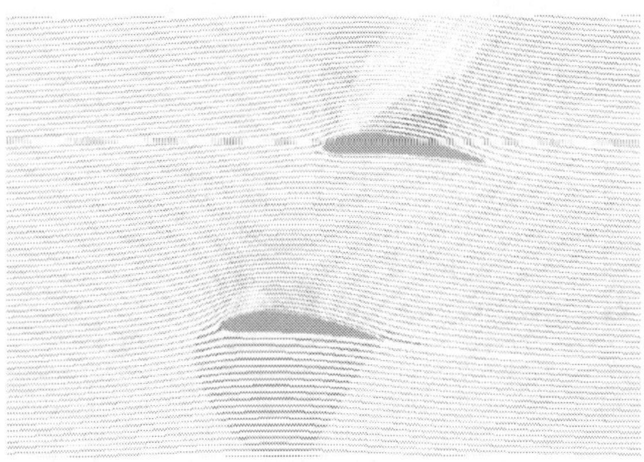

Figure 8. Streamlining of optimum setting.

Streamlining, as shown in Fig. 8 above, of the flow also supports the findings, particularly the downwash of each wing will not cause flow restrctions when combining for a long distance behind the aircraft. Thus, the drag will not be as great as otherwise expected. It is further supported with the response rate show for off-set in Fig. 5. When this is compared in conrast to the lowest lift but highest drag the influences compond on the wings to cause negatives in flow both about and between the wing positions [14]. Even though the hight is maximum the lift offered by twin-wings in this case is neglable in compasion, see Fig. 9 below. The sensitivity is also high from small changes to the stated inputs used.

Figure 9. Lowest lift to drag ratio pressure.

Taguchi's experimental arrays offer insights to soulutions with reduced experiements, not all are able to achieve useful results as they are statistically inderterminate. Thus, to finalise and achieve maximum settings further experiements are needed. Indeed, any variations allow for signal-to-noise ratios that identify sensitiivety and if variations are random or spurious.

5 Analysis

This research has shown that of the four inputs investigated; those of the wing heights and the stagger are significant. Speed and altitude less for lift, although each can increase the overall drag, from small changes within the medium to high settings on height and stagger. Likewise, increasing lift may have significant results with moderate changes from inputs.

What is not clear is if interactions exist between any of the settings and if those of speed and altitude are sensitive or just showing random not spurious variations. The results have been validated with a datum scenario and there is a high degree of confidence in the results.

6 Conclusions and future work

This research has shown that the classic way of using a twin-wing had many aerodynamic disadvantages, even though the negative stagger was necessary for operational use. There were four parameters investigated: Height, Stagger, Speed and Altitude. Height and Stagger are the two principal influences; those of speed and altitude were secondary. The results of this research suggest that high separation and positive stagger are critical. The Angle of Attack has not been included in any of these designs and their influence is needing to be known. Future research will also focus on the exact height separation and stagger. Additionally, how any angle of incident will complement the lift whist reducing drag to achieve the highest altitudes with minimum flight speed.

Future work will model in 3D with wing tip boundaries and fuselage drag. These constraints and parameters established here will focus the variabilities to determine robust settings and prediction models.

References

1. J. D. Anderson. Introduction to flight (McGraw Hill, 6th Ed. New York 2005).
2. Craig, Gale (2003). Introduction to Aerodynamics. Regenerative Press.
3. Flightglobal Archive. Aviation Histtory, 1940. Flight.
4. McAndrew I. R, Godsey O and Navarro E. (2014). Aerodynamics of drogue system refueling of unmanned aerial vehicles. Advanced Materials Research. 1016: 349- 353.
5. Lochner, Robert H., and Joseph E. Matar. *Designing for Quality: An Introduction to the Best of Taguchi and Western Methods of Statistical Experimental Design*. Milwaukee, WI: ASQC Quality Press, 1990.
6. J. Bertin & R. Cummings: Aerodynamics for Engineers (Prentice Hall, 5th ed. June 28, 2008)
7. Farnum, Nicholas R. *Modern Statistical Quality Control and Improvement*. New York: Duxbury Press, 1994
8. Federal Aviation Regulation: Aeronautical Information Manual, U.S Department of Transportation. Aviation Supplies & Academics, Inc. Newcastle, WA 98059-3153
9. Hirschel, Ernst H. (2004). Basics of Aerothermodynamics. Springer
10. Ross, P. J. *Taguchi Techniques for Quality Engineering*. New York: McGraw-Hill, 1988
11. McAndrew I. R, Navarro E and Witcher K. (2015). Drogue deflections in low speed unmanned aerial refueling. International Journal of research in Aeronautical Engineering & Mechanical Engineering, IJRAME. 3(1): 119-127.
12. Chanute, O *(1997). Progress in Flying Machines.* Dover Publications.
13. Hodge, B. K.; Koenig K. (1995). *Compressible Fluid Dynamics with Personal Computer Applications.* Prentice Hall
14. Bertin, J. J.; Smith, M. L. (2001). *Aerodynamics for Engineers (4th ed.).* Prentice Hall

Factorial Experiment on Route Structures Using SAAM on a Specified Area in ASEAN

S.M.Phyoe[1], J.W. Goh[2], R.Guo[3], Z.W. Zhong[4]

[1,2,3,4] *Air Traffic Management Research Institute, Nanyang Technological University, Singapore*

Abstract. Due to a rapid growth in the air transport industry, ASEAN (Association of Southeast Asia Nations) region has experienced a surge in the air traffic movement in recent years. Several forecasting papers foretell to have the continuous traffic growth for ASEAN region [1, 2]. The pilot study of ATS routes forecasts also conclude that there is a need to relook at the current route structures in selected areas in ASEAN [3]. The interaction between the structure of ATS routes and the number of intersections has been studied by using the number of conflicts [4]. This paper will serve as a continuation of the previous study, and the objective is to determine whether there is any consistency in the conclusion when the selected area is changed, using the methodology proposed.

1 Introduction

The exponential growth in the air traffic movement has left many researchers scrambling for solutions and improvements in the past decades. There are many proposed measures to counterbalance the acceleration, from straightforward solutions like expanding ground infrastructures to sophisticated concepts like the ground delay program [5-7]. The demand for each ATS route also increases. Optimization of en route flow was studied by Li et al [8]. Integer programming was employed for better traffic flow and follow the suitable ATS routes [8].

Our previous study has investigated the potential factors which might lead to an improvement in the current structure of some ATS routes by using the number of conflicts [4]. A conflict is when two or more aircraft experience a loss of minimum separation [9]. There are several studies on conflict resolution since this is directly related to aircraft safety.

Tomlin et al. created a methodology for two aircraft conflict resolution maneuvers based on reachable sets for hybrid systems with nonlinear dynamic within each discrete state [10]. Wang et al. established a mathematical model for resolving air traffic conflicts [11]. Ma et al. resolved the 3 flights conflict through combining the ADS-B reports with the track optimization and Genetic Algorithm [12]. According to Hong et al, the particle swarm optimization (PSO) was used to solve the nonlinear optimal conflict resolution and flow management problem in their study [13]. Han et al. proposed the potential field differential evolution (PFDE). The differential evolution algorithm was used to optimize the potential field (PF) and proved that PFDE performed better than PF [14]. Matsuno et al. proposed the near-optimal control method to resolve the aircraft conflict based on the polynomial chaos expansion and pseudospectral methods [15]. Due to the nature of conflicts reduction from our ATS routes improvement, it can also be one of the solutions for conflict resolutions.

However, the results and conclusions are restricted to that particular airspace and are not a representation of the general airspace. In this paper, using the concept of factorial experiment, we will investigate another area in ASEAN and determine if the same conclusion can be obtained.

2 Factorial experiment - factors and levels

There are a total of two factors and three levels for the setup of this factorial experiment. They are defined as follows:

1. Structure of the ATS routes (Figure 1)
   - 1.1. Bidirectional (Bi)
   - 1.2. Unidirectional clockwise (Uc)
   - 1.3. Unidirectional counterclockwise (Ucc)
2. Number of Intersections in the Region (Figure 2)
   - 2.1. Low (L): Fewer than 10 intersections in the area
   - 2.2. Medium (M): Between 10-20 intersections in the area
   - 2.3. High (H): More than 20 intersections in the area

3 Methodology

The methodology follows closely to that done in the previous study [4].

Firstly, an area with high traffic density was selected. This study chose the area of the experiment based on the ASEAN's traffic density which was calculated by using

SAAM (System of Traffic Assignment and Analysis at a Macroscopic Level) and 2016 June second day traffic data from INNOVATA.

As shown in Figure 3, the airspace was divided into small density cells and the color of the cells indicates the density of the traffic for that area. Since red color indicates the highest density, the particular area covered by red cells was chosen for this study.

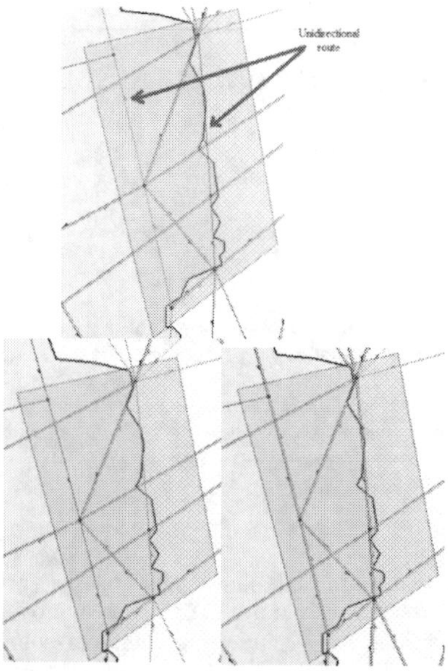

Figure 1. (From top, moving clockwise): Unidirectional clockwise route structure, bidirectional route structure and unidirectional counterclockwise route structure for the chosen ATS routes. The area of concern is shaded in orange.

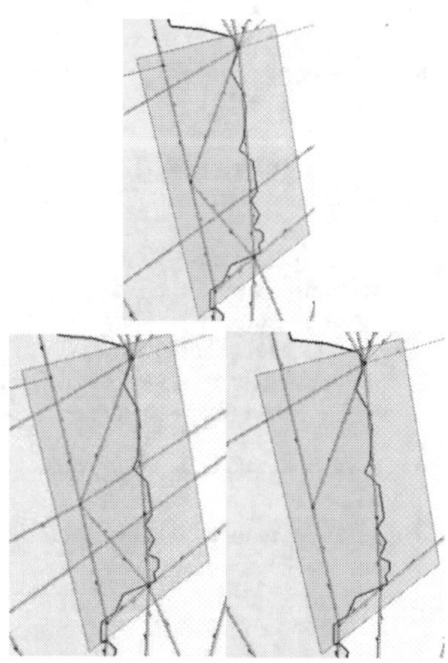

Figure 2. (From top, moving clockwise): Medium number of intersections (10 intersections), low number of intersections (5 intersections) and high number of intersections (16 intersections). The area of concern is shaded in orange.

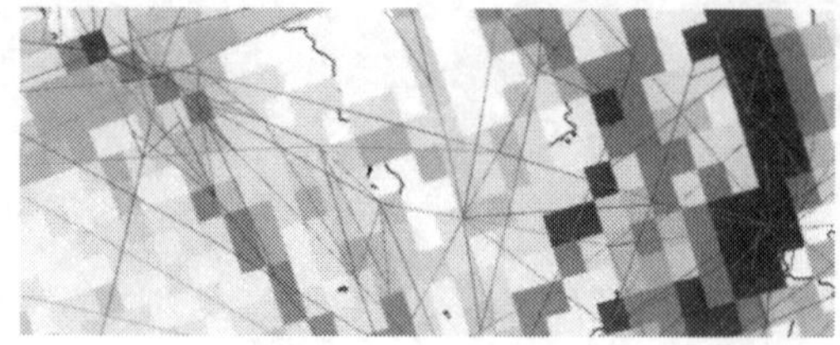

Figure 3. Example 3D Density based on 2016 June second day Traffic (INNOVATA)

Subsequently, two ATS routes were focused based on the extracted whole day information within the area. Following that, the structures of the ATS routes were modified as shown in Figure 1.

The introduction and removal of segments for the intersection modifications were done in a manner not to make the aircraft experience any major rerouting. Figure 2 demonstrates how certain segments of the ATS routes were added or removed.

Once the setup was completed, each scenario was ran thrice using SAAM. Then the number of conflicts was recorded. Finally, the interactions between the factors were studied, and the factorial ANOVA test was performed to determine if there was any difference in conflict rate due to the changes in route structures, the number of intersections and the interaction between route structures and intersections. The three different tests are shown below with their respective decision rules, tested at 5% level of significance.

1. H_0: $\mu_{Bi} = \mu_{Uc} = \mu_{Ucc}$
   H_1: $\mu_{Bi} \neq \mu_{Uc} \neq \mu_{Ucc}$
   Decision rule: F_α (2, 18) = 3.555
2. H_0: $\mu_H = \mu_M = \mu_L$
   H_1: $\mu_H \neq \mu_M \neq \mu_L$
   Decision rule: F_α (2, 18) = 3.555
3. H_0: interaction is absent
   H_1: interaction is present
   Decision rule: F_α (4, 18) = 2.928

4 Results and analysis

Table 1 records the number of conflicts after running each scenario thrice with SAAM.

Table 2 shows the values of the sum of squares (SS), degree of freedom (dF), mean squared (MS) and the F-value (F) based on the results in Table 1.

Table 1: The number of conflicts recorded after running each scenario thrice.

	Low	Medium	High
Bidirectional	10	19	15
	9	21	18
	17	41	48
Unidirectional clockwise	6	13	11
	11	23	11
	8	14	10
Unidirectional counterclockwise	16	32	34
	13	36	43
	4	7	28

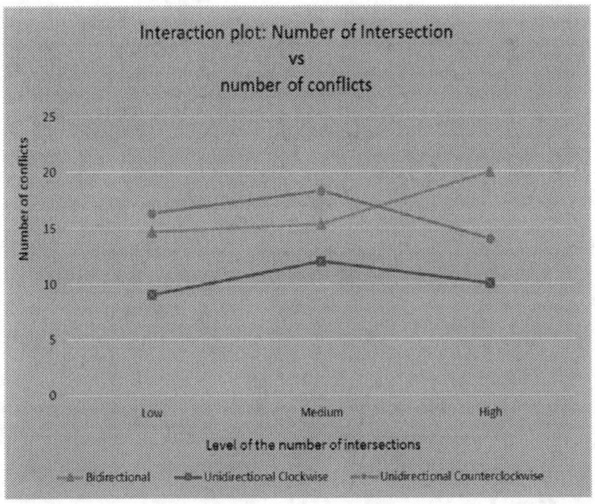

Figure 4. The interaction plot between intersections and conflicts

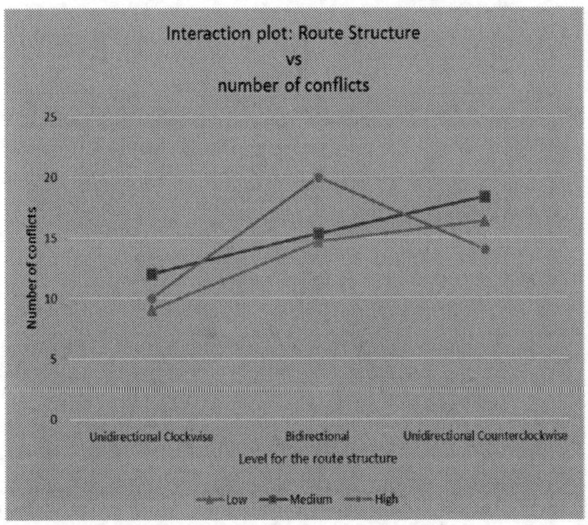

Figure 5. The interaction plot between route structures and the numbers of conflicts.

Figures 4 and 5 interpret that there is some form of interaction between the route structure and the number of intersections considering the plots are nonparallel. As shown in Figure 4, the plot lines for the unidirectional routes (both clockwise and counterclockwise) are more parallel comparing to the plot line for the bi-directional route, revealing that the effect of interaction is more significant for the bidirectional route structures.

As shown in Figure 5, the plot lines for intersections are nonparallel for all the three levels of intersections (low, medium & high). This observation supports the observation mentioned above for Figure 4, and thus it is deduced that the effect of the interaction is affected by the route.

Table 2. The numerical values for sum of squares, degree of freedom, mean squared and the F-value for each row

	SS	dF	MS	F-value
Structure of the ATS routes (RS)	731.19	2	365.59	3.62
Number of Intersections (NOI)	49.85	2	24.93	0.25
Interaction (RS×NOI)	1325.70	4	331.43	3.28
Error	1817.33	18	100.96	-
Total	3924.07	26	-	-

Numerically, Figure 5 reveals that when the number of interactions is high, employing the unidirectional route structure (both clockwise and counterclockwise) encounters less conflicts than the bidirectional route structure. Whereas, for low number of interactions, there are least conflicts by applying unidirectional clockwise route structure. In conclusion, among the three different structures, the unidirectional clockwise route structure yields the lowest conflicts for all levels. This is the different result since the bidirectional route structure was the lowest in our previous study. However, this information is insufficient to demonstrate that the change in the structure of the selected ATS routes indeed results in a change in a number of conflicts.

According to Table 2, since $F_{RS} \geq F_{\alpha}$, the null hypothesis is rejected and it is concluded that there is a difference in the number of conflicts when there is a change in the structure of the selected ATS routes. On the other hand, since $F_{NOI} \leq F_{\alpha}$, the null hypothesis is not rejected and it is concluded that there is no significant difference in the number of conflicts when there is a change in the number of intersections in a region. Likewise, the null hypothesis is rejected since F (RS × NOI) $\geq F_{\alpha}$, and it is concluded that the number of conflicts can be reduced with a modification of both the selected ATS routes structure and the number of intersections between routes in the targeted region.

The results of the factorial ANOVA test suggest that the number of intersections has no much influence on the number of conflicts. However the structure of the selected ATS routes, particularly their cruising directions, can be one of the primary factors that can improve the airspace or the route capacity of this selected area. In addition, the interaction between route structures and intersections can be another element to reduce conflicts and improve the airspace. It is critical to note that the above results are restricted to this particular area and may not apply to every case since the arrangement of the ATS routes differs in every country.

5 conclusion

From the results, the main conclusion is that the number of conflicts could be significantly affected by the direction of routes for this selected area and the interaction of the routes' direction and intersection. It can also conclude statistically that for this selected area, there is no significant difference in the number of conflicts if we change the number of intersections itself. In comparison with the previous result, the impact of the routes' direction remains the same, but the interaction of the routes' direction and the intersection is found to be the new influence factor. Therefore, we can assume that the effect on the number of conflicts also depend on the regions. By applying the methodology, the service providers can design the optimal route structures for their targeted airspace to achieve less number of conflicts. In future, more advanced and detailed study for the factorial ANOVA test will be conducted and the reason of different performance for different regions will be discussed.

Acknowledgment

This research was sponsored by the ATMRI of NTU and CAAS via ATMRI Project No. 2014-D2-ZHONG for Regional Airspace Capacity Enhancement - ASEAN Pilot.

References

1. Airbus S.A.S (2015), Airbus Global Market Forecast 2015-2034. [Online]. Available: http://www.airbus.com/company/market/forecast/
2. S.M. Phyoe, Y.X. Lee, Z.W. Zhong, Determining the Future Demand: Studies for Air Traffic Forecasting, International Journal of Technology and Engineering Studies, Vol **2**, No.3, pp. 83-86, (2016)
3. D. Raheja, Y.X. Lee, Z.W. Zhong, Utilizing Regression Algorithm for ATS Routes Forecasts , Singapore 34th International Conference on Engineering & Technology, Computer, Basic and Applied Sciences (ECBA-2017), Singapore, Jul (2017)
4. J.W. Goh, Y.X. Lee, S. Aneeka, S.M. Phyoe, Z.W. Zhong, Factorial Experiment on Route Structures leveraging on Current Air Traffic Management Technology on Selected Area in ASEAN, Singapore 28th International Conference on Engineering & Technology, Computer, Basic and Applied Sciences (ECBA-2017), Singapore, Feb (2017)
5. E. P. Gilbo, Optimizing airport capacity utilization in air traffic flow management subject to constraints at arrival and departure fixes, IEEE Transactions on Control Systems Technology, Vol **5**, pp. 490, (1997)
6. R.L. Hoffman, M.O. Ball, Measuring Ground Delay Program Effectiveness Using the Rate Control Index, Journal of Air Traffic Control, Vol **42**, No.2, pp. 19-23, (2000)
7. M. Ball, G. Lulli, Ground Delay Programs: Optimizing over the Included Flight Set Based on Distance, Air Traffic Control Quarterly, Vol **12**, No.1, pp. 1-25, (2004)
8. Q. Li, Y. Zhang and R. Su, "A Flow-based Flight Scheduler for En-route Air Traffic Management", IFAC-PapersOnLine, vol. **49**, no. 3, pp. 353-358, 2016.
9. J. Kuchar and L. Yang, "A review of conflict detection and resolution modeling methods", IEEE Transactions on Intelligent Transportation Systems, vol. **1**, no. 4, pp. 179-189, 2000.
10. C. Tomlin, G. Pappas and S. Sastry, "Conflict resolution for air traffic management: a study in multiagent hybrid systems", IEEE Transactions on Automatic Control, vol. **43**, no. 4, pp. 509-521, 1998.
11. K. Wang and R. Nie, "Research on Detection and Resolution Model of Air Traffic Conflict", in E-Product E-Service and E-Entertainment (ICEEE), Henan, China, 2010.
12. Y. Ma, Y. Ni and P. Liu, "Aircrafts Conflict Resolution Method Based on ADS-B and Genetic Algorithm," 2013 Sixth International Symposium on Computational Intelligence and Design, Hangzhou, 2013, pp. 121-124.
13. Y. Hong, B. Choi, G. Oh, K. Lee and Y. Kim, "Nonlinear Conflict Resolution and Flow Management Using Particle Swarm Optimization", IEEE Transactions on Intelligent Transportation Systems, pp. 1-10, 2017.
14. D. Han, X. Zhang and X. Li, "Aircrafts Conflict Resolution Using Differential Evolution", in 12th World Congress on Intelligent Control and Automation (WCICA), Guilin, China, 2016.
15. Y. Matsuno, T. Tsuchiya and N. Matayoshi, "Near-Optimal Control for Aircraft Conflict Resolution in the Presence of Uncertainty", Journal of Guidance, Control, and Dynamics, vol. **39**, no. 2, pp. 326-338, 2016.

Fluid Similarity Criterions in Aircraft Landing Gear

Liu Xiang-yong[1], Li Wan-li [1], Xu Zhi-qiang[2]

[1]*School of Mechnical Engineering, Tongji University, NO:4800 Caoan Highway, Shanghai, China.*
[2]*The Institute of Fishery Machinery and Equipment, NO:63 Chifeng Road, Shanghai, China.*

Abstract: Facing the design difficulty and risk of aircraft landing gear, the fluid similarity theory and model experiments are used to predict large prototype's property, which provide guidance for hydraulic designation and selection to reduce risk. Firstly, the matrix method is used to establish the fluid similarity relationship between the prototype and the model's luffing mechanism, and the above conclusions are used to design the prototype's hydraulic system. Secondly, the luffing mechanism's dynamical model with accumulator was established, the initial motion condition was set, and the pressure's similarity ratio is obtained by simulation method. Finally, the dynamic model and similarity ratio about time, velocity and pressure are verified by the prototype and model's experimental data.

1 Introduction

Aircraft landing gear is an important aircraft load bearing mechanism, and the aircraft landing gear include luffing cylinder and hoisting leg, as shown in Fig.1. [1] In history, there have been many design failures on the aircraft landing gear, and the mechanism can't bear load. In order to reduce the design difficulty and risk, it is urgent and necessary to use the shrunken model to predict the large prototype's performance. Besides, the similarity research has important theoretical and engineering value.

Figure.1 Aircraft landing gear

At present, scholars have done similarity research in some areas. Yang Jun-jie deduced the similarity theory and do the model experiment. Chen Xing-hua did the large structure test and comparision. Gezary did the micro dynamic similar model flight test. Li Gao did the similarity research of arch dam vibration. Cheng Su-qiu did the scaling laws application in underwater explosion models [2-9].

All of the above researches are based on the well-known equations, but the fluid's similarity research and its application in aircraft landing gear are rare. Our contribution is to establish the fluid similarity theory by computational fluid dynamics method and fluid similarity theory; The theory was verified by simulation and experiments.

2 Similarity criterions based on matrix method

The electro-hydraulic proportional directional valve controls the luffing cylinder's outstretch and retraction movement; In order to reduce the pump demand, an accumulator was added in the cylinder's large cavity, and the accumulator plays an energy-saving role [10-13].

2.1 Deduce the similarity criterions

The incompressible fluid was the research target, and the relationship among the pressure (P), oil viscosity (μ), acceleration (g), velocity (w), length (l), and density (ρ) were built.

$$\Pi = p^a \mu^b g^c w^d l^e \rho^f$$

The mass (M), length (L), time (T) were the basic factors, and the total relationship is:

$$\Pi = \left[ML^{-1}T^{-2}\right]^a \left[ML^{-1}T^{-1}\right]^b \left[LT^{-2}\right]^c \left[LT^{-1}\right]^d \left[L\right]^e \left[ML^{-3}\right]^f \quad (1)$$

Get the following power matrix form:

	a	b	c	d	e	f
	p	μ	g	w	l	ρ
M	1	1	0	0	0	1
L	-1	-1	1	1	1	-3
T	-2	-1	-2	-1	0	0

Since the M, L and T have no relationship, the following three equations are obtained from the power matrix:

$$\left.\begin{aligned} a+b+f&=0\\ -a-b+c+d+e-3f&=0\\ -2a-b-2c-d&=0 \end{aligned}\right\} \tag{2}$$

So the following three power groups were got:

$$\left.\begin{aligned} \Pi_1&=\mathrm{p}w^{-2}\rho^{-1}=\frac{p}{\rho w^2}\\ \Pi_1&=\mu w^{-1}l^{-1}\rho^{-1}=\frac{\mu}{\rho wl}\\ \Pi_3&=\mathrm{g}w^{-2}l=\frac{gl}{w^2} \end{aligned}\right\} \tag{3}$$

2.2 Deduce the similarity ratio and design selection

According to the formulas (5), (6), (7), when the length (*l*) ratio is 1: 4 and the acceleration (*g*) is equal, it is easy to get: the speed (*w*) ratio is 1: 2, the pressure (*p*) ratio is 1: 4, the oil viscosity (μ) ratio is 1: 8. According to the formula: $t = l/v$, when the model and the prototype have the same angle change, the time ratio is 1:2.

3 Similarity criterions based on dynamical model

3.1 Establish the dynamic model

The hydraulic system's flow distribution is shown in Fig.2. Q2 is the hydraulic valve's flow, Q1 is the accumulator's flow, and Q3 is the cylinder's flow. Besides the flow in the hydraulic system also includes the compressed and leaked oil parts.

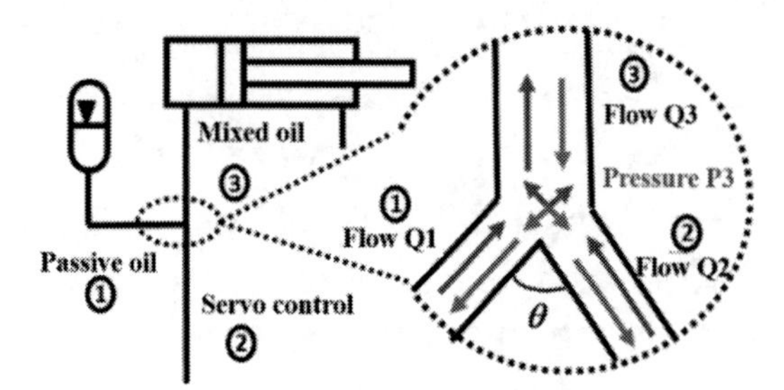

Figure.2 Mixture of accumulator oil

3.1.1 The accumulator's dynamical model

The capsule accumulator's structure and dynamical model are shown in Fig.3:

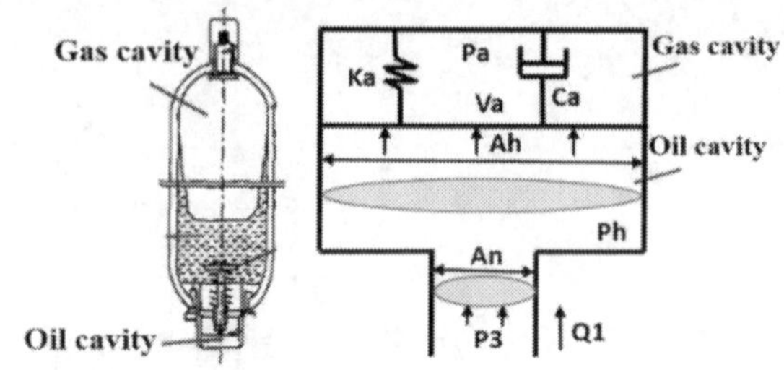

Figure.3 Accumulator and its model

In Fig.3, the accumulator's gas cavity equation can be described as followed:

$$(p_h - p_a)A_a = k_e\frac{V_a}{A_a} + C_e\frac{1}{A_a}\frac{dV_a}{dt} \tag{4}$$

In the formula (4), Ke is the gas stiffness coefficient, and Ce is the gas damping coefficient. The accumulator's oil cavity equation can be described as followed:

$$p_3A_n - p_hA_a = m_e\frac{d^2V_a}{dt^2}\frac{1}{A_a} + B_e\frac{dV_a}{dt}\frac{1}{A_a} \tag{5}$$

In the formula (5), Me is the oil's equivalent mass, and Be is the oil's damping coefficient. Set An = n.Aa, and from the above two equations, it is easy to get the accumulator's dynamical equation:

$$n.p_3 - p_a = \frac{1}{A_a^2}\left[m_e\frac{d^2V_a}{dt^2} + (B_e + C_e)\frac{dV_a}{dt} + k_eV_a\right] \tag{6}$$

The gas state equation is $p_{ao}V_{ao}^k = p_aV_a^k$, where k(1.14) is the gas variable index. Do the derivation on the gas state equation:

$$n.p_3 = \frac{1}{A_a^2}\left[m_e\frac{d^2V_a}{dt^2} + (B_e + C_e)\frac{dV_a}{dt} + \left(k_e - \frac{K.p_{ao}}{V_{ao}}\right)V_a\right] \tag{7}$$

In the formula (8), V0 is the accumulator's the initial volume, and the accumulator's flow Q1 satisfies the following formula:

$$Q1 = V0 - Va \tag{8}$$

Formula (7) can be expressed in Simulink shown in Fig4.

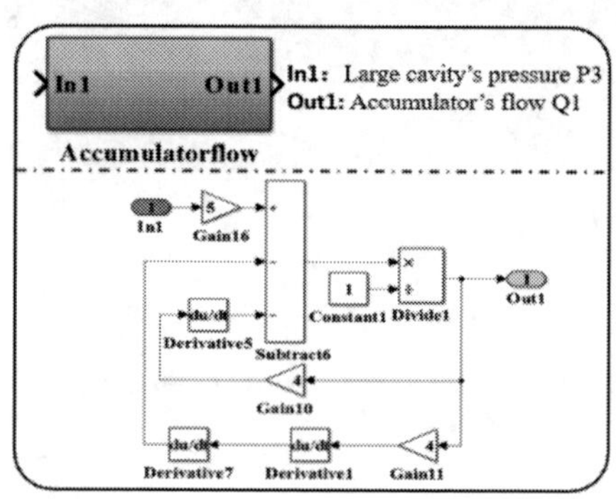

Figure.4 Accumulator's dynamic model

3.1.2 The Dynamical Model of Valve Control Cylinder

In the cylinder's outstretch stage, the working state is shown in Fig.5, and the flow distribution is: Q2=Q3-Q1

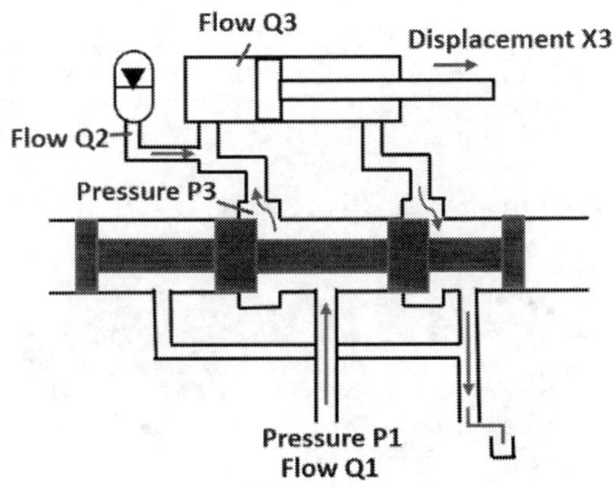

Figure.5 Cylinder's outstretch working state

The flow-continuity equation in the cylinder's outstretch stage:

$$C_{d2}.W_2.\Delta x_2.\sqrt{\frac{1}{\rho}(P_1-P_3)}=\left(\overset{\square}{A}_3\,\dot{x}_3+C_t.P_3+\frac{V_{t3}}{4\beta_e}.\dot{p}_3\right)-Q1 \tag{9}$$

$$P_3=\left[C_{d2}.W_2.\Delta x_2.\sqrt{\frac{1}{\rho}(P_1-P_3)}+Q1-\overset{\square}{A}_3\,\dot{x}_3-\frac{V_{t3}}{4\beta_e}.\dot{p}_3\right]\Big/C_t$$

The dynamical model in the cylinder's outstretch stage is shown in Fig.6

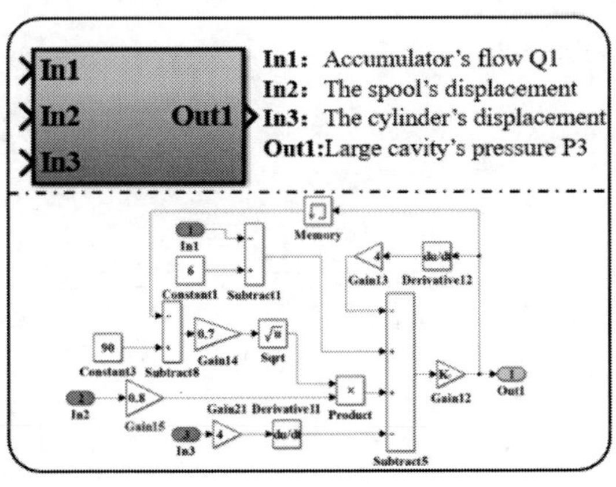

Figure.6 Cylinder's outstretch dynamic model

Combine the accumulator and the cylinder's dynamical model, and the whole simulation process is shown in Fig. 7.

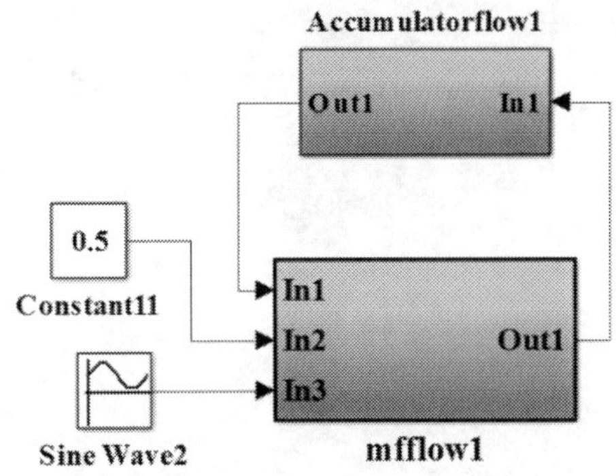

Figure.7 Hydraulic system's simulation process with accumulator

3.2 The similarity research based on the dynamical model's simulation

Model and prototype's cylinder outstretch movement are shown in Fig.8. Initial conditions: the length ratio is 1: 4, the used time ratio is 1: 2, so the average speed ratio is 1: 2. Besides, set the valve spool's diameter ratio 1: 4, valve spool's flow gradient ratio 1: 2, and the accumulator's volume ratio 1: 4. Do the simulation in Fig.10.

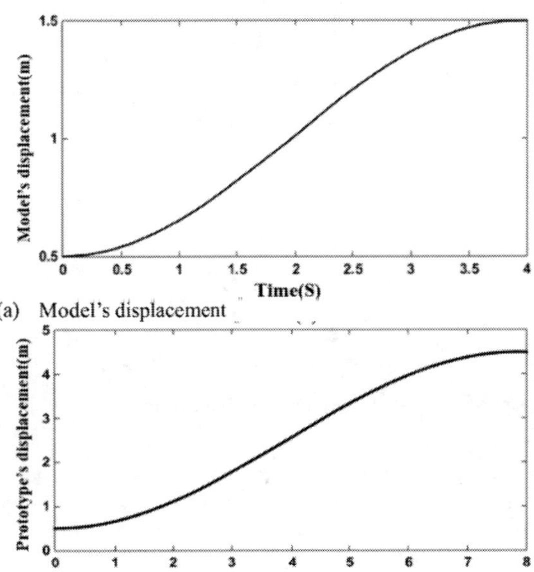

(a) Model's displacement

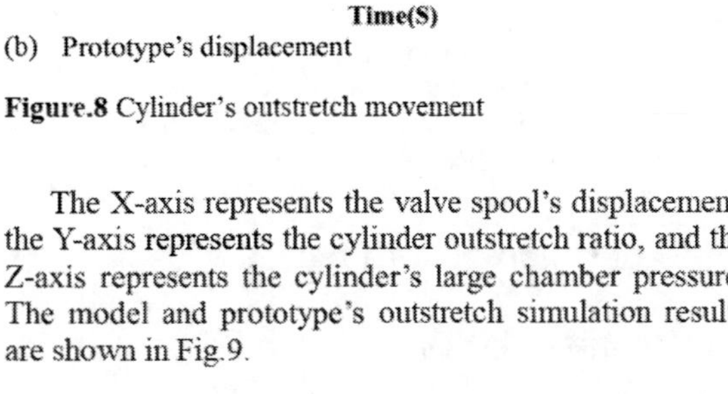

(b) Prototype's displacement

Figure.8 Cylinder's outstretch movement

The X-axis represents the valve spool's displacement, the Y-axis represents the cylinder outstretch ratio, and the Z-axis represents the cylinder's large chamber pressure. The model and prototype's outstretch simulation results are shown in Fig.9.

 https://doi.org/10.1051/matecconf/201815104007

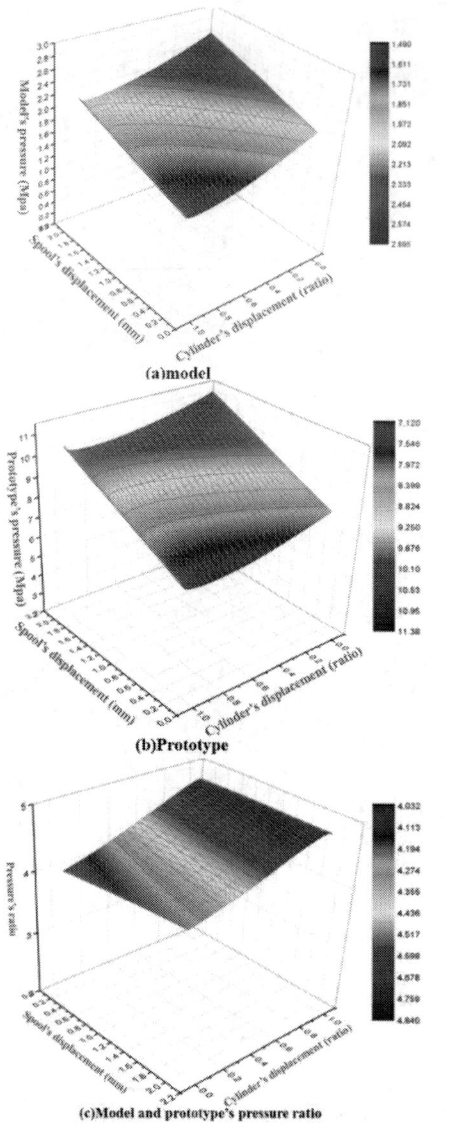

Figure.9 Cylinder's outstretch movement simulation

From the above simulation conditions, we can get that the prototype and model cylinder's length ratio is 1: 4, the speed ratio is 1: 2, so the calculated running-time ratio is 1: 2. In the cylinder outsretch stage: with the decrease of the valve spool's displacement and the increase of the cylinder's extension length, the cylinder's large chamber pressure is gradually reduced, the pressure ratio is about 1:4, and the pressure's decrease leads to the accumulator's oil discharge process.

4 Experimental verification

4.1 Luffing mechanism's Control and data acquisition

The prototype and model are shown in Fig.10. Model's luffing mechanism: ①The maximum displacement: 1m; ② Maximum working speed: 1 m / s; ③ Boom angle's range: -20 ° ~ 58 °. Prototype's luffing mechanism: ① Maximum displacement: 4m; ② Maximum working speed: 2 m / s; ③ Boom angle: -20 ° ~ 58 °;

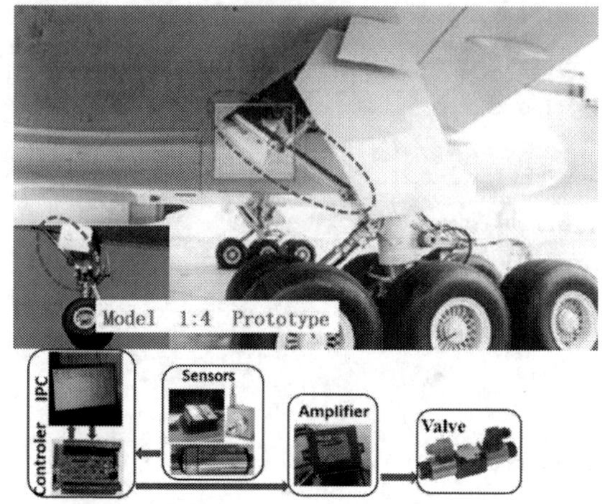

Figure.10 Model and prototype

4.2 Analysis of experimental data

(1)Set the appropriate outstretch step signal, so that the model cylinder's consumed time is 4S, prototype cylinder's consumed time is 8S. Model and prototype's outstretch movement are shown in Fig.11.

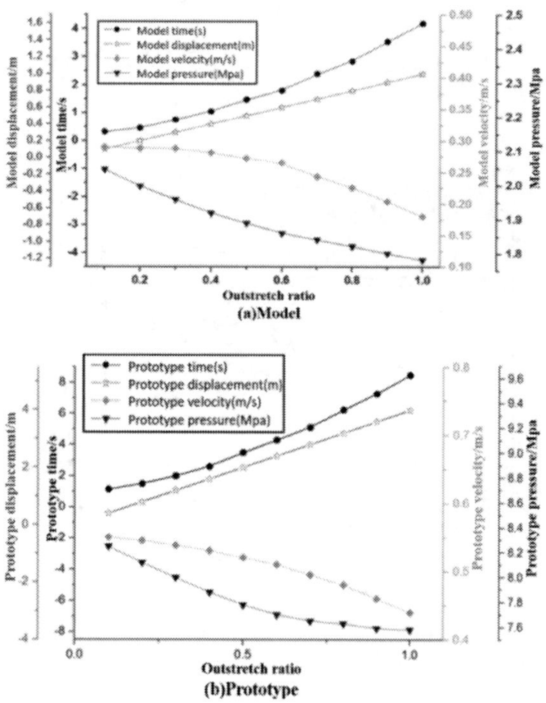

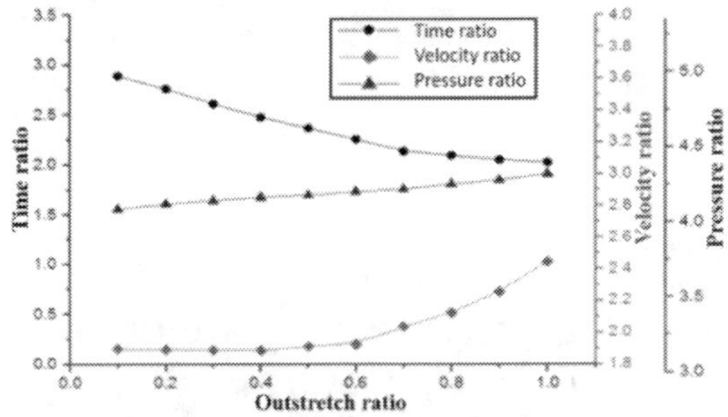

(c)Prototype and Model's ratio

Figure.11 Model and prototype's outstretch movement

Analysis: In the cylinder's outstretch stage, the accumulator discharge the oil; As the accumulator and pump provide oil together, the cylinder's speed increases firstly; When the accumulator has no oil to discharge, the pump is the only source to provide oil and the cylinder's speed decreases. Conclusion: when the model and prototype's cylinder are in the same extension ratio, the pressure ratio is about 1: 4, the time ratio is about 1: 2, the speed ratio is about 1: 2, and the conclusions are consistent with the similarity ratio obtained by the matrix method and the dynamic simulation method.

5 Conclusion

The aircraft landing gear faces the high cost and risk problems. Through the shrunken model's experiment and the deduced similarity criterions, it can provide a reliable approach for the prototype's hydraulic system designation.

The matrix method is based on the basic factor's equal method, the relationship were built, and the similarity criterions were deduced. The dynamical model is established to imitate the electro-hydraulic proportional control process. Through the dynamical model's simulation, the similarity ratios were calculated and the deduced conclusion were consistent with the matrix method's conclusion. The length ratio is 1: 4, the speed ratio is 1: 2, the time ratio is 1: 2, and the pressure ratio is 1: 4.

The model experiment and the similarity criterions provide good references for the prototype equipment, and it can reduce the cost and risk. In the future, the similarity should be verified by more ways.

Acknowledgements

This study was funded by Qingdao Ocean Science and Technology National Laboratory Aoshan Science and Technology Innovation Project (2016ASKJ10), and thanks to Tongji University and the Institute of Fishery Machinery and Equipment.

Reference

1. Nie Hong, Wei Xiaohui. Large-scale civil aircraft landing gear key technology [J]. Journal of Nanjing University of aeronautics and astronautics, 2008, 40(4): 427-432.
2. Chen C-S, Yan D-, Tian Z-C, dynamic similarity theory analysis to the model of Yueyang Dongting lake bridge [J]. Bridge construction,2002(1):48-51.
3. Pakdee , W. , Utaivorawit , N. and Hussein, A.K. Mathematical model in the form of vorticity-stream function for porous premixed combustion , Songklanakarin Journal of Science and Technology, Vol. 37, No. 5 , 2015, pp: 595-600.
4. Al-Kayiem , H. , Hussein, A.K. , Jaleel , J. and Hussain, S. Numerical Computation of 3-Dimensional Supersonic Flow Field Over Seamless Missiles , Indian Journal of Science and Technology , Vol. 7, No. 10 , 2014, pp: 1563-1572.
5. Zhang H, Lu Yaosi, Shi Lide. Similarity Analysis of Crane Model Test [J]. Hoisting and Transportation Machinery, 2000 (12): 27-29.
6. Li Zong-sen, YU Ping, HU Xiao-ping. Similarity theory's application in the design of shaking table model for concrete structure [J]. Foreign Building Materials Science and Technology, 2008, 29 (1): 55-57.
7. Wan Feng, Chen Xiaoyang, Min Yonglin, et al. Design of Simulation Experiment of Fire Gun Jet Trajectory Based on Similarity Theory [J]. Journal of Shanghai University, 2008, 14 (6): 629-632.
8. Sun Da-peng, CHENG Su-qiu, WANG Xue-song. Blasting Experiments of ship structure model based on similarity theory [J]. Blasting, 2012, 29 (2): 91-94.
9. Liu Yun, WANG Ai-lun. Study on complex system's similarity principle and similarity condition [J]. Journal of Systems Engineering, 2009,24 (3): 350-354.
10. Minxun Lu, Wanli Li. Fluid mechanics and Hydraulic Transmission [M]. Shanghai: Tongji University Press, 2006.
11. Analysis and Design of Electro-hydraulic Proportional Control System. Beijing: Mechanical Industry Press. 2005: 35-40.
12. Quan Lingxiao. Mathematical model and experimental study of bladder-type accumulator based on pipeline effect. [Yanshan University's Engineering and Master's degree thesis]. 2005
13. He Qiao-hai, Zhu Ji-li, etl. The hinged point location optimization to Telescopic boom and fork truck's luffing mechanism [J]. The journal of China central's science and technology university, 2011, 39 :423-429.

Comparing the Performance of Reference Trajectory Management and Controller Reconfiguration in Attitude Fault Tolerant Control

Rouzbeh Moradi[1], Alireza Alikhani[1] and Mohsen Fathi Jegarkandi[2]

[1]Aerospace Research Institute (Ministry of Science, Research and Technology), P. o. Box: 14665-834, Tehran, Iran
[2] Department of Aerospace Engineering, Sharif University of Technology, P. o. Box: 11365-11155, Tehran, Iran

Abstract. Reference trajectory management is a method to modify reference trajectories for the faulty system. The modified reference trajectories define new maneuvers for the system to retain its pre-fault dynamic performance. Controller reconfiguration is another method to handle faults in the system, for instance by adjusting the controller parameters (coefficients). Both of these two methods have been considered in the literature and are proven to be capable of handling various faults. However, the comparison of these two methods has not been considered sufficiently. In this paper, a controller reconfiguration mechanism and a reference trajectory management are proposed for the spacecraft attitude fault tolerant control problem. Then, these two methods are compared under the same conditions, and it is shown that the proposed controller reconfiguration has better performance than the proposed reference trajectory management. The reason is that the controller reconfiguration has more variables to modify the closed-loop system behavior.

1 Introduction

Fault occurrence in closed-loop system brings about one of the challenging situations in control problem. For severe faults, the conventional feedback controller may not show satisfactory performance [1]. Fault tolerant control (FTC) has been proposed as a method to handle faults in control systems. FTC is divided into two main parts: Active FTC (AFTC) and Passive FTC (PFTC). PFTC has a fixed structure and is designed to be robust against a pre-defined set of faults [1]. This approach has limited capabilities from a fault-tolerance point of view [1]. AFTC reconfigures the closed-loop system actively to recover the performance of the faulty system [1]. Essential components of this type of controller are fault detection and diagnosis (FDD), reconfigurable controller and a reconfiguration mechanism [1-3]. Almost an uncountable number of papers, books, and reports have considered AFTC. [4] and [5] are recently published review papers that have considered the FTC literature in aerospace and spacecraft attitude control problems.

One of the components of AFTC is the reference trajectory management (RTM) [1]. The main responsibility of RTM is to adjust/modify the reference trajectories to retain the pre-fault performance of the system [6]. There are several papers that have studied the effects of RTM on the performance and stability of the post-fault system [7-10]. According to these works, RTM has been able to deal with the actuator faults/failures efficiently.

According to the literature, AFTCs have been designed considering a variety of methods and disciplines. However, to the authors' best knowledge, no comparison is made between the performance of RTM and controller reconfiguration. This point is the subject of the present paper.

It is shown that under the same conditions, reconfiguring the controller (adjusting the controller parameters, in this paper) has a better performance than the RTM. The reason is that the proposed controller reconfiguration has more variables to modify the closed-loop system. This fact is verified via a simulation.

This paper consists of the following sections: Section 2 considers the spacecraft dynamics and control problem. The proposed controller reconfiguration and RTM are presented in section 3. Section 4 presents the numerical simulation and finally, the paper ends with a conclusion.

2 Dynamics and control

According to [11], if the components of $\mathbf{u}''$ are selected as:

$$u_1'' = \ddot{q}_{1,r} - k_{\dot{q}_1}\left(\dot{q}_1 - \dot{q}_{1,r}\right) - k_{q_1}\left(q_1 - q_{1,r}\right) \quad (1)$$

$$u_2'' = \ddot{q}_{2,r} - k_{\dot{q}_2}\left(\dot{q}_2 - \dot{q}_{2,r}\right) - k_{q_2}\left(q_2 - q_{2,r}\right) \quad (2)$$

$$u_3'' = \ddot{q}_{3,r} - k_{\dot{q}_3}\left(\dot{q}_3 - \dot{q}_{3,r}\right) - k_{q_3}\left(q_3 - q_{3,r}\right) \quad (3)$$

the closed loop system of the spacecraft will be asymptotically stable, provided the first and second order derivatives of the reference quaternion vector ($\mathbf{q}_r$) exist and controller coefficients (k_{q_1}, k_{q_2}, k_{q_3}, $k_{\dot{q}_1}$, $k_{\dot{q}_2}$, $k_{\dot{q}_3}$) are positive.

3 Controller reconfiguration and RTM

Depending on the method used (controller reconfiguration or RTM), the following cases will be considered:

3.1 Controller reconfiguration

In this case, the closed-loop system structure is shown in Fig. 1:

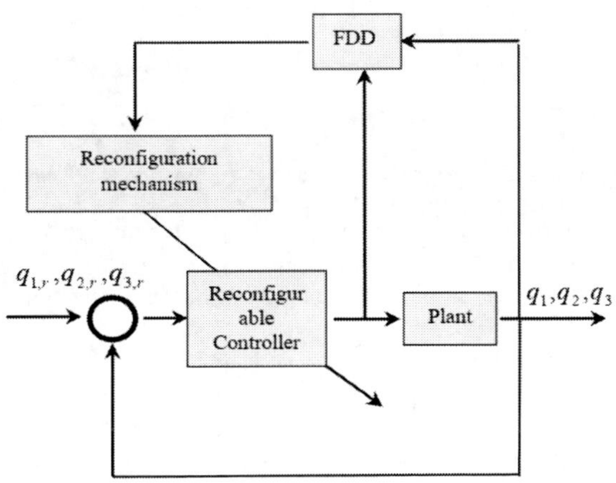

Figure 1. Closed-loop structure (controller reconfiguration).

The controller parameters are adjusted to compensate for the effect of faults. Block diagram of the reconfiguration mechanism is illustrated in Fig. 2.

The controller coefficients are determined at fixed Δt intervals. Then, a cubic spline [12] passes through these points to make the controller coefficients smooth functions of time. The simulation loop is repeated several times until the stopping criterion is satisfied. The subscript *sim* is used to confirm the simulation. The output of Fig. 2 is the smooth curves (controller coefficients) produced by cubic splines.

3.2 Reference trajectory management

In this case, the closed-loop system structure is shown in Fig. 3.

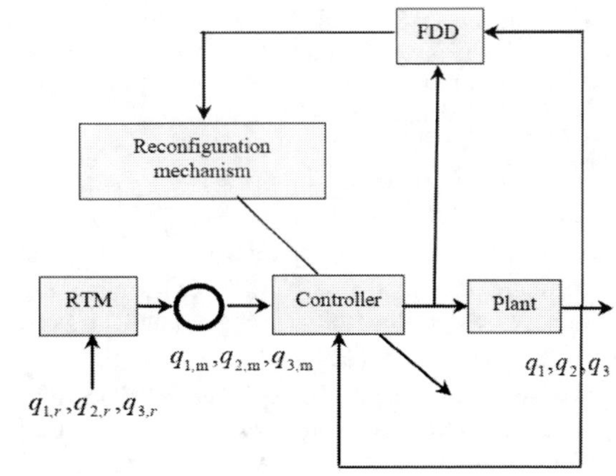

Figure 3. Closed-loop structure (RTM).

It should be noted that in this case, Eqs. 1, 2 and 3 are changed to the following form:

$$u_1'' = \ddot{q}_{1,m} - k_{\dot{q}_1}\left(\dot{q}_1 - \dot{q}_{1,m}\right) - k_{q_1}\left(q_1 - q_{1,m}\right) \quad (4)$$

$$u_2'' = \ddot{q}_{2,m} - k_{\dot{q}_2}\left(\dot{q}_2 - \dot{q}_{2,m}\right) - k_{q_2}\left(q_2 - q_{2,m}\right) \quad (5)$$

$$u_3'' = \ddot{q}_{3,m} - k_{\dot{q}_3}\left(\dot{q}_3 - \dot{q}_{3,m}\right) - k_{q_3}\left(q_3 - q_{3,m}\right) \quad (6)$$

The reference trajectories are adjusted to compensate for the effect of faults. Block diagram of the reconfiguration mechanism and RTM is illustrated in Fig. 4.

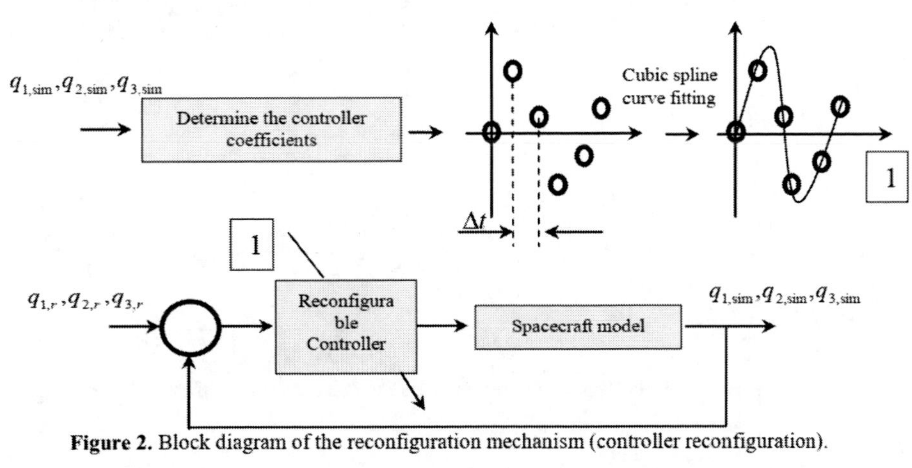

Figure 2. Block diagram of the reconfiguration mechanism (controller reconfiguration).

The modified reference trajectories are determined at fixed Δt intervals. Then, a cubic spline passes through these points to make the modified reference trajectories smooth functions of time. The simulation loop is repeated several times until the stopping criterion is satisfied. The output of Fig. 4 is the smooth curves (modified reference trajectories) produced by cubic splines.

3.3 Goal of the AFTC

The goal of the AFTC is defined as follows:

Determine the controller coefficients/modified reference trajectories such that the following cost function is minimized:

$$J = \int_{t_s}^{t_f} (\mathbf{q}_r - \mathbf{q})^T (\mathbf{q}_r - \mathbf{q}) dt \tag{7}$$

t_f and t_s are the final and settling times, respectively. Therefore, the goal of the AFTC is to minimize the distance between the faulty system response ($\mathbf{q}$) and the reference trajectories ($\mathbf{q}_r$).

In order to compare the performance of the proposed controller reconfiguration and RTM, a numerical example will be presented in the next section.

4 Simulation

The actuation system consists of six thrusters (without considering hardware redundancy), that are placed in opposite directions and each thruster is capable of producing maximum 50 N variable thrust. The effective moment arm of all thrusters is one meter along the principal body axis. However, the configuration of the thrusters is such that the thrust forces (T_1-T_2), (T_3-T_4) and (T_5-T_6) produce net moments about the first, second and third principal axes, respectively (Fig. 5). The direction of arrows shows the direction of the forces produced by the thrusters.

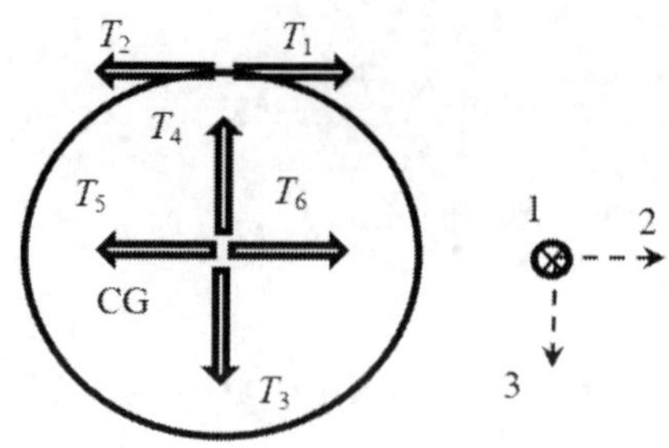

Figure 5. Thruster configuration (CG: Center of gravity).

Therefore, the relation between control torques and thrust forces can be obtained according to the following equations:

$$u_1^+ = T_1, u_1^- = -T_2 \tag{8}$$

$$u_2^+ = T_3, u_2^- = -T_4 \tag{9}$$

$$u_3^+ = T_5, u_3^- = -T_6 \tag{10}$$

The superscripts + and – show the positive and negative control moments, respectively.

It seems that the thrusters T_3, T_4, T_5 and T_6 pass through CG. However as stated in the previous paragraph, they have a moment arm of one meter along the first body axis.

Physical properties [13], initial conditions and controller parameters are presented in Table 1.

It is assumed that the faults occur at $t_{fault} = 0$, $t_s = 70$ s and $t_f = 100$ s. The reference trajectory has the following profile:

$$\mathbf{r} = [\phi_r(t), \theta_r(t), \psi_r(t)] = [0.1t, 0.1t, -0.1t] \tag{11}$$

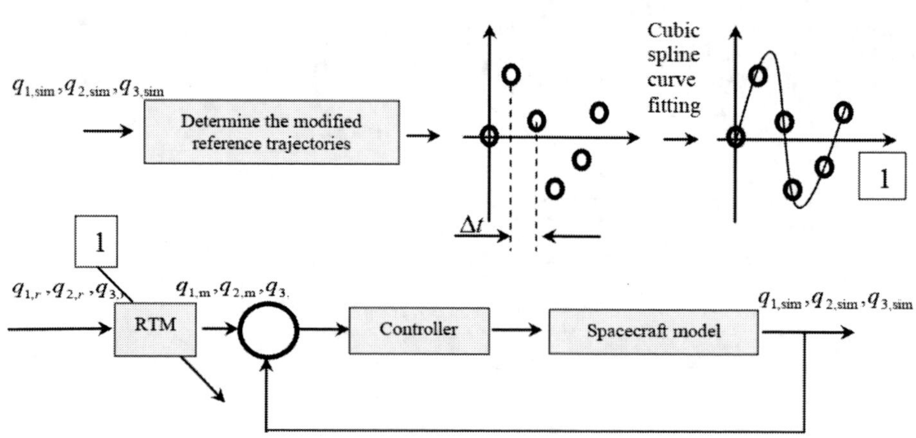

Figure 4. Block diagram of the reconfiguration mechanism and the RTM.

Table 1. Simulation parameters.

Moments of inertia (kg.m^2)	Initial condition s (deg)	Initial condition s	Controller coefficients	
$I_1 = 449.5$	$\phi(0) = 15$	$\omega_1(0) = 0$	$k_{q_1} = 1$	$k_{\dot{q}_1} = 0.1$
$I_2 = 264.6$	$\theta(0) = 10$	$\omega_2(0) = 0$	$k_{q_2} = 1$	$k_{\dot{q}_2} = 0.1$
$I_3 = 312.5$	$\psi(0) = -5$	$\omega_3(0) = 0$	$k_{q_3} = 1$	$k_{\dot{q}_3} = 0.1$

The fault scenario considered in this paper is the reduction in the actuator region [14]. The post-fault actuator region is assumed to be:

$$\text{sat}_n(u_i) = \begin{cases} u_i & \text{if } -u_{\max,n} \le u_i \le u_{\max,n} \\ u_{\max,n} & \text{if } u_i > u_{\max,n} \\ -u_{\max,n} & \text{if } u_i < -u_{\max,n} \end{cases} \quad (12)$$

$u_{\max,n}$ is the new actuator region obtained after the occurrence of actuator fault in the system.

The new actuator regions are determined according to the following relation:

$$u_{\max,n} = a u_{\max} \quad (13)$$

a is the actuator effectiveness coefficient and is a value between 0 and 1. Therefore, the maximum and minimum reductions in actuator region will be 100% (failure mode) and 0% (safe mode), respectively. It is assumed that after the fault, the values of actuator effectiveness coefficients are given in Table 2:

Table 2. Effectiveness coefficients.

a_1	a_2	a_3	a_4	a_5	a_6
1	1	1	1	0.1	0.1

Therefore, it is assumed that the thrusters T_5 and T_6 can produce 5 N-m moments about the third body axis.

Interior-point is used as the solver. The stopping criterion is considered to be 500 simulations. The parameters of this solver are the default values considered in MATLAB 2011a.

The values of cost functions obtained by the proposed methods are presented in Table 3:

Table 3. Cost functions.

	Controller reconfiguration	RTM
J	8.75E-11	7.44E-4

It is shown that in comparison to RTM, controller reconfiguration has made the cost function smaller and therefore, the faulty system response ($\mathbf{q}$) closer to the reference trajectories ($\mathbf{q}_r$).

The Euler angles corresponding to the controller reconfiguration and RTM are illustrated in Figs. 6 and 7, respectively.

A comparison of Figs. 6 and 7 shows the better performance (quality of response) of controller reconfiguration in comparison to RTM. The main reason is that the proposed controller reconfiguration has more variables to modify the closed-loop system behavior. Thrust profiles are also illustrated in Figs. 8 and 9.

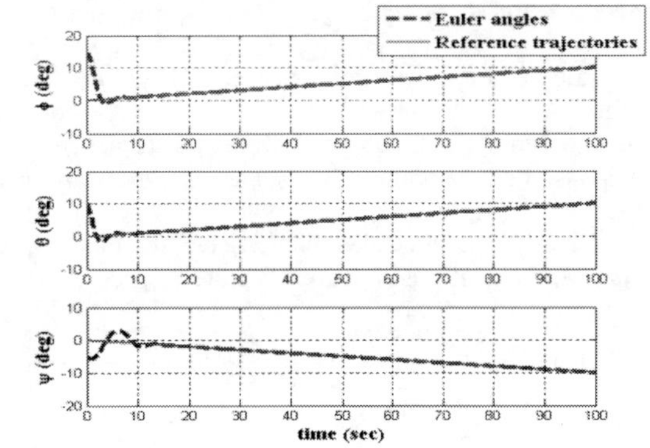

Figure 6. Time histories of Euler angles (controller reconfiguration).

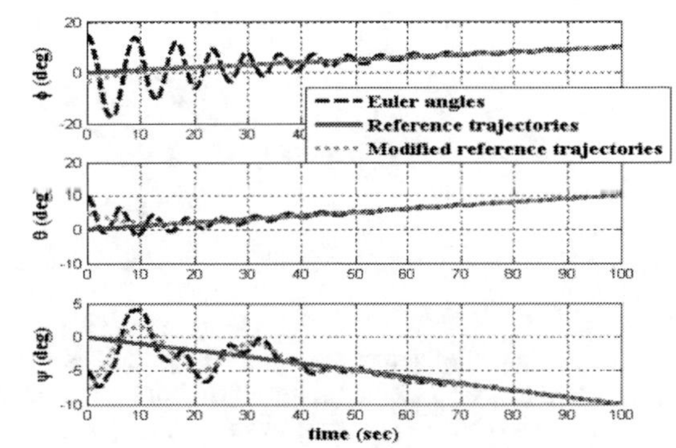

Figure 7. Time histories of Euler angles (RTM).

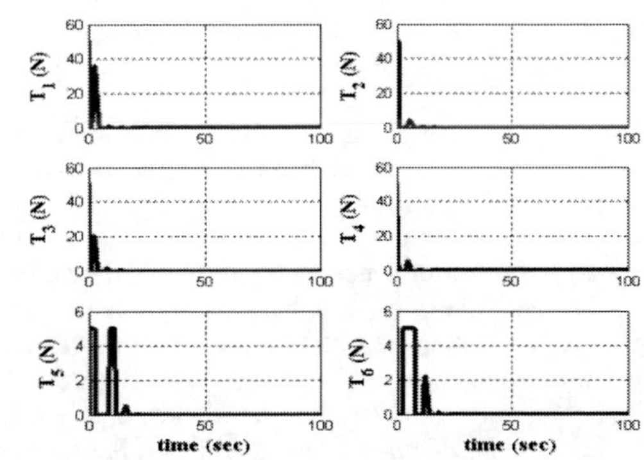

Figure 8. Thrust profiles (controller reconfiguration).

https://doi.org/10.1051/matecconf/201815104008

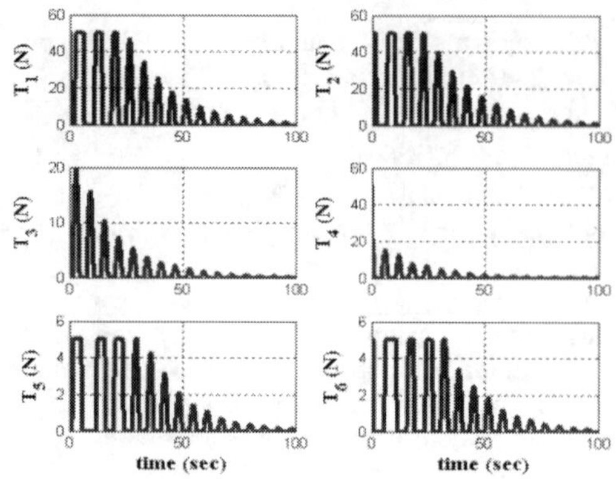

Figure 9. Thrust profiles (RTM).

Another important point is the energy consumption. Figs. 8 and 9 show that the proposed controller reconfiguration not only has a better performance from a response quality point of view but also from an energy consumption view point.

Finally, the controller coefficients for the proposed controller reconfiguration are illustrated in Fig. 10:

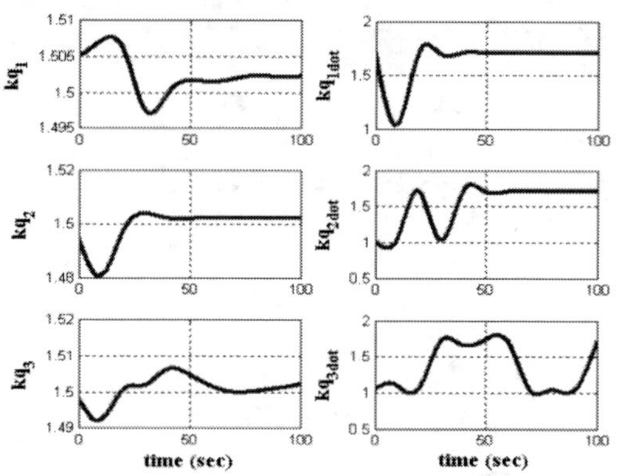

Figure 10. Time histories of controller coefficients (Controller reconfiguration).

It should be noted that the initial values considered for the controller coefficients (controller reconfiguration) are the same as Table 1. The initial values considered for the modified reference trajectories (RTM) are the same as the reference trajectories and $\Delta t = 10$ s.

5 Summary

Reference trajectory management and controller reconfiguration are two methods to design active fault tolerant controllers. Each method is capable of handling faults and retaining the pre-fault performance efficiently. Comparing the performance of these two methods, under the same conditions is an interesting problem. In this paper, a controller reconfiguration and a reference trajectory management were proposed to design active attitude fault tolerant controllers for a faulty spacecraft. It was shown that under the same conditions, the proposed controller reconfiguration shows better performance than the proposed reference trajectory management. The main reason is that the proposed controller reconfiguration has more variables to modify the closed-loop system behavior. As a future work, it is possible to consider analytical methods to have a better comparison between the two approaches.

References

1. Y. Zhang, J. Jiang, Annu Rev Control, *Bibliographical review on reconfigurable fault-tolerant control* **32**, 229-252 (2008)
2. J. Jiang, X. Yu, Annu Rev Control, *Fault tolerant control systems: a comparative study between active and passive approaches* **36**, 60-72 (2012)
3. J. Lunze, J. H. Richter, Eur J Control, Reconfigurable *Fault-tolerant Control: A Tutorial Introduction* **14**, 359-386 (2008)
4. A. Fekih, American Control Conference (ACC), *Fault* Diagnosis *and Fault Tolerant Control Design for Aerospace* Systems: *A Bibliographical Review* (2014)
5. Y. Shen, X. Bing, X. D. Steven, Z. Donghua, IEEE T Ind Electron, *A Review on Recent Development of* Spacecraft *Attitude Fault Tolerant Control System* **63**, 3311-3320 (2016)
6. E. Garone, S. D. Cairano, I. V. Kolmanovsky, Automatica, *Reference and command governors for systems with constraints: A survey on theory and applications* **75**, 306-328 (2017)
7. B. Boussaid, C. Aubrun and M. N. Abdelkrim, Conference on Control and Fault Tolerant Systems, *Fault adaptation based on reference governor*, 257–262 (2010)
8. B. Boussaid, C. Aubrun and M. N. Abdelkrim, 8th International multi-*Conference* on Systems, Signals and Devices (SSD), *Two-level active fault tolerant control approach* (2011)
9. B. Boussaid, C. Aubrun, J. Jiang, M. N. Abdelkrim, Int J *Robust* Nonlin, *FTC approach with actuator saturation avoidance based on reference* management **24**, 2724-2740 (2014)
10. FA. De. Almeida, J Guid Control Dynam, *Reference* management *for fault-tolerant model predictive control* **34**, 44-56 (2011)
11. R. Moradi, A. Alikhani and M. Fathi Jegarkandi, Aerosp *Sci* Technol, *Multi-objective optimization in graceful performance degradation and its application in spacecraft attitude fault-tolerant control* **69**, 465-473 (2017)
12. C. De. Boor, *A practical guide to splines, Applied* mathematical *science*, First edition (1978)
13. D. Wang, Y. Jia, L. Jin and S. Xu, Acta Astronautica, *Control analysis of an underactuated spacecraft under disturbance* **83**, 44–53 (2013)
14. T. Miksch, A. Gambier: Proceedings of 8th Asian conference (ASCC), Kaohsiung, Taiwan, *Fault-tolerant control by using lexicographic multi-objective optimization* (2011)

Design of a Symmetrical Quad-rotor Biplane Tail-Sitter Aircraft without Control Surfaces and Experimental Verification

Hongyu Wang[1], Xun Zhao[1], Hui Bai[2], Cunyue Lu[1], Baomin Zhang[1], Cong Li

[1]Department of Instrument Science and Engineering, Shanghai Jiao Tong University, Shanghai, China
[2]School of Information Science & Technology, Donghua University, Shanghai, China

Abstract. This paper presents the design of a symmetrical quad-rotor biplane tail-sitter VTOL UAV (Vertical Take-off and Landing Unmanned Aerial Vehicle) which is composed of four rotors and two symmetrically mounted fixed wings. This aircraft achieves high accuracy in the attitude control and smooth flight mode transition with four rotors rather than the conventional VTOL UAVs using control surfaces. The proposal of angled rotor mounting is adopted to address the issue of insufficient yaw control authority. The layout of symmetrically mounted fixed wings makes the aircraft have capability of rapid bidirectional flight mode transition to improve maneuverability. To validate the performance of the aircraft, simulation and flight experiments are both implemented. These results show that the aircraft has a rapid yaw response under condition of the stable attitude control. In comparative experiment, it is shown that the aircraft is more flexible than other similar configuration of aircrafts. This symmetrical quad-rotor biplane tail-sitter VTOL UAV will have a wide range of potential applications in the military and civilian areas due to its superior performance..

1 Introduction

In recent years, the potential uses for UAVs have increased drastically. These applications include but not limited to surveillance, rescue and reconnaissance, as well as medical supply, consumer package delivery and power line inspection [1]. The VTOL aircrafts can either take off and land vertically or fly in a high-speed cruise configuration [2]. At present, according to different ways to achieve flight mode transition, VTOL aircrafts may be categorized as tilt-rotor, tilt-wing, tail-sitter aircrafts, et al [3-6]. The simplest way is tail-sitting since it switches between level flight and hover flight mode by changing its pitch angle of the fuselage [7]. Tail-sitter aircrafts have been extensively studied for its simple and effective flight mode transition. Stone studied the T-Wing, which has a canard wing and tandem rotors [8]. Kita developed a simple tail-sitter VTOL UAV with a single propeller R/C airplane [9]. These tail-sitter aircrafts which are developed based on fixed-wing aircrafts, have an accuracy drawback in the attitude control for stationary hovering because of the use of control surfaces, such as aileron, elevator, and rudder. To overcome such a drawback, those tail-sitter aircrafts which are developed based on rotorcrafts have been a research hotspot. This strategy was proposed by Young at NASA Ames Research Centre in 2002 [10]. Oosedo designed a quad-rotor tail-sitter aircraft with one wing and succeeded in manual flight in 2013 [11, 12]. Bogdanowicz developed a quad-rotor biplane micro air vehicle in 2015 [13]. The cambered wing makes those aircrafts have different aerodynamic efficiency when they achieve bidirectional flight mode transition, which causes that they are inflexible. To address these issues above, a new symmetrical quad-rotor biplane configuration tail-sitter VTOL UAV is proposed. This aircraft realizes high accuracy in attitude control and smooth flight mode transition with four rotors. To solve the slow response of yaw and the insufficient yaw control authority problems, the proposal of angled rotor mounting is adopted. The configuration of symmetrical biplane enables the aircraft to have the same aerodynamic efficiency during bidirectional flight mode transition, which is a compromise between aerodynamic efficiency and maneuverability.

In order to verify the rationality of the design, the mathematical model is established and control system is designed. The flight experiments are carried out to evaluate the simulation results. A comparative experiment with quad-rotor monoplane and asymmetrical quad-rotor biplane is also implemented to evaluate the maneuverability of the aircraft.

2 Principle of flight

The symmetrical quad-rotor biplane tail-sitter VTOL UAV without Control surfaces has three flight modes: hovering mode, transitional flight mode and level flight mode, as illustrated in Fig. 1. In hovering mode, the force generated by rotors is mainly used to maintain the gravity. The aircraft has accurate attitude control with excess thrust in hovering mode. The transition between flight modes is attained through a pitching moment provided through RPM variation of the propellers. Varying RPM of propeller pairs allows for differential thrust which, when coupled with the installed moment arm, generates a sizeable pitching moment. Transition flight mode is between hovering mode and level flight mode. The configuration of symmetrical biplane makes the aircraft has the same aerodynamic efficiency to achieve a rapid bidirectional transition. It can be along with either the A path or the B path for transition, which improves the maneuverability of the aircraft, as illustrated in Fig.. 1. As the aircraft enters the level flight mode, at sufficiently high cruise speeds, the loads are then transferred to the wing. So, the force generated by rotors is mainly used to increase the cruise speed.

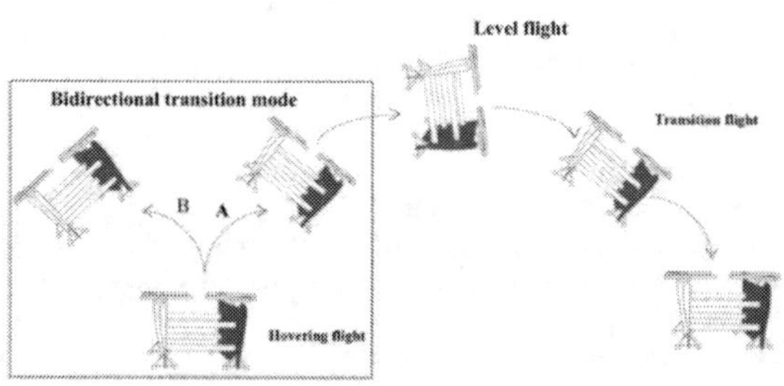

Figure 1. Concept of flight mode.

Table 1. Nomenclature

Symbol	Unit	Description
i		Index of each motor
j		Index of each axis
Ω_i	(rpm)	The ith motor speed
l	(m)	Distance of perpendicular form each motor to the center
C_T		Torque coefficient of each propeller
C_M		Lift coefficient of each propeller
a	(rad)	Tilt angle of each motor
α	(rad)	Angle of attack
m	(kg)	Mass of whole aircraft
g	(m/s^2)	Gravitational acceleration
τ_j	(N/m)	Torque of each axis
p	(rad/s)	Angular velocities of X axis
q	(rad/s)	Angular velocities of Y axis
r	(rad/s)	Angular velocities of Z axis
u	(m/s)	Translational velocities of X axis
v	(m/s)	Translational velocities of Y axis
w	(m/s)	Translational velocities of Z axis
I_j	$(kg*m^2)$	Fuselage moment of inertia of each axis
θ	(rad)	Pitch angle
ϕ	(rad)	Roll angle
ψ	(rad)	Yaw angle
L	(N)	Lift of wings
D	(N)	Drag of wings
$Q_{slipstream}$	(N/m)	Torque of propeller slipstream
J_r		propeller gyroscopic effect coefficient
M_w	(N/m)	Pitching moment of the longitudinal area
M_{lat}	(N/m)	Pitching inertial resistance of the lateral area

3 Configuration

3.1. Fuselage

An overview the aircraft is shown in Fig. 2. It consists of two pairs of counter rotating propellers arranged in a quad-rotor configuration. Two wings are affixed symmetrically to each pair of rotors through a set of rigid motor attachments. The wings are constructed from low density polyurethane foam using high lift, low Reynolds number airfoil. The battery and on-board microcontroller are incorporated in the centre of the vehicle. Each rotor shaft is tilted inward at an angle toward the mid-span of its respective wing.

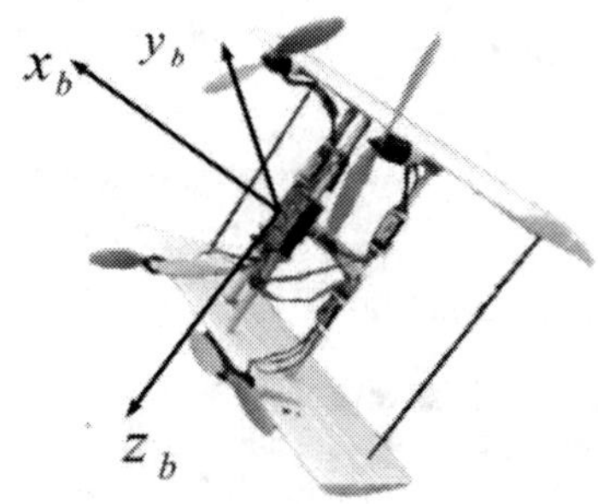

Figure 2. Overview of the aircraft.

3.2. Angled rotor mounts

The aircraft suffers from the insufficient yaw control authority, especially, when it enters the level flight mode, which results in the divergence of the yaw. The proposal of angled rotor mounting is adopted to address the issue, as show in Fig. 3. Each rotor shaft is tilted inward at an angle a, toward the mid-span of its respective wing. $F_Z$is the force generated by rotor. $F_V$is the vertical component of F_Z. F_H is the horizontal component of F_Z. These side-forces will provide a valuable menment which will compensate for yaw to attained sufficient yaw control authority, though, the vertical commopent of the force generated by rotor is decreased. Fig. 4 shows the direction and relative magnitude of the new forces which, when coupled with a moment arm, will generate a large yaw moment in the clockwise direction, which is the desired direction of rotation. The new yaw moment is given by Eq.1.

$$\tau_z = \left(C_M + \sqrt{2}C_T lsin\alpha\right)(\Omega_1^2 - \Omega_2^2 + \Omega_3^2 - \Omega_4^2) \quad (1)$$

Figure 3. Diagram of thrust.

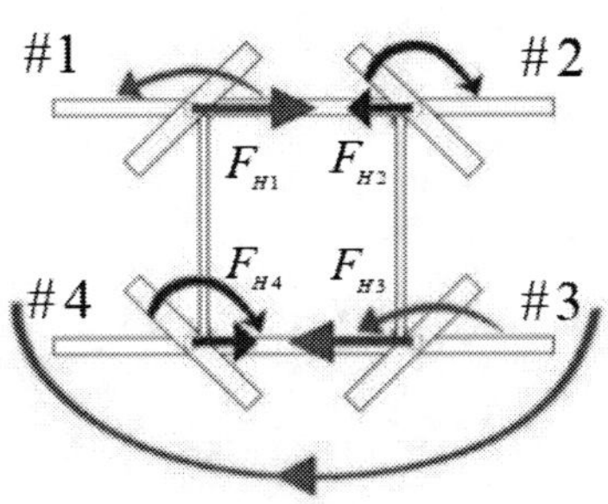

Figure 4. Diagram of thrust vectoring.

4 Modelling

4.1. Coordinate systems

The aircraft body coordinates (X_b, Y_b, Z_b) are defined as shown in Fig. 2. The earth-fixed coordinates define the X_i-axis as true north, the Y_i-axis as east and Z_i-axis as perpendicular downward. The X_b-axis of the aircraft coordinates coincides with the X_i-axis of the inertial coordinates when pitch and yaw angles are equal to 0.

4.2. Dynamics

The equations of translational and rotation motion of the aircraft are described by Eq.2-Eq.7. The effect of the slipstream on the wing is modelled as a yaw moment.

$$\dot{u} = \frac{rv - qw - Lcos\alpha - Dsin\alpha}{m} + gcos\theta cos\psi \quad (2)$$

$$\dot{v} = \frac{pw - ru}{m} + gcos\theta sin\psi \quad (3)$$

$$\dot{w} = \frac{qu - pv - Lsin\alpha - Dcos\alpha - T}{m} + gsin\theta \quad (4)$$

$$\dot{p} = \frac{(I_y - I_z) + \tau_x - J_r r \sum_{i=1}^{4}(-1)^i \Omega_i}{I_x} \quad (5)$$

$$\dot{q} = \frac{(I_z - I_x) + \tau_y + M_x + M_{lat} - J_r p \sum_{i=1}^{4}(-1)^i \Omega_i}{I_y} \quad (6)$$

$$\dot{r} = \frac{(I_x - I_y) + \tau_z + Q_{slipstream}}{I_z} \quad (7)$$

5 Control systems

5.1. Attitude and Altitude Control System

The attitude and altitude controller is designed based on cascaded PID controller, as shown in Fig. 5. The cascaded PID controller can improve the dynamic performance and robustness of the system. In order to address the issue of excessive overshot, which is caused by the frequent change of the reference angular velocity, the inner loop adopts differential forward PID controller. As shown in Fig. 6, the differential PID controller only differentiates the feedback signals rather than the reference signals. As shown in Fig. 5, firstly, the reference and current attitude are transformed into rotation matrices in avoidance of the "gimbal lock". Those rotation matrices are sent to the outer PID controller to get the reference angular velocity. Then, reference angular velocity, current angular velocity, reference altitude and current altitude are sent to the differential forward PID to get desired force and moments. Finally, the force and these moments are assigned to each rotor to achieve the attitude and altitude control.

5.2. Transition Control System

An important difference in aircraft dynamics between hovering and level flight mode is the lift generated by the wings. Transition flight, which refers to the flight phase between level flight and hovering, is supported by both thrust and lift. The aircraft is aerodynamically unstable during transition flight because a stall occurs at the fixed wings. The biggest problem is the change of the altitude. It is hard for a pilot to operate to get an optimal thrust at different speed and attack angel. So the transition control system is designed for constant altitude flight to attain the automatic transition, as shown in Fig. 7. The optimization problem of this transition strategy is solved by Levenberg-Marquardt algorithm. Under condition of constant altitude, optimal thrust will be obtained at every velocity and attack angel using Levenberg-Marquardt algorithm. For flight safety, the transition mode switch is set in the RC transmitter to ensure that the pilot has sufficient authority for the aircraft.

https://doi.org/10.1051/matecconf/201815104009

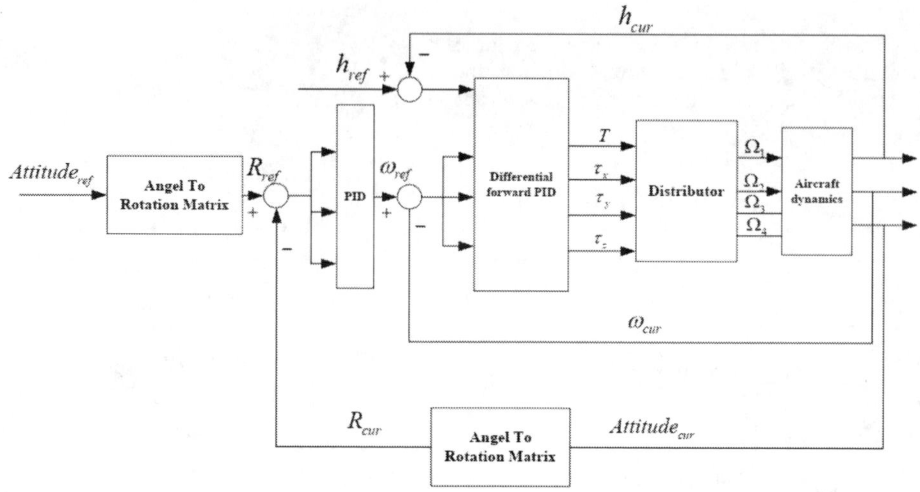

Figure 5. Attitude and altitude control system.

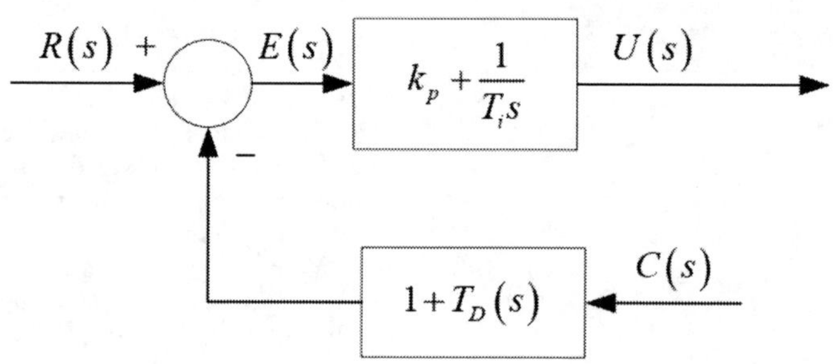

Figure 6. Differential forward PID controller.

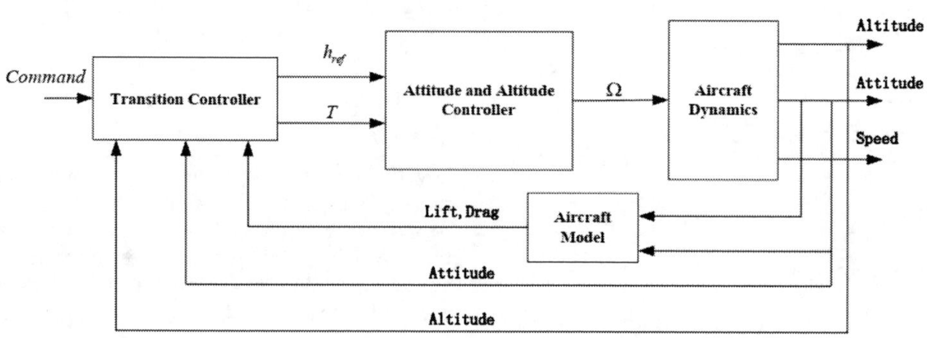

Figure 7. Transition control system.

6 Simulation and flight experiment

6.1. Simulation

The simulation experiments are composed of the aircraft taking off, the hovering mode and the transition mode at time of 5s. The simulation results are shown in Fig. 8. As can be seen in Fig. 8(a), the aircraft has rapid response under condition of high accuracy in the attitude control during either the hovering mode or the transition mode. The force generated by rotors is used to balance the gravity in hovering mode. At the beginning of transition mode, the force generated by wings is small due to small translational velocity while the vertical component of the force generated by rotors is still decreasing. For maintaining the altitude, the force generated by rotors needs to be increased. The lift generated by wings which is mainly used to maintain the gravity of the aircraft increases as the velocity increases. So, the force for lift which generated by rotors is continuously reduced, as shown in Fig. 8(b). The force generated by rotors is mainly used to increase the speed of the aircraft, as shown in Fig. 8(c). The cruise speed reaches up to 13m/s while the max speed of a quad-rotor aircraft is about 8m/s. Fig. 8(d) shows that the yaw is divergent before rotors tilting, especially during transition mode because of insufficient yaw control authority and influences of streamslip. The results that the yaw is convergent after rotors tilting validate that the proposal of angled rotor mounting is effective.

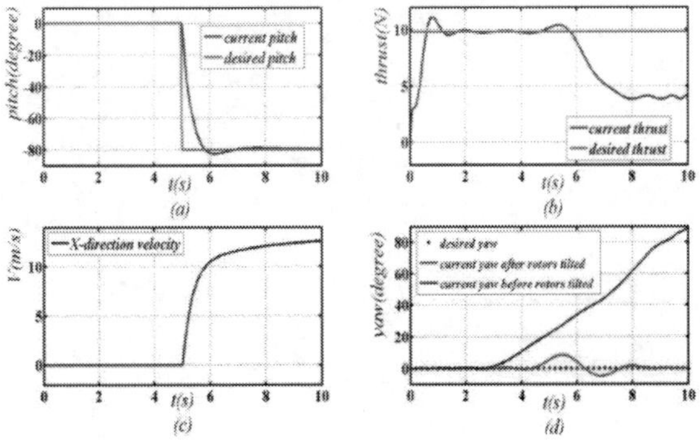

Figure 8. Results of simulation.

6.2. Flight experiment

The flight experiments are composed of evaluating the simulation results and comparative experiment with other configuration aircrafts. The results of the flight experiment as shown in Fig. 9. As can be seen in these figures, the results of flight experiment and simulation are basically consistent, except that the altitude change in flight experiment is more severe than it in simulation during transition flight mode. The reason is that the wind is not taken into consideration when the transition controller is designed. The comparative experiment is that a round-trip flight with a straight-line distance of 100m is carried out with three different configuration aircrafts. These aircrafts are the same except for the configuration. The results of comparative experiment with other aircrafts are listed in Table 2. As can be seen in Table 2, the configuration which is proposed by this paper has the minimal transition time and the minimal total time. So, the results of comparative experiment validate that the layout of symmetrical biplane improves the maneuverability of the aircraft.

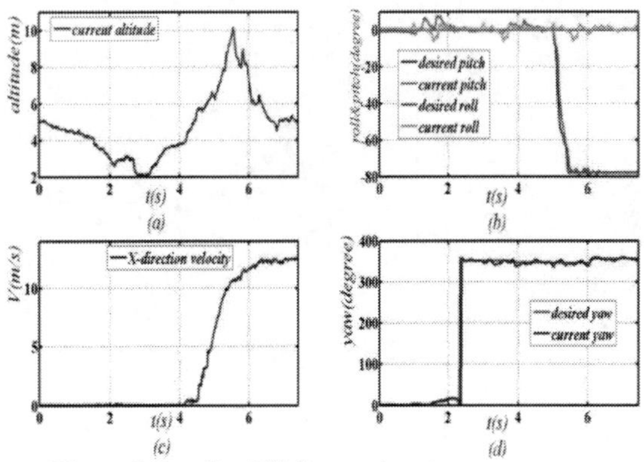

Figure 9. Results of flight experiment.

Table 2. Results of comparison experiment

Configuration	Symmetrical biplane	Asymmetric biplane	Monoplane
Max speed after forward X-direction transition	12.5(m/s)	13(m/s)	14(m/s)
Max speed after negative X-direction transition	12.5(m/s)	8(m/s)	9(m/s)
Time of mode transition	4.5(s)	4.2(s)	3.88(s)
Time of AB experiment	14.5(s)	17.8(s)	16.0(s)

7 Summary

In this work, a symmetrical quad-rotor biplane tail-sitter VTOL UAV without control surfaces is designed and mathematic model is established which is used for designing control system. Simulation and flight experiments are also carried out. The results of experiment verify the aircraft can achieve high accuracy in attitude control during hovering mode. The proposal of angled rotor mounting speeds up the response of yaw and increase the yaw control authority by compensating the moment of yaw. The results of the comparative flight experiment comfirm that the layout of symmetrical biplane improves the maneuverability of the aircraft. This symmetrical quad-rotor biplane tail-sitter VTOL UAV will have a wide range applications in the military and civilian areas due to it's superior performance. In the future, the work on how to enhance the robustness of the transition control system and how to realize hovering precisely in strong wind conditions will be carried out.

References

1. K.P.Valavanis, *Andvances in Unmanned Aerial Vehicle:State of Art and the Road to Autonomy*(2007)
2. Abdessameud A, Tayebi A, Automatica J.IFAC, **47**, 2383-2394(2011)
3. Green W E, OhPY, ICRA, 2164-2169(2006)
4. Maisel, Martin D., D. J. Giulianetti, D. C. Dugan. *The History of the XV-15 Tilt Rotor Research Aircraft from Concept to Flight* (2000)
5. Taylor, Frederick.M, *V/stol aircraft with variable tilt wing*(US, US 3666209 A, 1972)
6. Kubo D, Suzuki S, J.AIRCRAFT, **45.1**, 292-297(2008)
7. Stone R H, Anderson P, Hutchison C, J.AIRCRAFT, **45.2**, 673-685(2008)
8. Kita K, Konno A, Uchiyama M, IJRM,**21.2,**77(2009)
9. Young L, Aiken E, Johnson J, AIAA(2002)
10. Oosedo A, Abiko S, Konno A, ICRA, 317-322(2013)
11. Oosedo A, Abiko S, Konno A, *Autonomous Robots*, 1-17(2006)
12. Hrishikeshavan V, Bawek D, Rand O, ISMUR(2013)
13. Bogdanowicz C M, J.AIAA(2015)

MATEC Web of Conferences **151**, 04010 (2018) https://doi.org/10.1051/matecconf/201815104010
ACMAE 2017

Concept development of control system for perspective unmanned aerial vehicles

Vsevolod V. Koryanov[1], Tatiana V. Kokuytseva[1], Alexey G. Toporkov[2], Stepan N. Iljukhin[2], Il'ya O. Akimov[2], Noa Mohamado[2], Victoria Da-Poian[2]

[1] *Peoples Friendship University of Russia (RUDN University), 117198, Miklukho-Maklaya str.6, Moscow, Russian Federation*
[2] *Bauman Moscow State Technical University, 105005, 2-nd Baumanskaya, 5, b1, Moscow, Russian Federation*

Abstract. Presented actual aspects of the development of the control system of unmanned aerial vehicles (UAVs) in the example of perspective. Because the current and future UAV oriented to implementation of a wide range of tasks, taking into account the use of several types of payload, in this paper discusses the general principles of construction of onboard control complex, in turn, a hardware implementation of the automatic control system has been implemented in the microcontroller Arduino platform and the Raspberry Pi. In addition, in the paper presents the most common and promising way to ensure the smooth and reliable communication of the command post with the UAV as well as to the ways of parry considered and abnormal situations.

1 Introduction

Currently the problem of the creation of unmanned aerial vehicles (UAV) with different function and configuration is extremely urgent. Different applications areas can benefit of this new technology. Civilian UAVs are used for different purpose such as aerial photography, patrolling areas, inspections, delivery of goods, supply of communication with designated areas, control of airspace and many other areas.

Thanks to modern conditions, the practical component in the creation of a UAV design and its control systems is not anymore a complicated science-intensive process such as previously.

Though, the UAV control systems have identical morphology, but due to the various purpose of the aircraft mission, the architecture of their control system (governing bodies, structure of onboard systems, communication system, sensor equipment and so on) can be created in different way to fulfil specific objectives.

Therefore, it is possible to mark the following urgent tasks, which will be solved in the development process of management systems:

- simplicity and reliability of construction;
- simplicity of operation in service;
- low competitive cost;
- insignificant mass-dimensional characteristics;
- enough high universality which will provide a possibility of placement on other types of aircraft including piloted;
- high level of fail safety of system to emergency situations.

2 Appearance perspective UAVs

In Moscow State Technical University of N. E. Bauman, within research and development activities operation over various youth projects, one project is about the carried "Sub-orbital rocket plane" [1]. As an example, this project is especially interesting thanks to the management systems: it was created "from scratch", without use of expert estimation elements, in a type of absence of a sufficient array of the statistical data obtained in an analysis result of properties of the existing samples. It is also worth noting that, unlike the majority of amateur projects about the creation of UAV, this one focuses not only on the structuration and synthesis of the command, but also combined management systems were made.

The sub-orbital rocket plane represents the perspective reusable UAV that serves for deduction of the disposable loads on height of low orbits of artificial Earth satellites, carrying out research experiences in the conditions of microgravity, execution of functions of communication and remote sensing of Earth. One of the UAV configuration is shown in Figure 1.

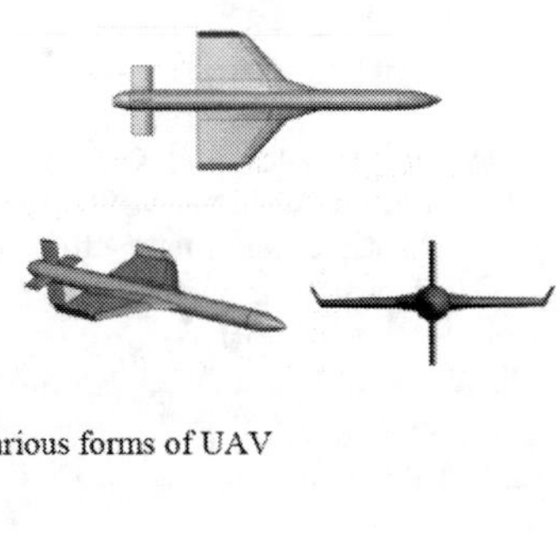

Figure 1. Various forms of UAV

https://doi.org/10.1051/matecconf/201815104010

The process of correcting the UAVs aerodynamic model from aerodynamic tests data was studied. The task of identifying the UAV aerodynamic model can be summarized as the correlation between the initial aerodynamic model and the data of aerodynamic tests [2]. This task consists in the optimization of the design solution for regularity criteria. The number of these criteria is equal to the number of calculated aerodynamic coefficients.

The task of assessing the completeness of the UAV aerodynamic model is to select its structure of an aerodynamic model with the simultaneous selection for each structure of optimal aerodynamic derivatives.

Experimental studies were preceded by analytical calculation and computer simulation. In this paper, the results of mathematical modeling for UAV subsonic flow are presented. Calculations were performed using specialized software: SolidWorks (Flow Simulation) and ANSYS CFX. Then, the results were compared with the results obtained in the analytical calculation.

An example of flow visualization used to determine the value of on UAS drag's coefficient is shown in Figure 2.

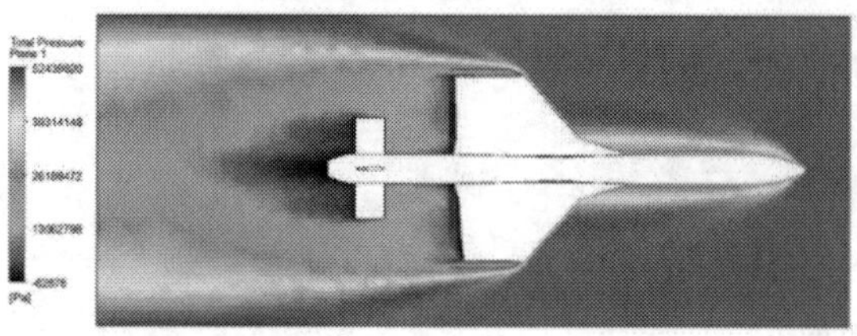

Figure 2. Example determining drag's coefficient

An analytical calculation was carried out according to the principle of the "UAV dismemberment" into isolated elements. Using this principle, the UAV was replaced by a designed scheme, in which the following elements were singled out: an isolated wing, an isolated aft body and an isolated hull.

The rocket plane, like any automatic aircraft, has a row of subsystems of the lowest hierarchical level as a part of the management system (see Figure 3). These includes: sensitive elements (MEMS sensors, temperature sensors, strain gages), electric power supplies of onboard systems, control moment gyroscope (two-degree force gyrostabilizers), the executive mechanisms (mechanism of disclosure of a wing, pyrosystem, servo actuators), and a basic element projected management systems - the onboard complex of control.

3 Emergency situations

As the UAVs have a long autonomy of flight and uses for a long time an autonomous mode without pilot tasks it can potentially constitute a threat for people near settlements, for infrastructure or for industrial facilities. Besides payload, the UAV capacity can be a special value, as well as the construction of the aircraft. For these reasons, the uncontrollable fall probability of the device is minimized, and the efficiency of the execution of the work is closer possible by all means and methods. Emergency situations have been studied in detail by NASA after communication issues [3], and unmanned missions emergency situations are still one main concern.

For the modern aircraft, the greatest danger is constituted by complex influence of operational factors, such as:

- failures of mechanical nodes of the aircraft;
- errors of operation of onboard systems at the level of algorithms and logic of functioning;
- the strong wind loads, in particular turbulence of the environment;
- rain, freezing and other meteorological factors;
- other factors.

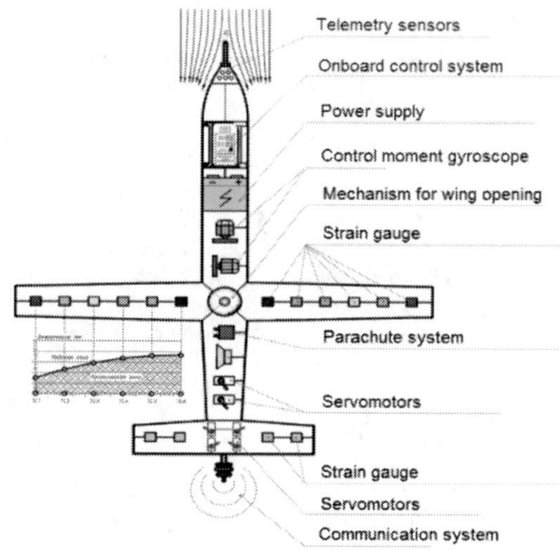

Figure 3. Composition of the UAVs onboard systems

Therefore, it is appropriate to consider an original problem of an emergency situation for two essentially different cases.

In cases of loss of communication with ground stations, the UAV belong to the first case. The normal mode is activated by implementing off-line control. In that case, special laws of control are used and the aircraft executes some implemented programs of control, and also realizes some algorithms in order to return the aircraft to a cover zone where signals of ground station control are provided.

In the second case, also with a loss of communication, certain emergency situations can emerge. It is known that compared to flight plan, these situations can be separated on considered and unconsidered emergency situations [4]. The considered emergency situations are those which probability of appearance was defined during flight planning stage. In this case, it is possible to write programs and algorithms of onboard systems operations in order to respond to these mergency situations.

The development of the instructions to avoid these considered emergency situations begins with compilation of the entire database of all aircraft subsystems, up to specific nodes and details. Then statistic analysis of probabilities and the reasons of their possible failures are carried out. Besides, development of techniques of detection of possible breakage and a technique of problematic situations removal are necessary.

During the flight of the perspective UAV, emergency situations will be removed in 3 steps. If operations at the current step do not allow to fix the problem, the process proceeds to the next step.:

1) Algorithms are guided by the indications of the sensors implemented in different systems. These sensors reveal the emergency modes and try to find instructions to eliminate the emergency situation found.

2) The algorithm removes the most likely causes of a problem. At the same time, the basis of the considered emergency situations ranked by probability of origin is used.

3) The algorithm adjusts to an emergency situation reducing as much as possible negative effect of its existence.

If the provided algorithms do not lead to the elimination of an emergency situation, it is then necessary to adapt the management system under the current conditions. In particular, to ensure the aircraft flight without using the faulty component, it is necessary to unify as much as possible the key algorithm control of the aircraft - a stabilization algorithm.

4 Management system movement and navigation

UAVs technology requires to analyse the relationship between the flight behaviour and all the technical parameters of the components the vehicle is made of [5]. Any effective system of automatic control represents a follow-up circuit. UAVs will be designed using optimal architecture system for future missions in order to benefit from advanced technologies and development facilities [6]. As a follow-up circuit, each such system possesses a certain set of measuring means. The system of angular stabilizing is based on the PID-regulator (Proportional Integral Derivative) which operation quality directly depends on the accuracy of its given coefficients. The perspective direction of development of autonomous systems of control are the adaptation of systems of control. For instance, under different weather conditions these coefficients will have different values. Also, if the UAV transports loads have different weight and overall dimensions, then these PID-regulator coefficients will also differ. In that sense, automated setup of "PID-coefficients" was applied. Such setup has to be made directly during the flight by the analysis of the UAV response to the available perturbations. This setup must be performed either at certain time intervals, either in case of essential change of the external environment parameters.

In order to deal with the task of navigation, information from the sensors GPS and GLONASS which is filtered and analyzed in the flight program are used. These navigation data are processed by an algorithm, and, according to the flight job, teams on governing bodies for aircraft orientation are formed. In a suborbital rocket plane MEMS – the following sensors represented in the Figure 4 are used: accelerometers, gyroscopes and compasses. The data obtained with these sensors are exposed to the linear filtering using Kallman's filter. To eliminate the accumulated error due to integration of data of a gyroscope, the complementary filter adding data of a gyroscope indexes of the accelerometer and calculating the true orientation angles of the aircraft is used. Also, the sensors located in different onboard systems and nodes, which allow to trace an aircraft time status at the moment are used. For instance, the strain gages set in a wing allow to trace its intense deformed status, and the differential pressure sensor (the pressure sensor of a filling flow allowing to determine aircraft speed) allows to adjust indications of the inertial MEMS - sensors and the GPS sensors monitor the aircraft orientation and its center of masses position.

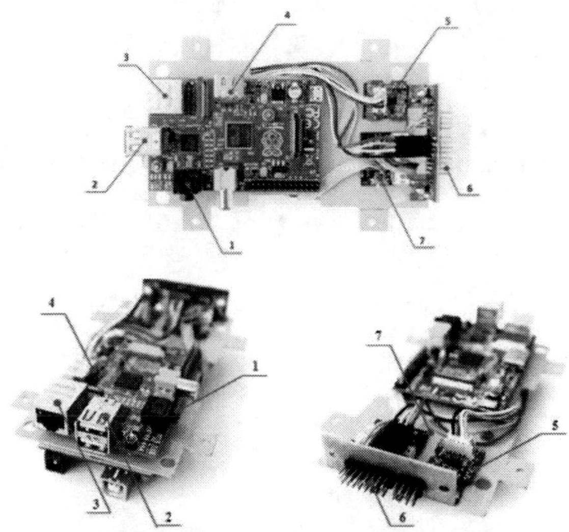

Figure 4. Experimental module of management system movement

1 – connector of the search and rescue device; 2 – USB; 3 – Ethernet; 4 – "Output" for debugging of the program of the onboard electro-computer; 5 – gyroscope; 6 – connector of control of the periphery; 7 – accelerometer.

In the case of hardware implementation of automatic control systems in this project, widely used microcontroller platforms, such as Arduino and Raspberry Pi are used. These platforms provide a fairly simple and intuitive interoperability layer with the programmer. Moreover, they have small weight, developed infrastructure and good productivity. In particular, Arduino [7] is used for the unit of stabilization system where the platform integrates sensors datas and, according to control box commands, gives the controlling impacts on executive bodies for orientation on the scheduled course. Functions of remaining units are performed, instantly, by Raspberry Pi as this platform has a processor with high clock rate and uses the Linux operating system.

The main characteristics of the developed module of management system movement are shown in the Figure 5.

Main characteristics

Supply voltage:	6..15 V
Power consumption (with sensors):	<150 ma
Dimensions:	147x77x47 mm
Weight (with servo drives):	180 g.

Inputs and outputs:

- 6 PPM channels for connection to the RC receiver
- 6 PPM output channels for connection to servo drives
- USB connector for PC connection
- SPI Connector
- I2C connector
- UART (GPS signal)
- Analog inputs

Figure 5. Main characteristics module of management system movement

Navigation of UAV. To navigate the information from the GPS / GLONASS sensors are used. Their data are further filtered and analyzed in the flight program. Information from the GPS-sensor is compared with the information obtained by integrating the readings of inertial sensors (Gyrometer sensor and accelerometer). The architecture of the navigation system is shown in Figure 6.

Also, when creating the navigation algorithm, the initial data of the NoFlyZone.org project was used and now, data from this portal can be downloaded to the base of the onboard control complex. Also in the flight program, the areas forbidden for flight can be set.

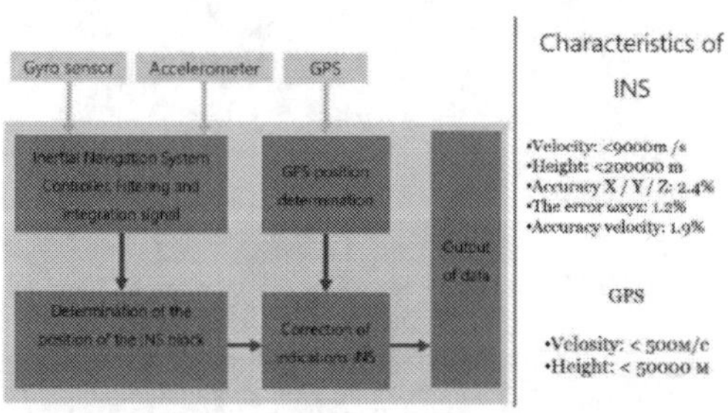

Figure 6. Navigation system

Stabilization of UAV. For the stabilization, inertial sensors are used: Gyro sensor and accelerometers. To correct the sensor readings, an incident pressure sensor is used to calculate the speed of the aircraft. The architecture of the stabilization system is shown in Figure 7.

Information from the sensors is processed by algorithms of control, which then send commands to the controls for the aircraft orientation at a given mission.

As mentioned above, we use MEMS sensors: accelerometers, gyroscopes and compasses (Figure 4). Data coming from MEMS sensors are processed by the Kalman filter (Figure 8). To eliminate the accumulated error of integrating the gyro data, a complementary filter that complements the gyro data with the accelerometer data is used and calculates the true angles of the aircraft orientation.

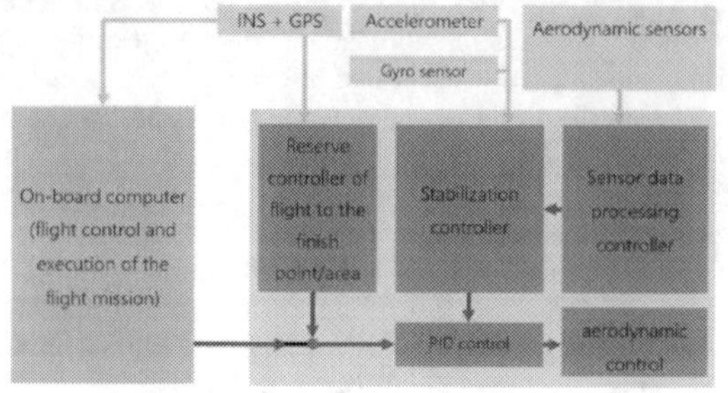

Figure 7. Stabilization system

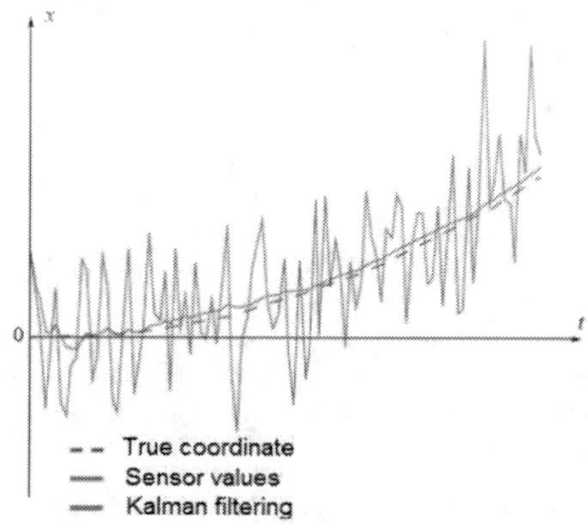

Figure 8. Kalman filter

Automated simulation stand. In order to check the software and the hardware of the UAVs automatic control system (ACS) and in order to timely detect errors, an automated simulation element was developed. This element is implemented on the basis of scripts operating in the MATLAB system running an IBM PC-compatible computer. This element is interfaced with the hardware module of the ACS via the USB interface.

Simulation element implements the process of controlling the model movement of the control object (UAV) under the control of the connected ACS. It is possible to specify various initial conditions and modeling scenarios, including simulating the occurrence of abnormal situations. At this stage, simulation provides: testing the operability of the hardware ACS modules; simulation of UAV motion with imitation of various abnormal situations and disturbing effects; exchange of information with the USB interface; reception, recording, processing and visualization of control signals generated by the ACS.

Simulation modeling and testing of the main systems and components of the experimental module was carried out using Matlab (see Figure 9). On the right of the figure, graphs of changes in the yaw, roll and pitch angles are illustrated.

These platforms have an additional advantage: during their existence, the big array of practical data and algorithms available to adaptation of the comparative analysis appeared.

https://doi.org/10.1051/matecconf/201815104010

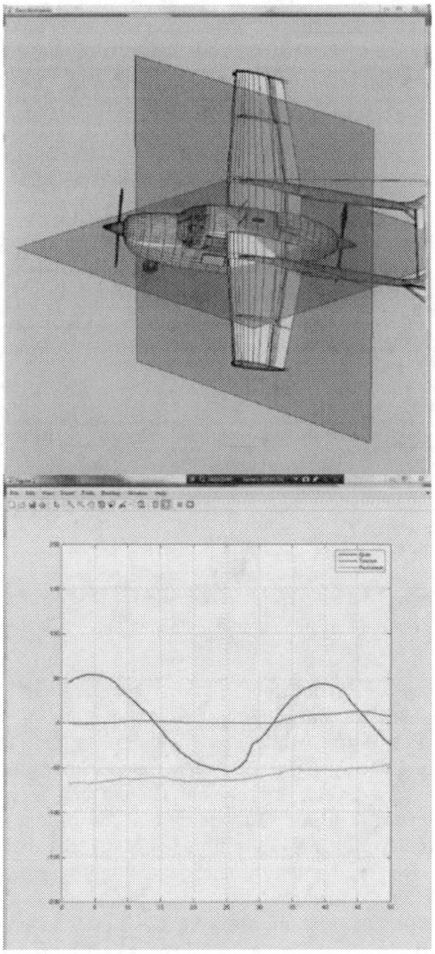

Figure 9. Simulation using Matlab

5 Conclusions

In operation, the main aspects of the UAV management systems for the example of a specific project are provided. At this stage of development, the management system has the following properties:

1. Quasiautonomy. Allows to minimize the involvement of the ground control station operator from take-off until the successful landing taking into account the origin of an emergency situation.

2. Modularity. The modular architecture allows to change and expand if necessary quickly a configuration and a functionality of system, thereby timely to increase fail safety of all control object in general, and also to reduce mass-dimensional and cost characteristics. The most important in case of operation of management system on different devices in different weather conditions taking into account requirements of the customer.

3. Universality. The system was developed for a sub-orbital rocket plane which could have different wing constructive implementations, however at this stage the developed system provides quasi-automatic control of the UAV weighing up to a 10 kilograms aircraft only with the fixed position of a wing for which flight tests were carried out. In the longer term regarding aircraft, the system will be able to control devices of helicopter type, multirotor systems, autogyros... The architecture and modularity of system will allow to test architectures, and system's modularity will allow to exercise further object managements not only terrestrial, but also maritime

References

1. Toporkov A.G., Mullin N.A., Pavlov N.G. *Innovatsionnyy proyekt «Suborbital'nyy raketoplan»* [*Innovation Project "Suborbital rocket plane"*]. Aktual'nyye problemy kosmonavtiki: Trudy XXXIX akademicheskikh chteniy po kosmonavtike, posvyashchennykh pamyati akademika S.P. Koroleva i drugikh vydayushchikhsya otechestvennykh uchenykh [Actual problems of space exploration: proceedings XXXIX Academic Space Conference dedicated to the memory of academician Korolev S.P. and other prominent russian scientists]. Moscow, BMSTU Publ., 2015, pp. 367–368 (in Russ).
2. Koryanov V.V., Nguen K.T., Nguen V.T. *Multiple Regression for UAV Aerodynamic Model Based on Experimental Data.* Vestn. Mosk. Gos. Tekh. Univ. im. N.E. Baumana, Mashinostr. [Herald of the Bauman Moscow State Tech. Univ., Mech. Eng.], 2016, no. 5, pp. 59–67. DOI: 10.18698/0236-3941-2016-5-59-67 (in Russ).
3. Manned spacecraft center NASA. *Results of the second US manned suborbital space flight,* 1961.
4. Solovyev V.A., Lysenko L.N., Lyubinskiy V.Ye. Upravleniye kosmicheskimi poletami. Chast' 2 [Control of space flights. Part 2]. Moscow, BMSTU Publ., 2010, part 2, 426 p. (in Russ).
5. Garcia Gallego, Gonzalo Augustin. *Development of a model-assisted approach for the requirement-driven design and optimization of a quadrotor unmanned aerial vehicle,* 2016.
6. Daniel Selva. *A concept for an Agile Mission Development Facililty for CubeSat and suborbital missions,* IEEE, 2016.
7. Jeremy Blum. *Exploring Arduino: Tools and Techniques for Engineering Wizardry,* Wiley, 2013, 384 p.

The publication was prepared with support of the «RUDN University Program 5-100».

UAV Delivery Monitoring System

Khin Thida San[1], Sun Ju Mun[1], Yeong Hun Choe[1] and Yoon Seok Chang[1]

[1]UTAC, School of Air Transportation and Logistics, Korea Aerospace University, Goyang City, Seoul 412-791, Republic of Korea (South Korea)

Abstract. UAV-based delivery systems are increasingly being used in the logistics field, particularly to achieve faster last-mile delivery. This study develops a UAV delivery system that manages delivery order assignments, autonomous flight operation, real time control for UAV flights, and delivery status tracking. To manage the delivery item assignments, we apply the concurrent scheduler approach with a genetic algorithm. The present paper describes real time flight data based on a micro air vehicle communication protocol (MAVLink). It also presents the detailed hardware components used for the field tests. Finally, we provide UAV component analysis to choose the suitable components for delivery in terms of battery capacity, flight time, payload weight and motor thrust ratio.

1 Introduction

The use of UAVs in urban areas is increasingly being considered to overcome ground delivery traffic issues. However, flight environments are subject to dynamic changes due to shifts in weather patterns and moving objects. To improve UAV flight safety, proposals have been offered to effectively monitor the entire UAV delivery process. This paper provides requirements that should be considered when building an effective ground control station (GCS). There are two main GCS types – ready-made open source GCS and customized GCS.

Ready-made GCS is typically open source software that can easily be downloaded from the internet and used for mission management and real-time flight operation. Lim et al. [1] analyzed eight open-source projects (OSPs) for UAVs. They discussed five autopilot components – processors, gyroscopes, accelerometers, magnetometers, and barometers – and introduced flight controllers with various functions to monitor and control UAVs. Moreover, additional functions can be added by extending the open source code. Mission Planner and Qgroundcontrol are representative open source software packages that deliver UAV flight automation. It enables UAVs to fly along predetermined waypoints and provides real time flight data. It also provides an automatic return function that engages when fuel or battery levels are low [2]. Ready-made GCS are convenient due to their wide functionality range. However, some functions of logistics purpose for delivery tracking in real time and history report of the previous flight missions are not provided. And, they are overly focused on flight functions, making it too difficult for normal users to apply for non-traditional tasks.

Customized GCS can be implemented according to project requirements. For example, Perez et al. [3] described the development of an efficient and comprehensive guidance system to monitor multiple UAVs in real time, while also discussing the main layout for a multi-UAV GCS. Generally, open source libraries are used to develop customized GCS. Dong et al. [4] developed a real-time software system for an unmanned helicopter with a two-layer framework, i.e. the data was transferred in the background while the visualization operated in the foreground.

The proposed monitoring system is a customized GCS inspired by FlightRadar24 (Live Air Traffic) [5]. FlightRadar24 facilitates the real time tracking of commercial flights by providing geographical locations and flight data such as flight numbers, speeds and headings. The monitoring system provides not only flight data but also delivery item management to improve logistics efficiency.

In this research, Pixhawk autopilot UAV is used with selected flight components to perform deliveries. MAVLink protocol [6-8] is used to receive the flight data and command of the UAV. Our main objectives for the monitoring system are (1) managing order assignment for a flight mission (item assignment and path planning), (2) controlling autonomous flight, (3) monitoring in real time, (4) tracking the UAV and delivery items, and finally, (5) dropping the parcel at a specific location autonomously (gripper servo movement).

2 System structure

Fig.1 depicts the required functions of the monitoring system for UAV delivery. The system is designed to combine useful UAV delivery system functions that are not normally supported by ready-made GCS.

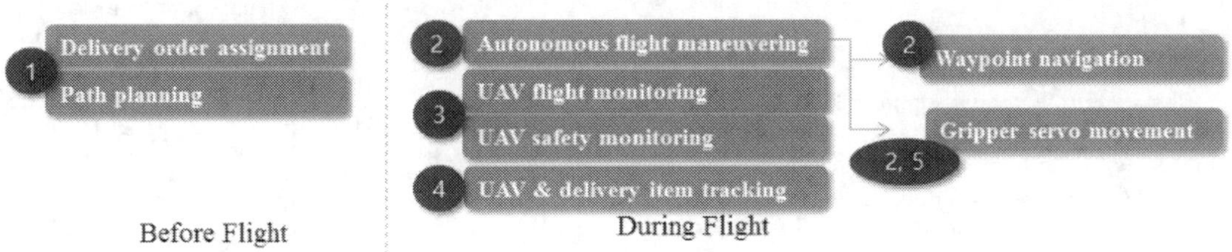

Figure 1. UAV delivery monitoring system functions

(1) The system combines order assignment and path planning functions together to deliver-items with minimal flight times. We use a concurrent scheduler approach together with a genetic algorithm (GA) to assign items to flight missions [9-10]. Accordingly, there are six parameters – item weight, order weight (for a collection of items), the UAV's sustainable payload, UAV flight time, delivery distance and UAV speed. All of the parameters must satisfy one another to achieve real flight. Therefore, GA with the concurrent scheduler approach is used to solve multi-parameters. Finally, the approach results in sets of missions with items, destinations, and selected UAV.

(2) The autonomous flight is controlled via MAVLink. The commands with detailed parameters for the required maneuvers are shown in Table 2.

(3) For real time monitoring and UAV flight safety, our monitoring system uses MAVLink messages to receive autopilot flight data, which are shown in Table 1.

(4) UAV and item tracking are shown in Fig. 2. Three main panels – UAV details, mission details, and real time data – are included. The UAV details panel shows detailed data regarding UAV flight time and sustainable payload weight. The mission details panel shows the current mission and ordered items information, as well as the home, destination, and waypoints. The real time data panel shows the real time data for the flights data, positions data, battery data, and telecommunication data.

(5) Finally, the specific destinations (latitude and longitude) are defined as waypoints to drop parcel. The system tracks UAV location during flight. UAV receives a command (MAV_CMD_DO_SET_SERVO [11-12]) to open the gripper servo when it reaches the proper altitude of the designated destination (MAVLink command, Table 2, No 6). The whole UAV delivery procedures can be seen in Fig. 3. The next steps of the delivery system are to implement emergency flight control with the help of the environment data and UAV components failure monitoring.

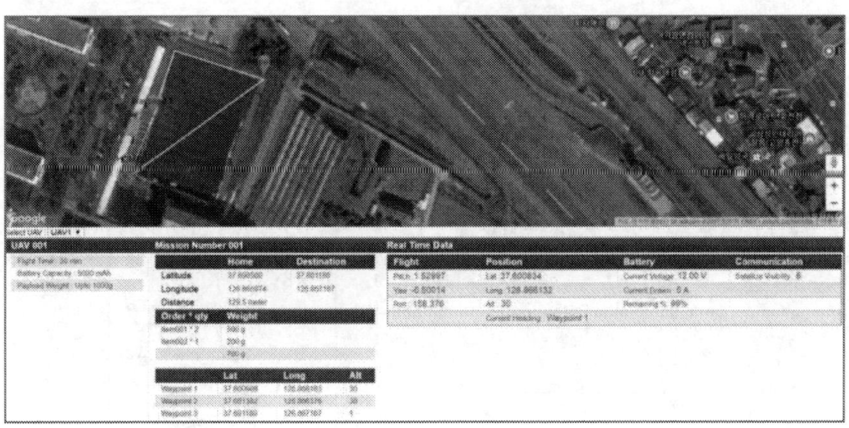

Figure 2. UAV delivery monitoring system

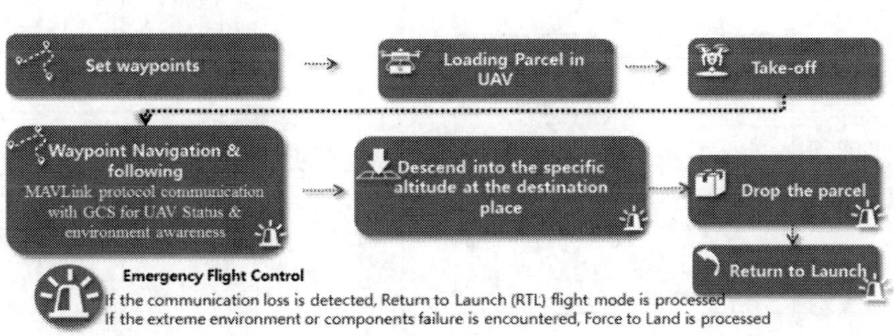

Figure 3. The whole UAV delivery step by step procedures

MATEC Web of Conferences **151**, 04011 (2018)
ACMAE 2017
https://doi.org/10.1051/matecconf/201815104011

3 Communication structure

In the system, 3DR telemetry (TX/RX) is used to communicate between the Pixhawk autopilot UAV and monitoring system. Pixhawk supports PX4 firmware and Ardupilotmega firmware. We use Ardupilotmega for our system. Specific system IDs are assigned to each autopilot and they are assumed as the UAV ID in MAVLink. The telemetry conveys MAVLink messages and data which contain the aircraft flight data and UAV GPS information with the respective system ID of the XML data file. At the same time, the monitoring system sends commands to the UAV to engage in functions such as take-offs, landings, and auxiliary servo movements. The commands are also sent to UAV via the command XML file. XML files are generated from the MAVLink C header library, which is decoded or encoded using Node.JS MAVLink of the JavaScript package manager [13]. We use two telemetries – TX and RX. One of the telemetries (Serial COM6) is used to send while the other (Serial COM5) continues to receive telemetry messages, as shown in Fig. 4.

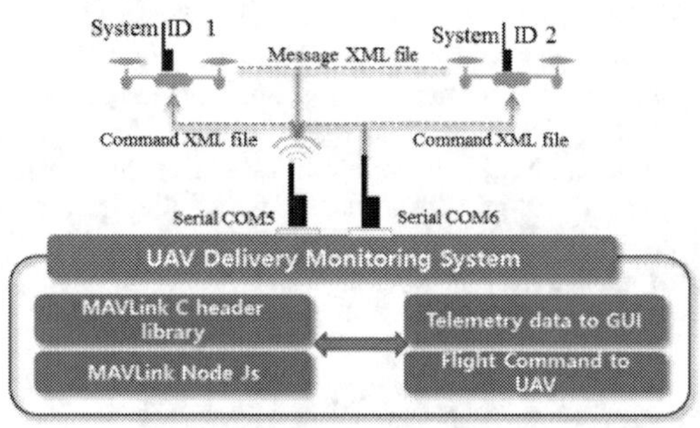

Figure 4. Communication for UAV delivery monitoring system

4 MAVLink communication protocol commands and messages

Tables 1 and 2 describe the MAVLink messages and commands used to implement the monitoring system. Table 3 describes the report status messages (failed or executed) that appear after the commands are sent.

Table 1. MAVLink Message

	Description/ Purpose/ Message ID	Field name and Remark
1	-Geographical location -UAV tracking GPS_RAW_INT [14]	lat (Latitude) long (Longitude) alt (Altitude)
2	-Satellite visibility -UAV communication GPS_RAW_INT [14]	satellites_visible (Number of satellites visible (0-20) – small visibility numbers indicate that communication between the UAV and GCS is weak) fix_type (status of GPS – No GPS, 3D GPS and 2D GPS)
3	-Current waypoint -UAV tracking MISSION_CURRENT [14]	seq (Number of sequence/waypoint to which the UAV is currently heading. We define sequence as a waypoint before actual flight)
4	-Battery monitoring data -UAV flight safety SYS_STATUS [14]	battery_voltage (currently used battery voltage in real time of flight) battery_current (drawn current in real time of flight) battery_remaining (remaining percentage in real time of flight)
5	-Flight data -UAV flight safety ATTITUDE [14]	pitch (pitch movement in real time of flight) yaw (yaw movement in real time of flight) roll (roll movement in real time of flight)
6	-Speed data -UAV flight safety GPS_RAW_INT[14]	vel (Ground speed - provided by GPS embedded on autopilot)
7	-Heading data -UAV flight safety GLOBAL_POSITION_INT [18] GPS_RAW_INT[14]	hdg (UAV heading yaw angle in degree) cog (Course over ground – direction of UAV movement between two waypoints with respect to the surface of the Earth)
8	-Flight data -UAV flight mode [15] HEARTBEAT [14]	base_mode custom_mode (The combination of base_mode and custom mode can change the flight mode details in Table 4)

Table 2. MAVLink Commands

	Description/Purpose/ Command ID	Remark and parameters
1	Defining waypoints for a mission/ flight To define waypoints for navigation MISSION_COUNT [14] MISSION_REQUEST [14] MISSION_ITEM [14] MAV_CMD_NAV_WAYPOINT [14]	To define waypoints into autopilot, a set of commands are used [16]. Sample waypoint assigning with MAVLink via Nodejs Javascript library can be seen in Fig. 5.

2	Mission clear
	To clear all waypoints of the autopilot
	MISSION_CLEAR_ALL [14]
		"MISSION_CLEAR_ALL",
		{
		'target_system': 1,
		// UAV ID
		'target_component': 1,
		// Autopilot
		'mission_type': 0
		//MAV_MISSION_TYPE = 0
		}

3	Mode control
	For UAV flight mode
	SET_MODE[14]
		"SET_MODE",
		{
		'target_system' : 1,
		'base_mode' : 89,
		'custom_mode' : 3
		}
		(The combination of base_mode and custom mode can change the flight mode , detail in Table 4)

4	Take off / Landing
	To take-off and land
	COMMAND_LONG [14]
	MAV_CMD_NAV_TAKEOFF [14]
	MAV_CMD_NAV_LAND [14]
		"COMMAND_LONG",
		{
		'target_system' : 1,
		'target_component' : 1,
		'command':22,
		//22 MAV_CMD_NAV_TAKEOFF
		//21 MAV_CMD_NAV_LAND
		'confirmation' : 0,
		'param1' : 0,
		'param2' : 0,
		'param3' : 0,
		'param4' : 2,
		'param5' : 37.600785,
		'param6' : 126.864739,
		'param7' : 2
		}

5	Servo control
	For the gripper servo movement in order to drop the parcel
	COMMAND_LONG[14]
	MAV_CMD_DO_SET_SERVO [12]
		"COMMAND_LONG",
		{
		'target_system': 1,
		'target_component': 1,
		'command' : 183,
		//MAV_CMD_DO_SET_SERVO
		'confirmation' : 0,
		'param1' : 10,
		///auxiliary 1 servo number
		'param2' : 1500,
		///PMW 1000 servo is closed
		///PMW 1500~1300 servo is opened
		'param3' : 0,
		'param4' : 0,
		'param5' : 0,
		'param6' : 0,
		'param7' : 0
		},

6	Arm/disarm
	To arm and disarm
	COMMAND_LONG
	MAV_CMD_COMPONENT_ARM_DISARM [14]
		"COMMAND_LONG",
		{
		'target_system' : 1,
		'target_component' : 1,
		'command' : 400,
		//MAV_CMD_COMPONENT_ARM_DISARM
		'confirmation' : 0,
		'param1' : 1,
		///1 arm //0 disarm
		'param2' : 0,
		'param3' : 0,
		'param4' : 0,
		'param5' : 0,
		'param6' : 0,
		'param7' : 0
		}

```
m.createMessage(
"MISSION_COUNT",
{
'target_system' : 1,
'target_component' : 1,
'count' : 2,
// For 2 waypoints setting
'mission_type' : 0
},

function(message) {port.write(message.buffer) ; } ) ;

m.on( 'MISSION_REQUEST', function(message , fields) {
console.log ("seq is " + fields.seq);
} ) ;
m.createMessage(
"MISSION_ITEM",
{
        'target_system' : 1, //UAV ID
        'target_component': 1, //autopilot
        'seq' : 0, // Waypoint 1
        'frame' : 3, // default value for quadrotor
        'command': 16, //default value, MAV_CMD_NAV_WAYPOINT
        'current' : 0, // 0: before mission /// 1: during mission
        'autocontinue' : 1, //default value
        'param1' : 0,
        'param2' : 0,
        'param3' : 0,
        'param4' : 0,
        'x' : 37.601057, //waypoint 1 latitude
        'y' : 126.864401, //waypoint 1 longitude
        'z' : 1, // altitude (meter)
        'mission_type' : 0 //default
 },

 function(message) {
  port.write(message.buffer);
 });
m.createMessage(
"MISSION_ITEM",
{
        'target_system' : 1, //UAV ID
        'target_component' : 1, //autopilot
        'seq' : 1, //waypoint 2
        'frame' : 3, // default value for quadrotor
        'command' : 16, //default value, MAV_CMD_NAV_WAYPOINT
        'current' : 0, // 0: before mission /// 1: during mission
        'autocontinue' : 1, //default value
        'param1' : 0,
        'param2' : 0,
        'param3' : 0,
        'param4' : 0,
        'x' : 37.600785, //waypoint 2 latitude
        'y' : 126.864739, //waypoint 2 longitude
        'z' : 2.68, // altitude (meter)
        'mission_type' : 0 //default
        },
function(message)
 { port.write(message.buffer);});

m.on ('MISSION_ACK', function(message, fields) {
        console.log("MISSION_ACK  is " + fields.type);
} ) ;
```

Figure 5. Assigning two waypoints with Node.JS MAVLink

Table 3. Acknowledgement message

	Description/Message ID	Field name and Remark
1	Command acknowledgement COMMAND_ACK [14]	result (MAV_RESULT) 0 - accepted 4 - failed This message is used after sending a command
2	Mission acknowledgement MISSION_ACK [14]	type (MAV_MISSION_RESULT) 0 - accepted 1 – failed This message is used after sending a series of sequences/ waypoints for a mission

Table 4. Flight Mode

Mode name	Base mode + Custom mode		Description
Stabilize	81	0	UAV can arm to fly only at the stabilize state
Alt Hold	89	3	UAV holds a specific altitude in order to drop the parcel from the air
RTL	89	6	Return to the start flying position after dropping the parcel
Auto	81	9	UAV changes into auto mode to navigate the predefined waypoints

5 Hardware analyses (flight time, weight and thrust force for motors)

For delivery purposes, the weight and flight time are correlated. Extra weight causes shorter flight times. Normally, flight time would be directly related to battery capacity. However, larger battery capacities may also result in heavier total payloads. Therefore, choosing an appropriate battery capacity often becomes a consideration. Normally, commercial UAVs can fly from 15 to 30 minutes. However, depending on customized components with advanced technologies and algorithms, some UAVs can fly for up to an hour.

5.1 Flight time estimation

Flight time estimation is not only effected by (i) the battery capacity but also (ii) the power consumption (current drawn) of the electronic components embedded in the UAV autopilot board. Moreover, for delivery purposes [17], (iii) the total all-up-weight (AUW), which includes the weight of products being delivered [17], and (iv) the thrust of the motor related to propeller size [18] are also considerable. A simple and an easy method of estimating flight time related to battery capacity and current draw is described in Eq.1.

$$\text{Flight time (minutes)} = (\text{Battery capacity}/\ \text{Current draw}) * 60 \tag{1}$$

For example, if the battery capacity is 5600mAh (LiPoly 60C) and the motor amp has a current draw of 25, Eq. 1 produces a flight time estimation of 13.44 minutes. However, the flight time is around 11 minutes with around 900 grams of UAV weight together with 500 grams of carry payload in the real experiment [17]. Choosing the propeller size in order to lift the entire UAV mass is described in the load calculation equations and its related calculations [18].

5.2 Thrust motors to carry a specific payload weight (Thrust to weight ratio)

As flight time varies according to payload, the thrust required to move forward related to the total mass of the UAV should be calculated. Eq. 2 represents a commonly used, easy equation to calculate motor thrust. The final total AUW is the total weight of the UAV, including components, i.e. the airframe, autopilot, sensors, battery, motors, ESC, propeller, camera, and delivery package [18-20].

Required thrust per motor = [**AUW** x 2 (or) 3] / number of motors (2)

2:1 thrust ratio is standard where AUW is ratio value of 2 for one/each motor thrust. For particularly acrobatic cases that use a 3:1 ratio to improve efficiency. If a copter is able to handle approximately 1 kilogram, according to the 2:1 equation ratio, a total thrust of 2 kilograms and 500 grams of thrust per motor are required. In the present study, the thrust to weight ratio is set as 2:1; this is a good starting point as it ensures that enough thrust exists for the desired performance, and it avoids overly agile UAV movements.

5.3 Choosing the correct Lipo battery capacity

The electrical power for a multi-rotor can be approximated using Eq. 3 [21-23]. The equation is related to the power estimation theory of a Lipo electric battery (discharge rate).

Power (P) Watts = Voltage (V) V· Current (I) A (3)

The electrical power (watts) used by a multi-rotor to hover is P, calculated by obtaining V and I from the onboard power module. The motor speed and weight can vary according to the current draw under various conditions such as the UAV heading with air resistance. In addition, the battery discharge rate is higher in hot and warm weather than in cold conditions. Therefore, the maximum electrical power or maximum sustained discharge rate must be defined. In the example with 25 °C with 3700 mAh Lipo, the maximum sustained discharge rate is 92.5 A, which can be calculated as I = 25*3700/1000 = 92.5 A. For the quad-copter with four motors, the current draw for each motor is 21 A with a maximum (21x4) of 84 A. The amount of electricity needed for flight operation is 92.5 A (92.5 > 84). Accordingly, the maximum sustained discharge rate of 92.5 A with 3700 mAh is suitable for flight operation.

6 UAV Delivery Flight Experiment

The gripper for the parcel is attached under the UAV frame, as shown in Fig. 6 (left). Two servo motors are used to operate the gripper, Fig. 6 (right-up). The parcel box requires a stick with a head blob for the gripper to grab onto, Fig. 6 (right-down). Table 5 shows our UAV's component specifications for the parcel delivery.

Figure 6. Quad-copter with the parcel (left), the gripper with two servos (middle), and the parcel (right)

Table 5. UAV component specifications

	Component	Specification	Weight
1	Frame and landing skid	S500 glass fiber quad-copter frame motor centers: 480mm height: 170mm	405 g
2	ESC	Hobby Wing X Rotor 20A wired type	
3	Propeller	Aerostar composite propeller/ 10x5.5cm	
4	Telemetry	3DR transceiver telemetry Radio 915mhz	300 g
5	Autopilot	Pixhawk 2.1 standard set	
6	GPS	3DR GPS	
7	Motor	LDPOWER MT2213-920 KV/ Brushless motor Max Power 302.4W Max Amps: 18A	221.2 g
8	Battery	3DRobotics / Lipo Battery , 11.1V 8C, 5100mAh	320 g
9	Two servos (Gripper) and UBEC	HS-7955 TH HV Digital servo/ UBEC model HobbyWing 3A,MAX 5A 2-6S Lipo	220 g
Total weight			1466.2 g
Parcel weight (payload)			700 g
Total weight			2166.2 g (2.1 kg)

g = gram, kg = kilogram

A flight test is performed to accomplish the continuous three waypoints, as shown in Fig. 2. The UAV starts from the home location with an altitude of 30 meters and changed into auto mode to follow the waypoints. It then descends to 1 meter when it reaches Waypoint 3. The monitoring system subsequently detects the UAV at a specific location (latitude, longitude and altitude). It sends a servo command to open the gripper to drop the parcel. UAV flight mode then changes to return-to-launch (RTL). This returns the UAV to its home position. The entire mission lasted a little over 2.5 minutes with a distance of 129.5 meters from home to the destination. The delivery process was achieved autonomously from take-off to landing.

6. Conclusions

UAVs are capable of safely delivering goods via UAV monitoring functions. These functions monitor the working conditions of essential UAV components in real time, including batteries, speeds, headings, and flight data. In addition, this approach also provides tracking delivery statuses such as upcoming waypoints, current items, and descriptions. Moreover, the system can send immediate manoeuvring commands to a UAV to avoid obstacles and engage a forced landing. In the present study, our system is implemented to fulfil a fully autonomous UAV delivery to overcome last-mile logistics challenges.

Acknowledgment

This research was supported by Ministry of Science and ICT (MSIT), Korea, under Information Technology Research Center (ITRC) support program (IITP-2017-2014-0-00678) supervised by Institute for Information & communications Technology Promotion (IITP).

References

1. H. Lim,J. Park D. Lee, H.J.Kim, "Build Your Own Quadrotor: Open-Source Projects on Unmanned Aerial Vehicles," *IEEE Robotics & Automation Magazine*, **19**,Issue. 3, 33-45 (2012).
2. A. Khalid, K. McFall, "Aerial Robotic Autonomous Patrol and Surveillance System," *14th AIAA Aviation Technology, Integration, and Operations Conference*, 1-10 (2014).
3. D. Perez, I. Maza, F. Caballero, D. Scarlatti, E. Casado, A. Ollero, "A ground Control Station for a Multi-UAV Surveillance System," *J Intelligent Robot System*, **69**, 119-130 (2013).
4. M. Dong, B.M. Chen, G. Cai, K. Peng, "Development of a Real-time Onboard and Ground Station Software System for a UAV Helicopter," *Journal of Aerospace Computing, Information and Communication*, **4**, 933-955 (2007).
5. https://www.flightradar24.com
6. http://qgroundcontrol.org/mavlink/start
7. S,n.d Balasubramanian, Information on "MAVLink *Tutorial* for Absolute Dummies (Part –I)," http://ardupilot.org/dev/docs/mavlink-commands.html
8. http://mavlink.org/messages/common
9. K.T. San, E.Y. Lee, Y.S. Chang, "The delivery *assignment* solution for swarms of UAVs dealing with Multi-dimensional Chromosome representation of Genetic Algorithm," *Ubiquitous Computing, Electronics & Mobile Communication Conference (UEMCON)*, IEEE Annual (2016).
10. K.T. San, Y.S. Chang, "Concurrent Scheduler Approach with Genetic Algorithm for Item Delivery by using swarm of UAVs," *The Proceedings of 21st International Symposium on Logistics (ISL 2016) Sustainable Transport and Supply Chain Innovation*, 510- 517 (2016).
11. D. Becker, N. Palumbo, et al. "HIGH ALTITUDE GLIDER," Final Design Report.
12. http://mavlink.org/messages/common#MAV_CMD_DO_SET_SERVO
13. https://www.npmjs.com/package/mavlink
14. http://mavlink.org/messages/common
15. http://ardupilot.org/copter/docs/flight-modes.html
16. http://qgroundcontrol.org/mavlink/waypoint_protocol
17. D. Robert, "Construction and Testing of a Quadcopter," A senior project presented to the Aerospace Engineering Department California Polytechnic State University, San Luis Obispo (2013).
18. Md. R Haque, M.Muhammad, D. Swarnaker, M. Arifuzzaman, "Autonomous Quadcopter for Product Home Delivery," *International Conference on Electrical Engineering and Information & Communication Technology (ICEEICT)*, (2014).
19. https://www.grc.nasa.gov/www/k-12/airplane/fwrat.html
20. https://aviation.stackexchange.com/questions/8316/how-to-calculate-the-required-motor-power-and-rotor-size-with-respect-to-the-wei
21. https://quadcopterproject.wordpress.com/static-thrust-calculation/
22. http://physics.stackexchange.com/questions/160435/relation-between-power-voltage-and-current-when-one-or-more-of-them-is-amplifie
23. C. Horne, "Expandable Open Source Autonomous Quadcopter," *The UNSW Canberra at ADFA Journal of Undergraduate Engineering Research*, **6**, No. 1, (2013).

Simulation and Effectiveness Analysis on One versus One Beyond Visual Range Air Combat

LIU Haoyu[1] , ZHANG Yunfei[1] and LI Shihao[1]

[1]*School of Aeronautic Science and Engineering, Beihang University, Beijing 100191, China*

Abstract. A kind of one versus one beyond visual range (BVR) air combat model has been established, which includes functional models of radar, missile and fighter and the process of several combat stages. Air combat effectiveness ratio (ACER) is defined to analyse the result. The 2^k factor design method is used to design combat test case and analyses the influence of three factors (fighter stealth character, missile range and flight height) on ACER. Simulation result reveals that when RCS of one fighter is reduced from 0dBm^2 to -10dBm^2 which cannot remarkably affect the opposition fighter's radar detection distance and missile launch distance, the RCS factor has small influence and the missile range factor has great influence. When RCS of one fighter is reduced from -10dBm^2 to -20dBm^2, the opposition fighter's radar detection distance will be reduced and lead the result of its missile launch distance be less than its missile range. Compared with the former case, the effect of RCS factor increases and the effect of missile range factor decreases. However, the effect of height is not significant.

1 Introduction

Modern air combat can be divided into within visual range (WVR) air combat and BVR air combat, and about 10km is the dividing line of two cases. BVR air combat is a kind of air combat model that two hostile fighters are outside the visual range, detecting target through the airborne detection equipment and attacking each other with middle/long-range missiles. With the development of sensors and weapon systems, BVR air combat has become the main form of modern air combat [1-3].

Modern BVR air combat research is divided into two methods. Some of the literature [1-5] uses effectiveness assessment as the basic means, which is a static assessment method that establishes the advantage function of some important parameters (such as entry angle, speed, height, etc.), and the function is used to judge the results of air combat. The shortcoming of this method is relying on experience, and cannot accurately descript the random process of the modern air combat. Part of the literature [6-8] simulates the air combat through computer, which establishes the model of fighters based on kinematics and dynamics equation. Their shortcoming is that it is rough for the radar detection and track model and the missile attack process, affecting the assessment accuracy of results.

This paper establishes a relatively complete one versus one BVR air combat model, including radar search, track target, launch and guide missiles and damage judgment. Using 2k factor design method [9] to count the simulation data. Provide reference for the tactical decision, through analysis of stealth performance (radar cross section, RCS), flight height, missile range and other parameters on the impact of air combat.

2 Radar detection and track model

The purpose of radar simulation is to calculate the probability of detecting and tracking targets. During the BVR air combat period, the detection situation of airborne radar may be divided into 3 cases of upward, horizontal sight and overlooking. For the Pulsed Doppler (PD) radar, in the situation of upward and horizontal sight, detection performance will be affected only by noise. In the situation of overlooking, the ground clutter will reduce the detection performance of radar [10].

The detection performance of HPRF PD radar may be expressed in critical radar cross section [11]

$$\sigma_{\rm cr}^{(c+n)} = \begin{cases} \sigma_{\rm cr}^{(n)} = \dfrac{R^4 L_\alpha}{C_{s1} F^4}, \\ \text{upward and horizontal sight,} \dfrac{|\Delta V|}{V_R} \geq 1; \\ \sigma_{\rm cr}^{(n)} \left\{ 1 + \dfrac{C_{s2}\gamma}{V_R H^2} \left[1 - \left(\dfrac{\Delta V}{V_R} \right)^2 \right] \right\}, \\ \text{overlooking,} \dfrac{|\Delta V|}{V_R} < 1; \end{cases} \tag{1}$$

R is the distance from the target, C_{s1} and C_{s2} is the 1st and 2nd radar system character constant, V_R and H is the speed and height of the aircraft. $(c+n)$ expresses the signal detection in condition of noise and clutter, and (n) expresses only in noise.

In the case of other parameters keeping constant, the signal noise ratio (SNR) is proportional to the RCS of the target, and is inversely proportional to the critical RCS of the target. Therefore, the SNR for a target with RCS of σ is

$$\frac{S}{N} = \frac{\sigma}{\sigma_{cr}^{(c+n)}} \tag{2}$$

According to the relationship of detection probability with SNR, the detection probability to a target may be gotten for a given false alarm probability $\hat{P}_{fa}$.

When the radar has detected a signal, it can't be confirmed by one or twice scans if the signal is a target or not. The radar has to continuously scan the target to get several signals, avoiding these cases of detecting the signal of irrelevant target or environmental false target. The requirement is in n times of scan, the radar has to detect the target signal no fewer than m (m<n) times. And the relation between m and n is

$$m = 1.5\sqrt{n} \tag{3}$$

Here m and n is respectively 3 and 5. It means that signal has to be detected no fewer than 3 times within five times scan, then it can be said that the target is being tracked.

3 One versus one air combat process

One versus one air combat model was built on basis of following assumptions:

• Simplifying the fighter as a particle, and red fighter and blue fighter are flying in parallel flight path at a constant supersonic speed. After launching the missile, the fighter's speed decreases to subsonic velocity.

• When the distance between the target and the missile is less than terminal guidance distance, the missile enters the autonomous terminal guidance. Nevertheless, the terminal guidance is not simulated here; a constant kill probability is given for this stage.

• The air-to-air missile is also considered as a particle, which is with a constant speed.

3.1 Simulation and data statistics

In this simulation, the fighter tactical decisions of both sides are the same, which include searching, detection, tracking, launching missile and missile terminal guidance and target killed, etc.

If one side missile has killed the opposition fighter, but the latter has not launched its missile or guided its missile into the terminal guidance, it is judged that the former is alive and the latter is dead.

Air combat is a random process and it will be influenced by the probability of these combat process such as detection, tracking, hitting and killing, etc. For the same combat case, the simulation has to be executed for many times, and simulation results should be counted up to give the integrated results such as success probability of the red side, success probability of the blue side, success probability of the two sides and failure probability of the two sides.

The whole simulation flowchart of air combat is show in Figure 1, and the flowchart of fighter against module is shown in Figure 2

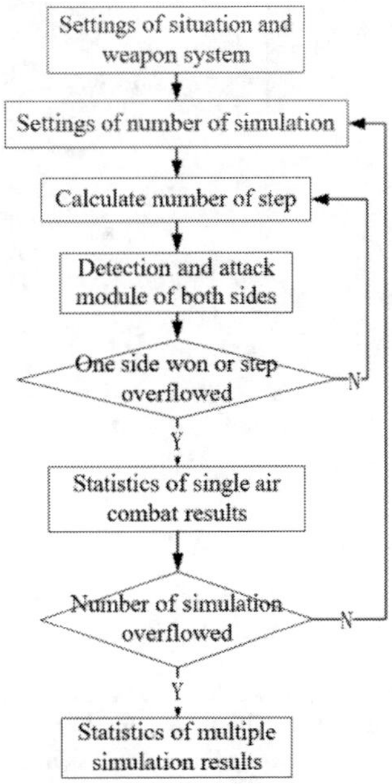

Figure 1. Whole simulation flowchart of air combat

3.2 Radar search

After the radar module calculates the detection probability P_d, the fighter should determine whether find the target or not using Monte Carlo method: firstly produce a random number α of [0, 1], and then if $\alpha < P_d$, it is considered that the fighter has found the target [12].

3.3 Track target

If the fighter has found the target, it need turn to the target and try to track it using pre-tracking method shown in Figure 3. O is the fighter; T is the target, and the flight direction is along vector $\boldsymbol{V_T}$; the fighter does not flight to the target directly, but add a pre-tracking vector $\boldsymbol{kV_Tt}$ (k is pre-tracking coefficient, t is the radar sweep time), and then flights along vector OT'.

In the tracking state, if the detection result is not meet the discovery condition referred in part 2, it means the fighter has lost the target and need return the radar search state.

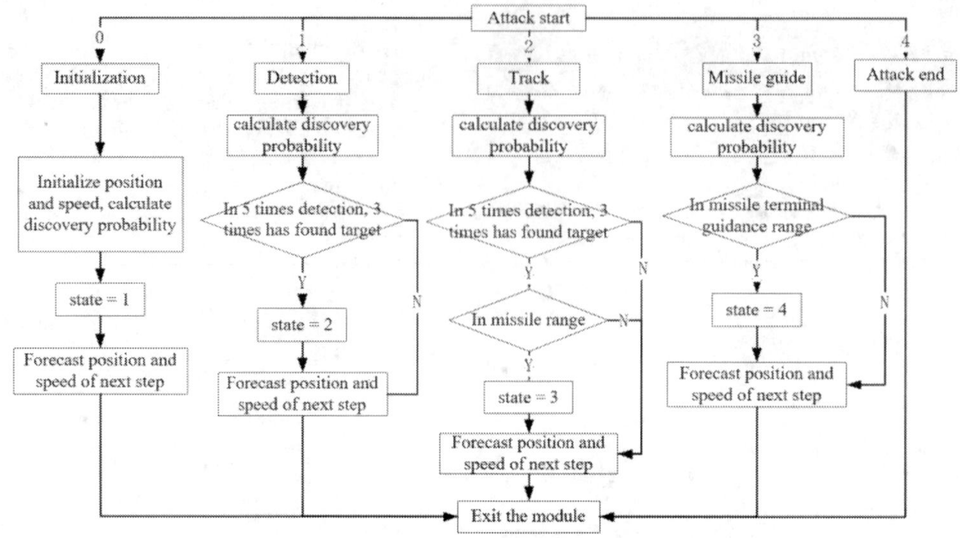

Figure 2. Flowchart of fighter against module

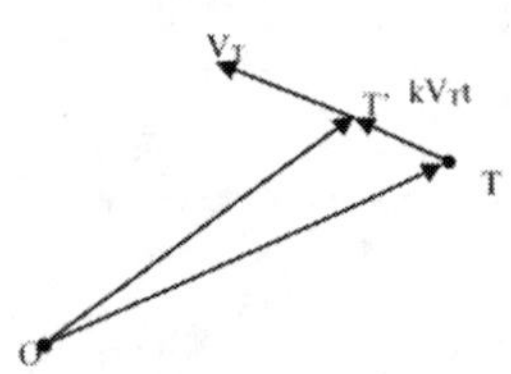

Figure 3. Pre-tracking method

3.4 Launch and guide missiles

When the fighter tracks the target, the fighter need launch missile if the distance between them is less than the range of missile. At this time, the airborne radar need continuously track the target and guide the missile flight to it. And the fighter should make the edge of scanning range turn to the target to avoid attack from enemy, which is shown in Figure 4. V_O is the speed vector of the fighter

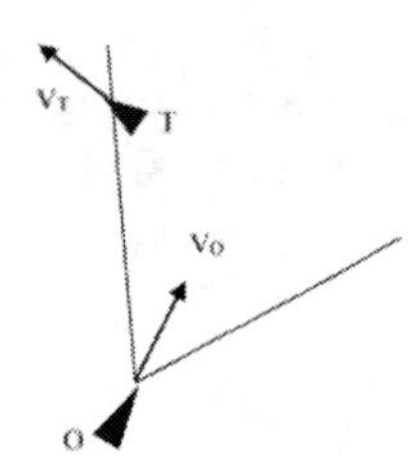

Figure 4. Missile guidance method

3.5 Damage judgment

After the missile enters terminal guidance, the fighter should maneuver to avoid the threat. Using Monte Carlo method to judge the result of air combat.

4 Data statistics and analysis

4.1 2^k factor design method

2k factor design method need select two values for each effect factor, and simulate all possible combinations of all effect factors and all values. If k=3 (3 factors), there are 8 combinations (shown in Table 1). The sign factor s_{ij} (+, –) represent two values of factor J_i; response value R_j is the value of simulation result.

Table 1. Combinations of factors

Combination j	Factor J_1 sign s_{1j}	Factor J_2 sign s_{2j}	Factor J_3 sign s_{3j}	Response value R_j
1	-	-	-	R_1
2	+	-	-	R_2
3	-	+	-	R_3
4	+	+	-	R_4
5	-	-	+	R_5
6	+	-	+	R_6
7	-	+	+	R_7
8	+	+	+	R_8

The main effect is the result R_j's average increment that the value of factor J_i changes from – to + in the case of other factors unchanged, which is J_i's effect to result. And the calculation formula of main effect e_i is

$$e_i = \frac{1}{4}\sum_{j=1}^{8} s_{ij}R_j \left(i = 1,2,3\right) \tag{4}$$

The interactive effect of two factors is one factor's influence on the other factor's main effect, which is the relevance of the two factors. And the calculation formula of e_{12} is.

$$\begin{aligned} e_{12} &= \frac{1}{2}\left[\frac{1}{2}\left(R_8 - R_6 + R_4 - R_2\right) - \frac{1}{2}\left(R_7 - R_5 + R_3 - R_1\right)\right] \\ &= \frac{1}{4}\left(R_1 - R_2 - R_3 + R_4 + R_5 - R_6 - R_7 + R_8\right) \end{aligned} \tag{5}$$

The first part in the square bracket is the main effect of I_2 when I_1's value is +, and the second part is the main effect of I_2 when I_1's value is -. So e_{12} is the increment of I_2's main effect when the value of factor I_i changes from – to +, and it is easy to prove that $e_{12} = e_{21}$. Similarly,

$$e_{13} = \frac{1}{4}\left(R_1 - R_2 + R_3 - R_4 - R_5 + R_6 - R_7 + R_8\right) \qquad (6)$$
$$e_{23} = \frac{1}{4}\left(R_1 + R_2 - R_3 - R_4 - R_5 - R_6 + R_7 + R_8\right)$$

4.2 Definition of air combat effectiveness ratio

Define the air combat effectiveness ratio (ACER) of Red/Blue is

$$R_{ex}\left(r/b\right) = \frac{P_r}{P_b} = \frac{P\left(\bar{r},b\right) + P\left(\bar{r},\bar{b}\right)}{P\left(r,\bar{b}\right) + P\left(\bar{r},\bar{b}\right)} \qquad (7)$$

$P(\bar{r},b)$ is the probability that red fighter is alive and blue fighter is dead; $P(r,\bar{b})$ is the probability that red fighter is dead and blue fighter is alive; $P(\bar{r},\bar{b})$ is the probability that both sides are alive.

The ACER of Red/Blue is the advantage of red fighter against the blue fighter. If the ratio is more than 1, it means red fighter is in advantage; and if the ratio is less than 1, blue fighter is in advantage.

4.3 Results and analysis

In order to avoiding the uncertainty brought by the change of RCS with the azimuth, we select sphere as the aircraft model. Air combat scene is that red fighter and blue fighter are flying head to head from the distance of 200km, and each simulation is repeated 10000 times. Some parameters are list in Table 2. Both sides use the same airborne radar, and
Table **3** shows the radar detection distance for targets with different RCS. RCS, flight height and missile range are chosen as the three factors, and the ACER of Red/Blue is chosen as response value.

Table 2. Fighter and missile speed

fighter speed during searching[km·s^{-1}]	0.3
fighter speed after launching missile[km·s^{-1}]	0.2
missile speed[km·s^{-1}]	1.0

Table 3. Radar detection distance

RCS [dBm^2]	Upward and parallel sight [km]	Overlooking [km]
0	134	96
-10	75	54
-20	42	30

4.3.1 Example 1

For the air-combat scene settings shown in Table 4, using the method above calculate the ACER of Red/Blue and the effect (shown in Table 5).

According to the results, as to the main effect, missile range's is the largest, and height's is least. When the missile range increases from 50km to 100km, the ACER increases 2.42 times; when RCS of Red/Blue increases from 0/(-10) dBm2 to 0/0 dBm2, the ACER increases only 0.16 times; when the height of Red/Blue increases from 5/5 (km) to 10/5 (km), the ACER decreases 0.03 times.

Table 4. Settings and result of example 1

Combat case	Red/Blue			ACER
	RCS [dBm^2]	*missile range R[km]*	*height H[km]*	
1	0/-10	50/50	5/5	0.80
2	0/0	50/50	5/5	1.01
3	0/-10	100/50	5/5	3.28
4	0/0	100/50	5/5	3.39
5	0/-10	50/50	10/5	0.80
6	0/0	50/50	10/5	0.98
7	0/-10	100/50	10/5	3.24
8	0/0	100/50	10/5	3.36

Table 5. Main effect and interactive effect

Main effect	*RCS*	0.16
	R	2.42
	H	-0.03
Interactive effect	*RCS-R*	-0.04
	RCS-H	-0.01
	R-H	-0.01

The reason that the main effect of RCS is small is: when the RCS of blue fighter decreases from 0 dBm2 to -10 dBm2 and the missile range is 50km(combination 1 and 2), detection range for red radar is more than 50km, so it doesn't affect launch distance of red missile, but will affect the ACER for the difference of RCS; when the missile range is 100km(combination 3 and 4), the variety of RCS will affect launch distance of red missile, but red fighter's launch distance is far greater than blue fighter's, so the ACER is not significantly affected by RCS.

4.3.2 Example 2

For the air combat scene settings shown in Table 6, using the method above calculate the ACER of Red/Blue and the effect (shown in Table 7).

According to the results, as to the main effect, RCS's is the largest, and height's is least, and the impact of RCS on ACER is significantly increased.
For cases that the red missile range is 50km, when blue fighter's RCS is -10 dBm^2, the red radar's detection range of upward/overlook is 75/54km, which is larger than missile range, indicating that launch distance is not affected; when blue fighter's RCS is -20 dBm^2, detection range of upward/overlook is 42/30km, and the launch increases by 10~20km. Moreover, the low RCS worsens

https://doi.org/10.1051/matecconf/201815105001

target tracking conditions of red radar, so the exchange ratio of Red/Blue reduces by about 0.7.

Table 6. Settings and result of example 2

Combat case	Red/Blue			ACER
	RCS [dBm²]	*missile range R[km]*	*height H[km]*	
1	-10/-20	50/50	5/5	0.31
2	-10/-10	50/50	5/5	1.01
3	-10/-20	100/50	5/5	0.31
4	-10/-10	100/50	5/5	3.24
5	-10/-20	50/50	10/5	0.31
6	-10/-10	50/50	10/5	0.94
7	-10/-20	100/50	10/5	0.31
8	-10/-10	100/50	10/5	3.23

Table 7. Main effect and interactive effect

Main effect	*RCS*	1.80
	R	1.14
	H	-0.13
Interactive effect	*RCS-R*	1.14
	RCS-H	-0.01
	R-H	0.02

For the 100km-range missile of red side, the launch distance entirely depends on the radar detection range. So when the RCS of blue fighter reduced from -10 dBm^2 to -20 dBm^2, detection range of upward/overlook reduced from 75/54km to 42/30km, and the exchange ratio reduces by about 2.9.

Therefore, the reason that the main effect of RCS is the largest is: the decrease of RCS reduces the radar detection distance, resulting in decrease of missile launch distance, as well as the low RCS worsens target tracking conditions, so the ACER reduced significantly.

The reason that the main effect of missile range is less than example 1 is: when blue fighter RCS is -20dBm^2, regardless of missile range, red fighter has to wait until detection range within 42/30km, so the two kinds of missile range has little impact on ACER; when blue fighter RCS is -10dBm^2, for the 100km-range missile, red fighter can attack when get the target (75/54km), but for the 50km-range missile, red fighter has to wait even if it has found the target. So the combined effects of missile range are reduced. Therefore, the decrease of RCS reduced radar detection distance, so that missile range cannot fully play a role.

5 Conclusion

A kind of method of one versus one BVR air combat has been established. It is built for the method of radar searching, detection probability calculation and target tracking, and missile guidance and control. As a result, it is formed for the fighting simulation process on two opposition sides attacking alternately. The combat success probability and ACER are given by amount of combat simulations for several supposed combat scenarios.

According to simulation result, when RCS of one side fighter is reduced from 0dBm^2 to -10dBm^2, therefore for the opposite side, the radar detection distance is affected slightly so that the missile launch distance is reduced unobviously. In such case the effect of RCS is small and the effect of missile range is great. When RCS of one side fighter is reduced from -10dBm^2 to -20dBm^2 that for the opposite side the radar detection distance is significantly reduced and the missile launch distance is shorter than missile range. Compared with the former case, the effect of RCS is increased and the effect of missile range is decreased. However, the effect of height is not significant.

It is revealed that the RCS reduction of one side fighter should make the opposite radar detection distance shorter than the missile range, so the opposite missile cannot play its full role, resulting that the effect of RCS on the combat success probability and ACER become significant.

References

1. ZHANG Hongbo, LI Guiying, Kong fane. Research on Situation Assessment in BVR Air Combat. EOC, **17**, 4 (2010), 9-13
2. DONG Yanfei, WANG Liyuan, ZHANG Hengxi. Synthesized Index Model for Fighter Plane Air Combat Effectiveness. AAAS, **27**, 6 (2006), 1085-1087
3. GAO Yong, XIANG Jinwu. New Threat Assessment Non-Parameter Model in Beyong-Visual-Range Air Combat. JSS, **18**, 9 (2006), 2570-2572
4. ZHOU Siyu, WU Wenhai, ZHU Meng. Analysis of Air Combat Situation Assessment Based on Nonparametric Methods. ACT, **41**, 4 (2011), 13-16
5. JamesS.McGrew, JonathanP.How. Air-Combat Strategy Using Approximate Dynamic Programming. AIAA-2008-6796
6. ZHENG Chang, DONG Wenhong, NIU Qinggong. Air Combat Construction and Emulation Design of Fighter Beyond Horizon. SEE, **29**, 12 (2009), 135-137
7. JU Chuanwen, YANG Xiuzhen. Study on Simulation of the Attacking Ability of Air to Air Missiles for Beyond Visual Range Targets. JSS, **16**, 9 (2004), 2048-2051
8. KONG Xiangjun, SHANG Chongyang, GAO Zhenghong. Research on Simulation Model of Fighter Maneuver. JNPU, **25**, 3 (2007), 331-335
9. DONG Xiaolong, SUN Jinbiao, SUN Naixiang. Simulation Experiment Analysis of Effectiveness Index of Air-combat Beyond-visual-range. JSS, **23**, 11 (2011), 2321-2326
10. ZHANG Kao, ZHANG Yunfei, MA Dongli. *Calculation and Analysis of Radar Stealth Performance of Aircraft* (NDIP, Beijing, 1997)
11. ZHANG Yunfei. *Radar Stealth Performance Calculation and Software Integration System for Aircraft* (Beihang University, Beijing, 1998)
12. LI Jianping, WU Hong, WANG Weiping. Metamodel Method for Radar System to Detect and Track Target. JSS, **19**, 3 (2007), 639-642

An Airborne Target Localization Algorithm Based on UKF and Optimal Maneuver Strategy

Chunxiao Zhang[1], Xiaofeng Shi[1]

[1] School of Electronics and Information Engineering, Beihang University Beijing, China

Abstract.:The airborne bearing-only passive target localization performance is relevant with the specific filter and maneuver model. This paper presents a combination of Unscented Kalman Filtering (UKF) with control inputs and optimal route planning algorithm to improve the performance. Taking the minimum trace of unbiased UKF estimation covariance matrix as criteria, single-step heading traverse method is applied to reach every optimal solution. A single flight path simulation and Monte Carlo analysis validate that the optimal strategy can improve both the convergence and localization accuracy.

1 Introduction

Benefitting its flexibility and concealment, airborne single-observer localization is applied in interference investigation, radio rescue and military reconnaissance, etc. The bearing-only algorithm has been a long-term and most interesting topic in passive localization area [1]. Even now, in some certain circumstances, passive single-station equipment can only detect the bearing of the target radiation source, which makes the bearing-only localization remain its significance [2-3].

Based on the airborne direction-finding system, a couple of emitter bearing angles relative to the aircraft, are collected at different points of the flight trajectory firstly. Then, the bearing-only algorithm estimates the emitter location by simple triangulation principle [4]. There are two problems in this estimation processing. At first, noises are inevitably introduced into the bearing measurements, which are propagated to final location estimation. Additionally, the mathematical model of bearing-based localization is non-linear. Filter is employed to solve the problems above, such as Extended Kalman Filtering (EKF) [5], Unscented Kalman Filtering (UKF) [6]. However, removing high-order parts in the Taylor series expansion, EKF suffers linearizing errors and degrades localization accuracy, even leads to the divergence[7]. Since no linearization is required, UKF is more suitable for states estimation in nonlinear systems, and attracts more attentions [8].

Another issue to be considered is that better observability should be maintained to ensure accurate bearing angles as possible. Besides the direction-finder equipment, the aircraft maneuver mode is critical for the angle measurement [9]. As a result, the localization accuracy and time are closely related to the specific aircraft maneuvering mode, which is investigated by lots of scholars already. The most applied, optimal criteria are Fisher Information Matrix (FIM) and its inverse matrix CRLB [10-14]. The posterior CRLB of the discrete-time nonlinear filtering problem to reduce the computational complexity was used in Ref.[15]. Steven C. Nardone proposed that the Hamilton-jacobi equation in the optimal control theory be used to find the optimal trajectory of the observation platform by maximizing the Fisher's information matrix lower bound[16]. Oshman and Davidson used differential inclusion to solve the resulting optimal control problem based on maximizing the determinant of the FIM[11]. In Ref. [17] and Ref.[18], adopting maximizing rate of change of the azimuth as the optimization rule, the optimization of single-observer trajectory of passive location was analyzed. In order to achieve the best estimation accuracy at each second of localization process, an adaptive observer maneuver control approach based on minimizing the GDOP(Geometrical Dilution of Precision) of target localization was presented in Ref.[19].

In this paper, the combination of an UKF filter and an optimal maneuver strategy is presented to improve convergence and accuracy of the airborne single-observer localization algorithm. In this strategy, taking the trace of covariance as the control criterion, UKF determines the optimal maneuvering path for better observability, resulting improved emitter localization.

2 Mathematical model

In general, the state and the measurement equations in the nonlinear filter model are

$$\begin{cases} X_{k+1} = \mathrm{f}(k, X_k) + w_k \\ Z_k = \mathrm{h}(k, X_k) + v_k \end{cases} \tag{1}$$

Where, the process noise w_k is zero mean Gaussian white noise; the measurement noise v_k is also zero mean Gaussian white noise, but independent with w_k.

As is shown in Fig. 1, source target T is located at (x_T, y_T). The aircraft moves from the origin as a straight motion with the velocity $u=[u_x, u_y]$ to measure the bearing of source target. ϕ_k is the target bearing of the aircraft detecting at the kth observing points (x_{ok}, y_{ok}), the measurement equation for airborne single observer localization system is

$$\begin{aligned}\phi_k &= \arctan\left(\frac{y_T - y_{ok}}{x_T - x_{ok}}\right) + v_k \\ &= \arctan\frac{y_k}{x_k} + v_k\end{aligned} \tag{2}$$

Where, $[x_k, y_k] = [x_T - x_{ok}, y_T - y_{ok}] = X_k$ is relative position vector of airborne observer and target.

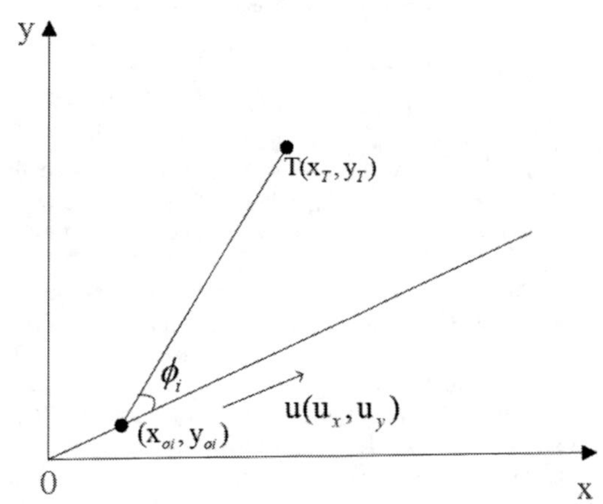

Figure 1. Airborne single observer passive of bearing-only method

The aircraft is moving fast and the measuring period is short, therefore slow moving or static targets can approximately be regarded as fixed. Taking aircraft speed u as a control input, the system state equation is

$$X_k = AX_{k-1} + Bu_{k-1} + w_{k-1} \tag{3}$$

Where, $A \in R^{n\times n}$ is coefficient matrix of state vector; $B \in R^{n\times l}$ is coefficient matrix of control input vector.

3 Algorithm based on ukf with control inputs and optimal maneuver

3.1 UKF with control inputs

The system equation of UKF with control inputs is

$$\begin{cases} X_{k+1} = AX_k + Bu_k + w_k \\ \phi_k = h(k, X_k) + v_k \end{cases} \tag{4}$$

Where, $h(k, X_k) = \arctan\frac{y_k}{x_k}$.

The processing steps in UKF with control inputs are as follows:

1) Set initial state solution $\hat{X}_{0|0}$ and initial state error convariance matrix $P_{0|0}$:

$$\begin{cases} \hat{X}_{0|0} = E[X_0] \\ P_{0|0} = E[\tilde{X}_{0|0}\tilde{X}_{0|0}^T] = E[(X_0 - \hat{X}_{0|0})(X_0 - \hat{X}_{0|0})^T] \end{cases} \tag{5}$$

2) $k \in [1, N]$ (iteration number)

(1) Calculate the sigma points $\chi_i(k|k)$

$$\chi_i(k|k) = F(\hat{X}_{k|k}, P_{k|k}) = \begin{cases} \hat{X}_{k|k} & i = 0 \\ \hat{X}_{k|k} + \left(\sqrt{(L+\lambda)P_{k|k}}\right) & i = 1, 2, \cdots, L \\ \hat{X}_{k|k} - \left(\sqrt{(L+\lambda)P_{k|k}}\right) & i = 1+L, 2+L, \cdots, 2L+1 \end{cases} \tag{6}$$

Where, $\hat{X}_{k|k}$ is state solution at the kth observing point, $P_{k|k}$ is state error convariance matrix solution, $\lambda = \alpha^2(L+\eta) - L$ is a scaling parameter. α is determines the spread of the sigma points around the mean state and is usually set to a small positive value. η is a secondary scaling parameter (usually set to 0); L is the dimension of the state vector.

(2) Update equations

$$\chi_i(k+1|k) = A\chi_i(k|k) + Bu \tag{7}$$

$$\hat{X}_{k+1|k} = \sum_0^{2L} W_i^{(m)} \chi_i(k+1|k) \tag{8}$$

$$P_{k+1|k} = \sum_{i=0}^{2L+1} W_i^{(c)} \Delta X_{ik+1|k} \Delta X_{ik+1|k}^T + Q_k \tag{9}$$

$$\xi_i(k+1, k) = h(k+1, \chi_i(k+1)) \tag{10}$$

$$\hat{\phi}_{k+1|k} = \sum_0^{2L} W_i^{(m)} \xi_i(k+1|k) \tag{11}$$

$$P_{\phi\phi} = \sum_{i=0}^{2L+1} W_i^{(c)} \Delta\Phi_{ik+1|k} \Delta\Phi_{ik+1|k}^T + R_k \tag{12}$$

$$P_{X\phi} = \sum_{i=0}^{2L+1} W_i^{(c)} \Delta X_{ik+1|k} \Delta\Phi_{ik+1|k}^T \tag{13}$$

$$K_{k+1} = P_{X\phi} P_{X\phi}^{-1} \tag{14}$$

$$X_{k+1} = \hat{X}_{k+1|k} + K_{k+1}(\phi_{k+1} - \hat{\phi}_{k+1|k}) \tag{15}$$

$$P_{k+1|k+1} = P_{k+1|k} - K_{k+1} P_{\phi\phi} K_{k+1}^T \tag{16}$$

Where,

$$\Delta X_{ik+1|k} = \chi_i(k+1|k) - \hat{X}_{k+1|k} \tag{17}$$

$$\begin{cases} W_0^{(m)} = \dfrac{\lambda}{L+\lambda} & i=0 \\ W_0^{(m)} = \dfrac{\lambda}{L+\lambda} + (1-\alpha^2+\beta) & i=0 \\ W_i^{(m)} = W_i^{(m)} = \dfrac{\lambda}{2(L+\lambda)} & i=1,2,\cdots,2L \end{cases} \quad (18)$$

β incorporates priori knowledge of distribution of the state (the optimal value is 2); $\xi_i(k+1,k), i=0,1,2,\cdots 2L$ are the sigma points of the one-step predictive measurement $\phi_{k+1|k}$; $\Delta\Phi_{ik+1|k}=\xi_i(k+1|k)$-$\hat{\phi}_{k+1|k}$, Q_k is processing noise variance, R_k is measurement noise variance.

3.2 Optimal maneuver path planning

The variance lower bound of the unbiased estimator is CRLB. As an unbiased estimation method, the mean square error of UKF with control inputs estimator is $P_{k|k}$. According to the statistical lemma [20],

$$\begin{aligned} P_{k|k} &= \mathrm{tr}(\mathrm{Cov}(\hat{X}_{k|k})) + \left\| \mathrm{E}\hat{X}_{k|k} - X_k \right\|^2 \\ &= \mathrm{trCov}(\hat{X}_{k|k}) = \mathrm{tr}(P_{k|k}) \end{aligned} \quad (19)$$

Therefore, $\mathrm{tr}(P_{k|k})$ can be used as optimal maneuvering optimization criteria. The $\mathrm{tr}(\bullet)$ represents the trace of the matrix; $\mathrm{Cov}(\hat{X}_{k|k})$ is the covariance matrix. The problem discussed in this section is turn to solving the following equation:

$$\begin{cases} \min tr(P_{k|k}) = tr(P_{k|k\text{-}1} - K_k P_{\phi\phi} K_k^T) \\ \hat{X}_{k|k} = A\hat{X}_{k\text{-}1|k\text{-}1} + Bu_{k\text{-}1} \end{cases} \quad (20)$$

The solution of the equation (20) is actually a global optimization process. In order to make it easier to be solved and more practical for engineering, this paper presents a single-step optimal method. The method exhausts all the headings of the airborne observation platform at every step with the minimum $\mathrm{tr}(P_{k|k})$ as the evaluation criterion, and obtains the optimal solution of each step. Finally, a local optimal solution is achieved.

The steps of one-step optimal control are described as follows:

1) At time k, the relative position of the platform and target position $\hat{x}_{k|k}$ and $\hat{X}_T$ are estimated via UKF based on the last k-1 bearings of target and observation platform positions.

2) Traverse the N flight directions of the observation platform with a certain step length.

(1) In the flight direction i ($i \in [0, N-1]$) of the observation platform, use the UKF method to estimate possible position of observing platform and target source at next step and get the next possible state error convariance matrix $P_{k|k}$.

(2) Calculate $\mathrm{tr}(P_{k|k})$.

(3) Repeat (1) and (2) until all possible flight directions are traversed, choose the direction with minimum $\mathrm{tr}(P_{k|k})$ for the actual flight at next step.

3.3 Algorithm steps

Assuming the aircraft velocity flies with a constant speed and only the heading can be changed among $\pm 60°$, the emitter position is estimated as follows:
1) Set initial value of the target source position and covariance.
2) Calculate all possible UKF predictions at time k and all predicted $\mathrm{tr}(P_{k|k})$, find the direction with minimum $\mathrm{tr}(P_{k|k})$ as the next direction of flight.
3) Update u , the control input of UKF. Where, $u = [|u|\cos(\theta), |u|\sin(\theta)]$, θ is the angle between the direction of the aircraft and the x-axis.
4) Repeat with 2), 3) until the end of positioning.

4 Simulation and analysis

Experiment One: Optimal path planning validation. The simulation environment is set as: the position of target source is (100km, 100km), the initial position of the aircraft is (0km, 0km), its velocity vector is (200 m/s, 0 m/s). The observation interval is 1 sec; the bearing error is Gaussian white noise with the standard deviation of 1 degree. The heading change in each step in the optimal maneuvering strategy is 5 degrees, and the number of observations is 3000 times. The results of the trajectory simulation are shown in Fig. 2. Fig. 2 (a) shows the motion trajectory of the UKF with input controls and optimal maneuvering strategy. Fig. 2 (b) is the motion trajectory which does not take the optimal maneuvering strategy into account. The blue curve is the trajectory of the airborne observation platform. The "+" is the position estimation of the target and the red "O" is the target real position.

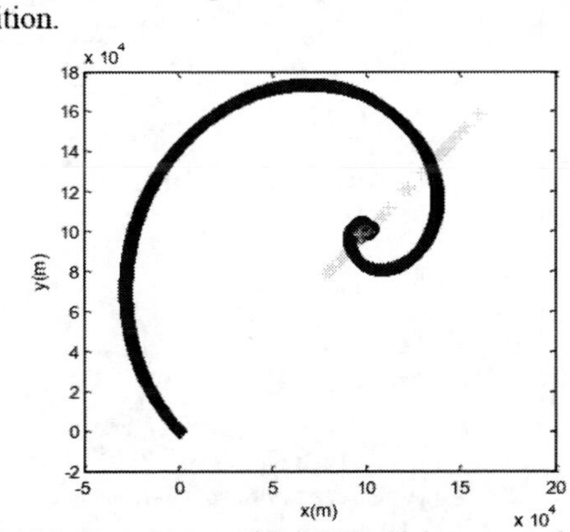

(a) Motion trajectory of the UKF with input controls and optimal maneuvering strategy

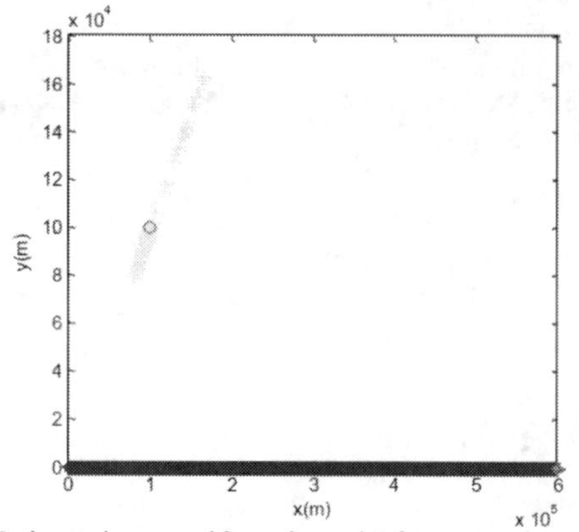

(b) Motion trajectory without the optimal maneuvering strategy.

Figure 2. Trajectories comparison

The comparison shows that under the influence of the optimal control algorithm, the trajectory of the observation platform is no longer a straight line, but a spiral motion. When the observation platform performs the optimal maneuver selection in each step, the heading of airborne platform has been changed. This results in changes of control inputs to ensure the observation platform observability and achieves the localization error convergence.

Experiment Two: Comparison of algorithms performance. The simulation parameters are same as Experiment One, except the number of observations is set 200. The number of Monte Carlo experiments is set to 50. Fig. 3 shows the simulation results of the UKF algorithm both with and with no the optimal maneuver. The abscissa represents the number of observation while the ordinate indicates the error. In this paper, the relative localization error is used as the evaluation criterion of the algorithm precision, and the convergence threshold is set as 5%. The relative error is defined as

$$r = \frac{\left\| \hat{X}_{k|k} - X_k \right\|}{\left\| X_T - X_0 \right\|} \times 100\% \qquad (21)$$

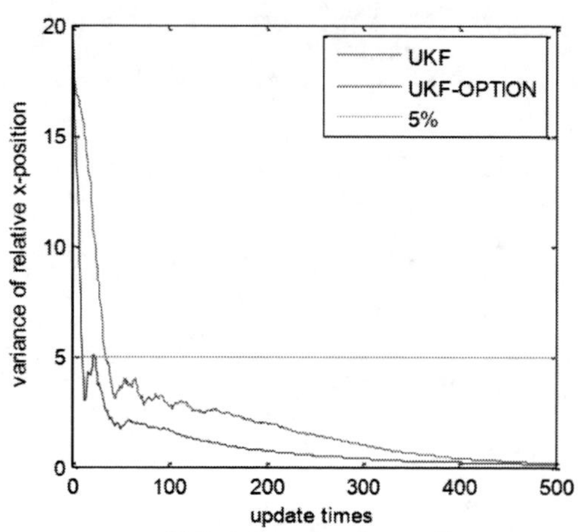

(a) Relative error of x-coordinate

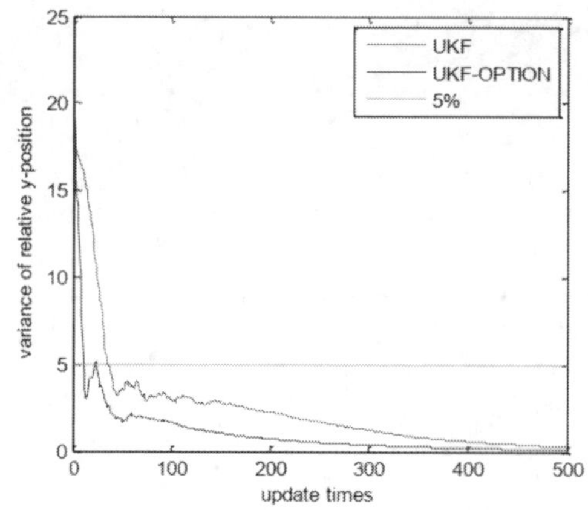

(b) Relative error of y-coordinate

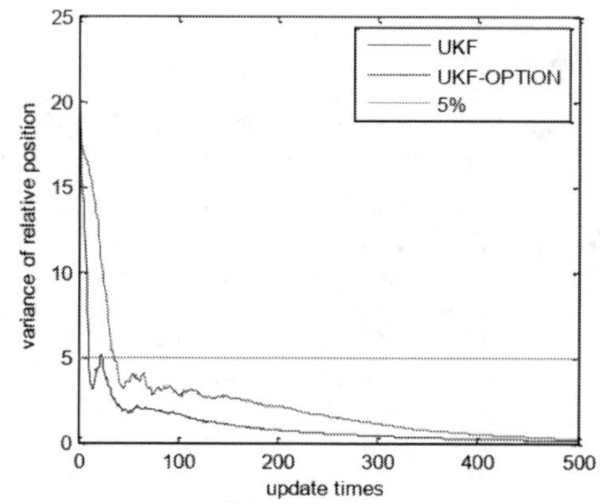

(c) Relative mean square error

Figure 3. Algorithm performance comparison of the UKF with and with no optimal maneuver

From the comparison among (a), (b) and (c) in Fig. 3, it can be concluded that with the same iteration times, the error of UKF algorithm with the optimal maneuver is smaller than the UKF algorithm only. It proves that the optimal maneuvering algorithm improves the localization accuracy. Moreover, after processing with the algorithm base on UKF and optimal maneuver, the relative error curve becomes smoother and can reach below the convergence threshold (5%) in less iterations. It demonstrates that the optimal maneuver improves both the stability and convergence of the UKF localization algorithm.

5 Conclusion

The UKF algorithm with control inputs is suitable for solving the problem in bearing-only airborne passive single observer localization solution. In the UKF method, the control input has a certain influence on the convergence speed and localization accuracy which can be improved by optimizing control inputs. In this paper, a single station localization algorithm based on UKF with optimal inputs is proposed, where the minimum trace of state estimation covariance is taken as the criterion to optimize the maneuvering direction. The simulation experiments verify that the algorithm improves both the stability and convergence. The paper has built the bearings-only mathematic model under the condition that the target is fixed or moving slowly. If the target is also maneuvering, the optimal maneuvering of the observer should be further studied. Simultaneously, the calculation cost of the UKF algorithm is large. Other filtering methods to reduce the computational load remain to be further studied, too.

References

1. Y. Shan, Z. Sun, K. Huang. Asrpspace Electronics Warfare,**6**,4 (2001)
2. O. N. Gharehshiran,V. Krishamurthy.IEEE T SIGNAL PROCES,58(2010).
3. S. Chen, H. Wu, H. Zhang. *Computer Science & Service System*,(Nanjing, China,2012).
4. W. Su, G.Wang,P. Zhong, J. Xiu. Journal of Naval Aeronautical and Astronautical University,**24**(2009)
5. M.J. Moorman, T.E. Bullock. *Aerospace and Electronics Conference* (NAECON ,1992)
6. S. Julier, J. Uhlmann, H. F. Durrant-Whyte. IEEE T AUTOMAT CONTR, **45**(2000).
7. N. J. Gordon, D.J Salmond, A.F.M. Smith. IEE Proceedings-F, **140**(1993)
8. C. Liu. *Passive location and tracking* (Xidian University Press,Xian,China,2011)
9. Z. Xu, A. Sheng, Z. Guo.*Proceedings of IEEE International Conference on Networking, Sensing and Control* (Sanya, China, 2008).
10. P. T. Liu. An optimum approach in target tracking with bearings measurements.Journal of Optimization Theory and Applications,1987, 56(2): 205 – 214.
11. Y. Oshman, P. Davidson. IEEE T AERO ELEC SYS, **35**(1999)
12. J M Passerieux , D Van Cappel. IEEE T AERO ELEC SYS, **34**(1998)
13. J P Helferty,D R Mudgett. *Proceedings of the 32nd Conference on Decision and Control* (San Antonlo, Texas, 1993).
14. W Zhang, M Liu, Z Wang etal. *Second International Conference on Information and Computing Science* ,(2009)
15. P.Tichavsky, C. H. Muravchik, A. Nehorai. IEEE T SIGNAL PROCES, **46**(1998)
16. S. C. Nardone, V. J. Aidala. IEEE T AERO ELEC SYS,**17** (1981)
17. Y Zhang. Fire Control Radar Technology, **33**, 5 (2004)
18. Z Dai, X Deng. Aerospace Electronic Warfare, **4** (2007).
19. X Deng, Y Zhou, Q Lu. Acta Electronica Sinica,**4** (2007)
20. G Wang, M Chen, L Chen.*Linear Statistical Model* (Higher Education Press, Beijing China,1999).

Construction and application research of knowledge graph in aviation risk field

Qian Zhao[1], Qing Li[1] and Jingqian Wen[2]

[1] *School of Mechanical Engineering and Automation, Beihang University, Beijing 100191, China*
[2] *School of Mechanical Engineering, Beijing Institute of Technology, Beijing 100081, China*

Abstract. Since the causes of aviation accidents and risks are complicated, concealed, unpredictable and difficult to be investigated, in order to achieve the efficient organization and knowledge sharing of the historical cases of aviation risk events, this paper put forward the method of constructing vertical knowledge graph for aviation risk field. Firstly, the data-driven incremental construction technology is used to build aviation risk event ontology model. Secondly, the pattern-based knowledge mapping mechanism, which transform structured data into RDF (Resource Description Framework) data for storage, is proposed. And then the application, update and maintenance of the knowledge graph are described. Finally, knowledge graph construction system in aviation risk field is developed; and the data from American Aviation Safety Reporting System (ASRS) is used as an example to verify the rationality and validity of the knowledge graph construction method. Practice has proved that the construction of knowledge graph has a guiding significance for the case information organization and sharing on the field of aviation risk.

1 Introduction

With the rapid development of the air transport industry, the modern society's demands for the aviation safety have been continually improved [1]. The contingency and huge losses of aviation accidents have made people pay close attention to the analysis and summary of historical cases. In the 1970s, nearly ten countries and regions, led by the United States, established the Aviation Safety Reporting System [2] to record air risks and accident cases in the form of reports. The data of reports are from pilot, air traffic controller, dispatcher, cabin crew, maintenance technicians and others. ASRS encodes case reports from the aspects of aircraft, environment and incident evaluation, and store the valuable data in relational database to realize the structured organization and persistent storage of the aviation risk and accident case information. As the world's largest information sources in aviation safety, ASRS in America has formed more than 1.3 million reports after nearly 40 years of accumulation. Especially in recent years, up to an average of 1,774 per week rapid growth.

However, the flat storage form based on relational database can't intuitively show the complex network relations between aviation cases, and also the keywords-based retrieval method is not easy for us to query and use the multi-dimensional information, so that the historical cases accumulated in ASRS system can't be well utilized, and the knowledge contained is difficult to be effectively explored. That is why there are many historical cases, but failed to be fully used to avoid similar incidents. We take the ice accident as an example, an air police-200 early warning aircraft from the Chinese Air Force, which is modified by the Yun-8AEW, crashed in the eastern part of Anhui province on June 3, 2006 due to aircraft wing icing [3]. The situation exists a similarity with the US Eagle Air Flight 4184 air crash [4] on October 31, 1994. This shows that ASRS system has laid a good foundation for information collection and sharing to avoid aviation risks and analyse aviation accidents. While how to make more effective use of the large number of historical cases accumulated by ASRS, how to gain experience and lessons from it and how to improve security rules to provide auxiliary analysis support to avoid similar risks have been the key issue which need to be solved urgently.

2 Related work and technology

As an effective tool for describing massive knowledge entities, attributes and relationships between them, knowledge graph has been widely used by academics and industries since its successful application to search engines by Google in 2012 [5]. The knowledge graph can be divided into general field and vertical field according to its knowledge range. General field knowledge graph is characterized by large amount of data, strong versatility and wide range, but vertical field knowledge graph has the characteristics of relatively small amount of data, high quality, high density, high concentration, high authority and high knowledge accuracy [6].

At present, the research of general knowledge graph at home and abroad are developing rapidly. There are

Probase [7], DBpedia [8], Freebase [9], Baidu Zhixin, Sogou knowCube, CN-DBPedia [10] and so on. Foreign vertical field knowledge graph involved the GeoNames in geographical field, DBLife [11] in academic field, UniProtKb [12] in biological field and so on, but the domestic vertical field knowledge graph is limited to academic research. For example, Ruan Tong et al. [13] proposed a data-driven incremental vertical knowledge graph construction method; Li Wenpeng et al. [14] proposed a software knowledge graph construction method for open source software projects; Ge Bin et al. [15] proposed a method and a computational framework for military knowledge graph construction.

In the field of aviation risk, the researchers mostly analyse the impact of a factor on aviation accidents, or use an algorithmic theory to analyse aspects of aviation risk [16, 17]. However, aviation risk and accidents are caused by results of multiple factors, so we need to consider the mutually influences of aircraft, environment, human factors and other aspects in the process of case analysis. Knowledge graph presents a great advantage in building a knowledge network. It provides a new means of acquiring, storing, organizing, managing, updating and displaying [18] for knowledge which contains complex relationships and it also can supply a more knowledgeable approach to cognitive habits.

3 Construction of aviation risk knowledge graph

As a complex event, aviation accident may be influenced by the aircraft performance, crew contingency, bad weather, geographical conditions and other factors. The various influencing factors and the event itself constitute a complex network relationship for each aviation risk event, and the same event set make up a huge knowledge network system, so two-dimensional relationship table can't show the complex relationships at all. Knowledge graph, as a direct representation of the relationship, is more conducive to our effective analysis of the specific potential risks existing in those complex relationships.

In order to fully organize and share the aviation risk knowledge, this paper combines the characteristics of aviation risk domain and constructs the aviation risk knowledge graph, the graph means a knowledge semantic network which is used to describe the whole event, it is composed of entities that make up the aviation risk event and the associations between these knowledge entities. Among them, the knowledge entity refers to the fact that is distinguishable, identifiable and owing some definite semantic meanings in the aviation risk event. The associations between knowledge entities refer to a certain type of binary relationship between knowledge entities.

The construction process of aviation risk knowledge graph includes domain ontology modeling, instance-to-ontology mapping, visualization analysis and application maintenance. Firstly, establish the ontology model of aviation risk event by combing the top-down and bottom-up methods, which is used as a data model to define the knowledge graph and describe the concepts and relationships between them. Secondly, transform the relational data of ASRS into knowledge entities, attributes and relationships in the knowledge graph by using the pattern-based knowledge mapping mechanism. Finally, carry out the analysis of data relations, and achieve the visualization and maintenance of knowledge graph. The construction process is shown in Figure 1.

Based on the above mentioned construction process, the main contents of this paper are divided into three parts: representation of knowledge, data mapping, management and application of knowledge graph.

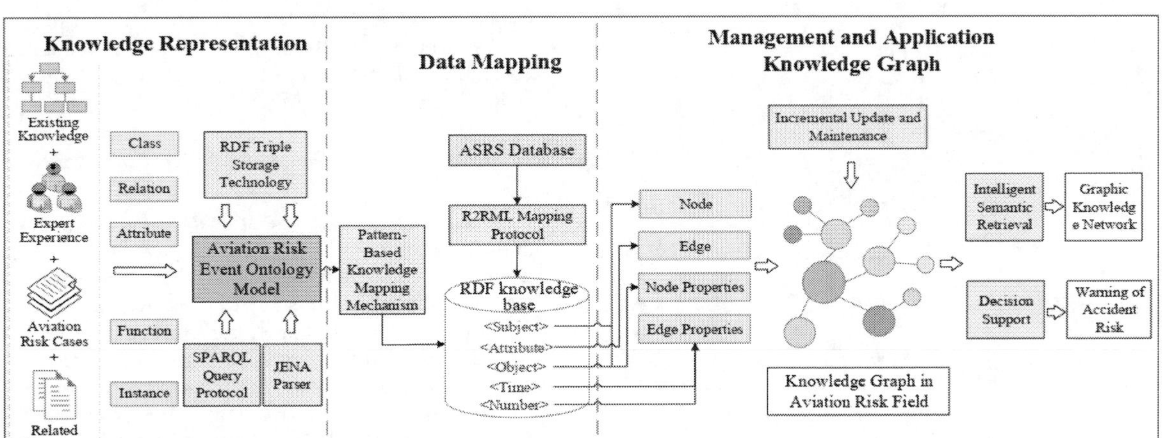

Figure 1. Overall structure diagram of the knowledge graph construction in aviation risk domain.

3.1 Knowledge Representation

Knowledge in aviation risk cases has rich hierarchical structure and complex logical relationships. The premise of knowledge graph construction in aviation risk domain is to classify, summarize and standardize the knowledge, and to construct an effective and extensible knowledge model. The knowledge representation method based on ontology [19] which works as a clear and normative field concepts, the level and logical relationships between concepts, the attributes and constraints of concepts, etc. The method can guarantee uniqueness and no ambiguity of understanding in the transfer and sharing process of knowledge, so as to become the important technical method for product domain knowledge modeling.

Because of the complexity of aircraft structure and the diversity of aviation risk events, the existing aviation accident knowledge base is not really suitable in China.

Thus this paper focuses on aviation risk case and builds the aviation risk knowledge ontology framework. It use data-driven incremental ontology modeling method to further extend the concepts and relationships of the module details. First, the aviation risk event ontology model is defined as O = <C, R, A, I, F> quintuple, the meaning of each letter in the fomular is as shown in Table 1.

Table 1. Meaning of ontology model definition.

Letter	Meaning of letter
O	Ontology - Refers to unified and standardized expression of the aviation risk knowledge.
C	Class - Refers to a collection of entities.
R	Relation - Refers to a logical relationship between concepts, including composition, inheritance and other business relationships.
A	Attribute - Refers to the concept (or instance) attribute relationship.
I	Instance - refers to the basic elements or the specific entity of the class.
F	Function - A function or axiom used to represent an association or constraint that exists between relationships or functions.

According to the above definition, the knowledge of aviation risk cases is systematically classified and organized to form a network of knowledge structure. Each concept contains one or more instances. The instance inherits all attributes of the concept. There is cross concepts or association relationships between instances. Figure 2 shows a partial of the aviation risk event ontology model.

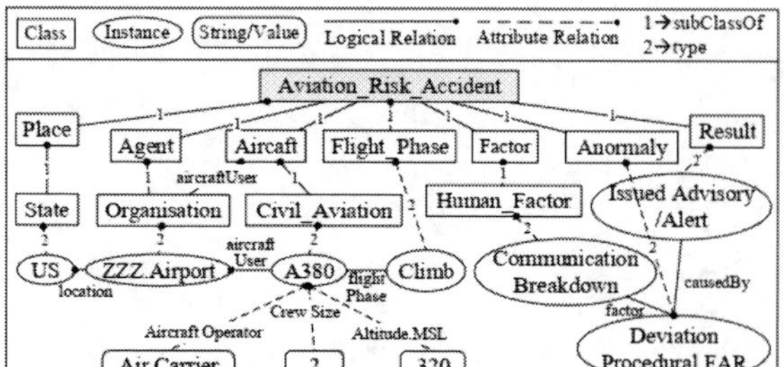

Figure 2. Example of partial aviation risk event ontology model.

After establishing the ontology model, the next step is to store the knowledge in the database. In this paper, we use Jena, a ontology parsing java toolkit, to transform the ontology metadata into the resource description framework RDF [20], then store and query knowledge in the form of <Subject-Attribute-Object>.

3.2 Data Mapping

The knowledge source of this paper is the structural data of the American Aviation Safety Reporting System (ASRS), whose data have high quality, high reliability and very analytical value. This paper stores the relevant data collected in the relational database and uses the pattern-based data mapping mechanism to complete the transformation structured knowledge, that is, RDB2RDF data conversion process. W3C introduced two mapping language standards in 2012 [21]: Direct and R2RML, where Direct Mapping uses the mapping mechanism that directly output the relational database table structure to RDF graph, but R2RML achieve the transformation through a custom vocabulary table, the latter approach is more customizable and more flexible. Therefore, based on the pre-built aviation risk event ontology model, this paper uses the custom R2RML mapping language to build up pattern-based data mapping.

R2RML mapping [22] design a logical table, it retrieve data from the relational database. We will define a SQL query of table in relational database as a logical table. Each logical table is converted into RDF data by a triples map, that is, each row of instance data in the logical table is mapped to several RDF triples. R2RML mapping mechanism expression is:

$$triplesMap = \{logicalMap, subMap, preobjMAP\} \quad (1)$$

Three elements in the formula are: (a) logicalMap: mainly describe the name of database table. (b) subMap: the common subject of all the RDF triples corresponding to one logical row, is used to generate the subject of RDF triples. (c) preobjMap: each mapping consists of a predicateMap and an objectMap or a valueMap, which is used to make the predicate and object of RDF triples.

The whole R2RML mappings of a relational database constitute a mapping document, which consists of a series of RDF triples. It can be described as:

$$mapdoc = \{tripleMap1, tripleMap2, K\ tripleMapN\} \quad (2)$$

According to logical relationships between concepts in the aviation risk ontology model, the R2RML mapping document is written based on this expression. The mapping principle is illustrated by Figure 3. It contains three parts, the the upper part means the relational database table (wing_icing1 table) and the corresponding attribute columns and tuple row data. The middle part is the definition of logical relations. The lower part is the RDF data generated by the mapping, expressed as a triple.

On the basis of this mapping, this paper adds <Time> (record the time of event occurrence to facilitate the timing analysis) and < Number > (record number of occurrence to make it easy for data analysis) in each triple. Finally the data storage pattern becomes the form of <Subject-Attribute-Object-Time-Number> quintuple.

Input <Relational database tables>

ACN	*Organization*	*Aircraft*	*Flight Phase*	*Anormaly*	*Handling*
873164	ZZZ.Airport	A380	Final Approach	Deviation-Procedural Published Material/Policy	Air Traffic Control Separated Traffic
955328	SFO.Tower	A350	Taxi	Inflight Event	Air Traffic Control Issued New Clearance

R2RML mapping document

```
@prefix map: <#>.
@prefix rr: <http://www.w3.org/ns/r2rml#>.
@prefix rdf: <http://www.w3.org/1999/02/22-rdf-syntax-ns#>
@prefix rdfs: <http://www.w3.org/2000/01/rdf-schema#>.
@prefix owl: <http://www.w3.org/2002/07/owl#>.
@prefix base:<http://www.owl-ontologies.com/2017/ASRS.owl#:
map:handle
   rr:logicalTable [ rr:tableName '"wing_icing1"'; ];
   rr:subjectMap[
     rr:template 'base:{"Anormaly"}';
     rr:class base:Anormaly;
   ];
   rr:predicateObjectMap [
     rr:predicate base:handlingMethod;
     rr:objectMap [ rr:column '{"Handling"}' ]
   ].
```

```
map:aircraftloc
   rr:logicalTable [ rr:tableName '"wing_icing1"'; ];
   rr:subjectMap[
     rr:template 'base:{"Aircraft"}';
     rr:class base:Aircraft;
   ];
   rr:predicateObjectMap [
     rr:predicate base:aircaftUser;
     rr:objectMap [ rr:column '{"Organization"}' ]
   ];
   rr:predicateObjectMap [
     rr:predicate base:flightPhase;
     rr:objectMap [ rr:column '{"Flight Phase"}' ]
   ].
```

Output <RDF Triple>

base: A380	base: institution	base: ZZZ.Airport
base: A380	base: flightPhase	base: Final Approach
base: Deviation-Procedural Published Material/Policy	base: handlingMethod	base: Air Traffic Control Separated Traffic

Figure 3. Partial R2RML mapping schematic of entities

3.3 Management and application of knowledge graph

The essence of knowledge graph is a semantic network of entities and relationships between them. The ontology modeling in 3.1 defines the basic data pattern for knowledge graph. 3.2 greatly enriches the relationships between entities by realizing the mapping from structured data to RDF. So the knowledge are finally stored in RDF triples and quintuple, respectively.

The RDF storage pattern includes two types of tuple from the ontology mapping and data mapping. We extract the <Subject> and <Object> as entities in the knowledge graph, and extract <Attribute>, <Time>, and <Number> as the attribute and the association in the knowledge graph, so we can construct the knowledge graph.

In the construction of general field knowledge graph, there is a need for knowledge fusion, that is entity links and knowledge merging to eliminate the concept ambiguity and misconceptions. In contrast with it, the knowledge graph construction in the aviation risk field does not require subsequent processing, because the data source is pre-processed structured data and the manually constructed ontology model also ensures the high quality and uniqueness of the knowledge. With the increasing number of data, this paper proposes a data-driven incremental ontology modeling technology, it expand concepts and instances on the basis of the original ontology. Concept refers to a class with the same entities, the changes of concepts and attributes will not be particularly frequent, so we mainly pay more attention to the automatic update of instances. The richness of relationships between entities depends on the pattern-based data mapping mechanism. Due to the consistency of the database table structure, the pattern rules can be reused without frequent modification after being defined.

The application of knowledge graph in aviation risk field is mainly in two aspects: (1) intelligent semantic retrieval. The knowledge graph makes the large number of hidden knowledge and associated data to be materialized and explicit, it organizes and stores all concepts, entities, attributes and relationships in the form of graph structure. We analyze the user query statement by using knowledge graph, get the keywords and map it to one concept or a group of concepts of knowledge graph. Then show users the graphic knowledge network about keywords. (2) decision support. It is critical to use the historical experience data on aviation risk events efficiently to obtain valuable information. In contrast to the flattened information, the graphic structure of knowledge will focus on the effective organization and association of the entire event at the semantic level, that is more direct to guide personnel and more conducive to follow-up knowledge reasoning, thus supply deeper level of decision support for the enterprise or government.

4 Implementation of aviation risk knowledge map

Based on the construction method of aviation risk field knowledge graph proposed in this paper, the effectiveness of the method is realized and verified by concrete practice.

In the first place, the model of the aviation risk event ontology is established by using the ontology construction tool Protégé. The ontology model regulates the data pattern of the knowledge graph, which contains 230 concepts, 65 logical relations between concepts and 85 attribute relations between concepts.

Secondly, obtain structured data from the [ASRS Database Online] in ASRS. Use "icing" event as an example to illustrate the data acquisition method. We input "icing" in the field of [Narrative] and [Synopsis], and click [Run Search] button, then we can get a total of 3028 cases. Then export the results to excel tables and import them into Oracle database as the data source of the knowledge graph. In the end, based on the rules defined in the manually defined R2RML mapping document, use the open source r2rml-parser tool to implement the mapping from relational data to RDF quintuple.

Finally, develop the aviation risk field knowledge graph construction system by using the java development tool Eclipse IDE and the show middleware Dorado7. The system consists of two parts: knowledge graph visualization module and knowledge graph update module. The knowledge graph visualization module uses ontology-based RDF query and Gojs foreground visualization javascript plugin to model the knowledge graph, and it can achieve the classification modeling for key concepts. In the meantime, it use ontology-based word segmentation technology and SPARQL-based semantic retrieval technique to realize intelligent search function of knowledge graph for natural query statement. The concrete implementation of this module is shown in Figure 4. The user enters the query statement in the search box, the system will pick out the keywords by using the word segmentation technique based on the ontology dictionary, the keywords refer to key knowledge entities with domain significance in the query statement. Then the system return to the user with graphical knowledge graph related to keywords. For example, the input word is "Icing event occurred in Washington DC ", the system automatically segments the sentence and gets keywords as "Icing", "event" and " Washington DC ", visualize the entities related to these two key words in the form of knowledge graph and display personnel, hardware facilities, environment and influencing factors of the ice accident in Washington DC. Known from the analysis of figure 4, three top airports in Washington DC which exist icing accidents are: DCA Airport, IAD Airport and ZDC Airport and the main reason leading to the icing accident in Washington DC is aircraft equipment failure. So as to remind relevant maintenance technicians that they must be responsible for equipment maintenance, test and security work to avoid the occurrence of similar incidents before the flight departure.

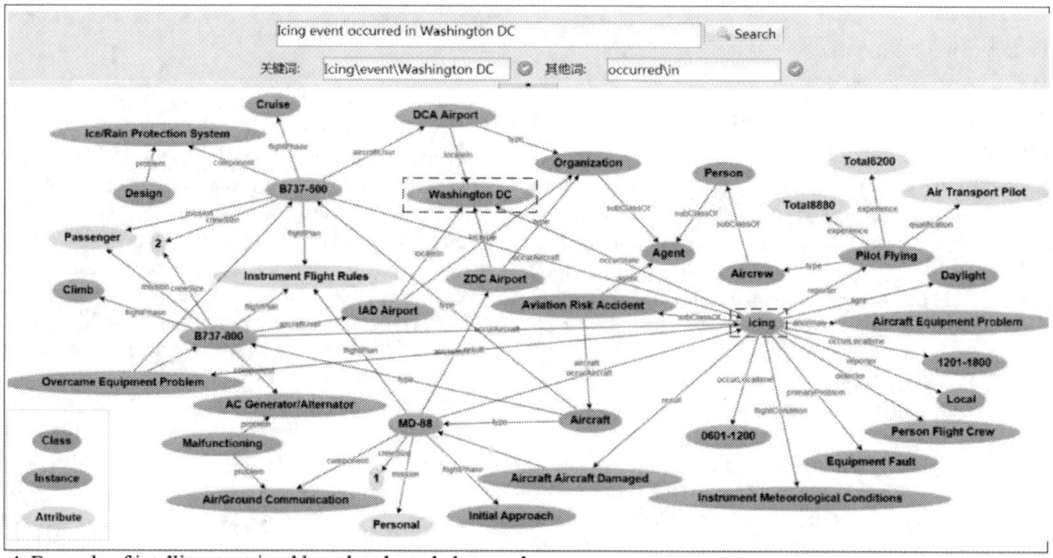

Figure 4. Example of intelligent retrieval based on knowledge graph.

The knowledge graph update module is to realize the update and maintenance of ontology concepts and conceptual attributes. This module includes the ontology hierarchy tree management, ontology file import and export management, ontology data statistics, ontology knowledge editing and knowledge attribute management.

5 Conclusion

Aim to organize and share aviation risk history cases effectively, this paper put forward the knowledge graph construction method in the field of aviation risk, constructed the aviation risk ontology model, proposed the knowledge acquisition method based on structured data, realized the mapping mechanism from the relation data to ontology, developed an aviation risk knowledge map construction system and verified the effectiveness and rationality of the method by certain cases. The application of aviation risk field knowledge graph can effectively improve the efficiency of knowledge retrieval, improve the way of knowledge organization, and help users to make decision through knowledge relationships.

It also has high application value in the knowledge mining and knowledge sharing.

References

1. J. Liu. Civil Aviation University of China. 7(2007)
2. J. Tian. Civil Aviation University of China. 10(2010)
3. J. Zhang. China Air Force 6·3 air crash: airplane ice caused the plane crash dead[DB/OL]
4. Wikipedia, the US Airways flight 4184 air crash[DB/OL].https://en.wikipedia.org/wiki/American_Eagle_Flight_4184. 2017.06.20
5. F. Hu. East China University Of Science And Technology. 2(2015)
6. T. Li, C. Wang, H. Li. Journal of Nanjing University of Science and Technology. **41**, 30 (2017)
7. W. Wu, H. Li, H. Wang, KQ. Zhu. ACM. 481-492(2012)
8. J. Lehmann. Semantic Web. **6**(2),188(2015)

9. K. Bollacker, C. Evans, P. Paritosh, T. Sturge, J. Taylor. *SIGMOD 2008*. 1247(2008)
10. B. Xu, Y. Xu, J. Liang, C. Xie, B. Liang, W. Cui, Y. Xiao. CN-DBpedia. *Advances in Artificial Intelligence: From Theroy to Practice*. 430(2017)
11. P. Derose, W. Shen, F. Chen,Y. Lee, D. Burdick, A. Doan, R. Ramakrishnan. *CIDR 2007*. 169(2007)
12. U.P. Consortium. Nucleic Acids Research. 37(2009), (Database issue):D190
13. T. Ruan, M. Wang, H. Wang, H. Hu. *KM Forum*. **3**(2016)
14. W. Li, J. Wang, Z. Lin, J. Zhao, Y. Zou, B. Xie. Journal of Frontiers of Computer Science & Technology. **11**(6), 851(2017)
15. B. Ge, Z. Tan, C. Zhang, W. Xiao. Journal of Command and Control. **2**(4), 302(2016)
16. K.P. Das, A.K. Dey. *JSM*. (2016)
17. W.R. Knecht. Accident Analysis and Prevention. **60**(60C), 50(2013)
18. Y. Wu, A. Yin, K. Lin, X. Yu, G. Lai. Journal of Fuzhou University(Natural Science Edition) . **45**(3) 329(2017)
19. M. Zhao, Y. Du, H. She, J. Zhang, H. Wang, Y. Chen. Transactions of the Chinese Society for Agricultural Machinery, **47**(9), 278(2016)
20. L. Xiang. Wuhan University of Science and Technology. 9(2013)
21. Q. Liu, Y. Li, H. Duan, Y. Liu, Z. Qin. Journal of Computer Research and Development. **53**(3), 583(2016)
22. S. Zhou, Z. Xu. "http://www.paper.edu.cn". (2013)

Adaptive Command Filtered Integrated Guidance and Control for Hypersonic Vehicle with Magnitude, Rate and Bandwidth Constraints

Liang Wang, Weihua Zhang, Ke Peng and Donghui Wang

National University of Defense Technology, College of Aeronautics and Astronautics, 410073 Changsha, China

Abstract. This paper proposes a novel integrated guidance and control (IGC) method for hypersonic vehicle in terminal phase. Firstly, the system model is developed with a second order actuator dynamics. Then the back-stepping controller is designed hierarchically with command filters, where the first order command filters are implemented to construct the virtual control input with ideal states predicted by an adaptive estimator, and the nonlinear command filter is designed to produce magnitude, rate and bandwidth limited control surface deflection finally tracked by a terminal sliding mode controller with finite convergence time. Through a series of 6-DOF numerical simulations, it's indicated that the proposed method successfully cancels out the large aerodynamics coefficient uncertainties and disturbances in hypersonic flight under limited control surface deflection. The contribution of this paper lies in the application and determination of nonlinear integrated design of guidance and control system for hypersonic vehicle.

1 Introduction

Hypersonic vehicle has the potential to fulfill the role of fast-speed and long-range aircraft capable to reach anywhere in a few time. The typical configuration is a wedge-shaped lifting body with a scramjet engine, which produces severe coupling of aerodynamics forces. So the hypersonic flight control is always a challenging research area. The guidance and control system needs to deal with large flight envelop, aerodynamics coefficients uncertainties, as well as flight states coupling (see [1, 2]).

The conventional design of the guidance and control system contains two hierarchical loops. In this manner, the inner autopilot loop is designed as a solution to track the angle of attack (AOA) or acceleration command generated by the outer guidance loop. The inner and outer control laws are independently designed based on frequency separating approximation. However, the approximation does not hold in hypersonic flight due to the rapid change of flight states. The conventional design may cause instability and increase the steady error. The integrated approach can eliminate the time lag existing inevitably in the conventional design, and exploit fully the synergistic relationships between guidance and control system. On one hand, the design process and cost can be significantly reduced. On the other hand, the performance of the flight control system is potential to improve. In [3], the term "integrated" means that the nonlinear flight dynamics is built without states separation and the full-state feedback generates directly the fin deflections. In this case, the guidance law and autopilot control law is designed together. The term also means the guidance law is independently designed with full-state feedback. In [4], it was shown that the performances of both two full-state feedback designs were superior to separated design. In [5, 6], a two-loop controller structure of partial integrated guidance and control is proposed. The outer loop is designed to generate the pitch angle rate command not AOA or acceleration command, and the inner loop tracks the outer loop command.

For cascade nonlinear integrated guidance and control, it's very useful to recursively design indirect (or pseudo, virtual) control input for lower dimension subsystems step by step so as to deal with the un-matched uncertainties and disturbances in different channels. The studies and examples of this back-stepping technique in the case of integrated guidance and control are found in [7, 8]. However the back-stepping controllers suffer from the problem of repeated differentiation of the virtual control input, which results in signal chattering. Hou [7] employ low-pass command filter to avoid directly differentiating to each virtual control input for missile integrated guidance and control system. In [7-11], the employment of command filter in back-stepping control is presented as 'dynamic surface control' (DSC) technique. It is widely extended in hypersonic flight control issues. In [12], the second-order command filters instead of the first-order filters are applied in DSC. With the help of command filters, the performance of back-stepping controller is significantly improved in stability and steady-state tracking accuracy, and the detail analysis is made in [13]. In [14], the stability of command filtered back-stepping control is further improved by composite learning. In [15, 16], a linear second order command filter is designed to impose magnitude and rate limitations on one of the virtual control input. In [17], the global uniform ultimate boundlessness of the tracking errors in DSC is guaranteed in the presence of the input constraints. Deep discussion about control input constraint, finite time convergence in IGC design arises rapidly in [18-20]. In [20], first order auxiliary dynamics is developed to deal with the input constraints.

As opposed to non-adaptive control method, adaptive control allows for performance over a wider flight envelope and uncertain operational environment such as actuator failure or unexpected changes of flight dynamics. In [21], L1 adaptive control architecture (as shown in Figure 2) is introduced, where the command filter $C(s)$ offers great benefit in filtering the plant uncertainties. The features have been verified in a large number of flight tests in recent papers. It is a powerful and low-cost controller in the presence of large uncertainties, un-modeled dynamics, and disturbances. In [22, 23], output feedback L1 adaptive control is designed and applied to missile guidance and control system. It demonstrates robustness to significant changes in the missile dynamics up to $\pm 50\%$ change in the missile longitude model parameters. In [24], the application of L1 adaptive control for hypersonic re-entry is presented. The reference AOA command is generated by a slower outer-loop flight path angle PID controller. Several scenarios including degradation of control surface efficiency, reduced longitudinal stability margins and controller processing delays are simulated without controller reconfiguration to test the robustness. In [25-28], it mainly addresses the application of L1 adaptive control in IGC design for hypersonic glider. A dynamics pole placement controller is implemented as the baseline controller, and it is augmented with an L1 adaptive controller to cancel out the matched and unmatched uncertainties.

Motivated by the reviews and discussions on the integrated guidance and control design, the adaptive command filtered controller is proposed for IGC of hypersonic vehicle in terminal phase. This paper is organized as follows: In Section 2, the nonlinear IGC model is built with uncertain aerodynamics parameters and un-matched time-varying disturbances. In Section 3, the adaptive IGC controller based on command filter and back-stepping technique is introduced, the control law is derived. In Section 4, the numerical simulations of different scenarios are presented. The system dynamics of more general types including time and state dependent nonlinearities are tested. In Section 5, the paper is concluded.

2 Problem Formations

In this section, the nonlinear IGC model is developed. The aerodynamics coefficients are taken from a scaled down version of NASA's CAV (see Appendix A). The planar geometry in terminal phase is placed in Figure 1. The inertial coordinate system OXY is built without considering the earth rotation, OX axis is local horizontal. R is the distance between mass center and landing point. q, α, ω_z are respectively the LOS (Line of Sight) angle, AOA and pitch angle rate. The angle and angle rate directions shown in Figure 1 are positive.

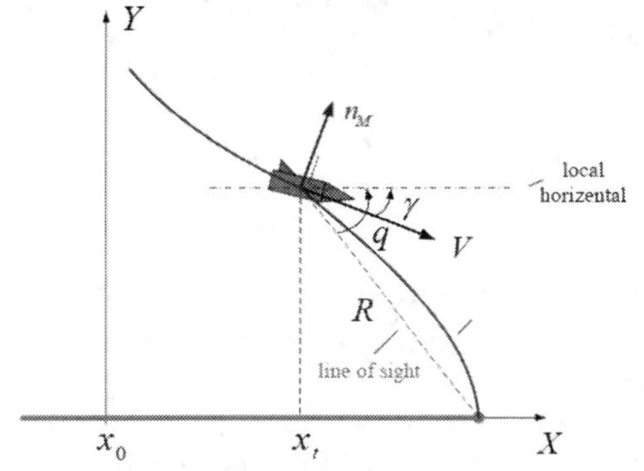

Figure 1. The planar geometry in terminal phase

2.1 Longitudinal Dynamics

The nonlinear longitudinal flight dynamics is presented as follows:

$$\dot{R} = -V\cos(q-\gamma) \tag{1a}$$

$$\dot{q} = \frac{V}{R}\sin(q-\gamma) \tag{1b}$$

$$\dot{\gamma} = \bar{L}_o + \bar{L}_\alpha \alpha - \frac{g}{V}\cos\gamma \tag{1c}$$

$$\dot{\omega}_z = \bar{M}_o + \bar{M}_\delta \delta_z \tag{1d}$$

when the flight velocity is high, the lift force coefficient can be expressed as $C_L = C_L^\alpha \alpha + C_{Lo}$. C_L^α is the lift coefficient derivative with respect to AOA, S_{ref} is the aerodynamics reference size, V is approximated as constant velocity, m, g are constant mass and gravity acceleration. Then the lift consists of two parts: $L = L_\alpha \alpha + L_o$.

$$\bar{L}_\alpha = \frac{L_\alpha}{mV} = \frac{1}{2}\rho V S_{ref} C_L^\alpha,\ \bar{L}_o = \frac{L_o}{mV} = \frac{1}{2}\rho V S_{ref} C_{Lo} \tag{2}$$

where L_α is the main contribution of AOA to lift, and L_o means the other factors contribution to lift such as control surface and body. Then the two parts of lift are unified by momentum.

The aerodynamics moment is also described in two parts: $M_z = M_{\delta_z}\delta_z + M_o$, M_{δ_z} is the moment coefficient derivative with respect to control surface deflection, and M_o means the other factors contribution to moment. Both are unified by the moment of inertia with respect to z-axis I_z .

$$\bar{M}_o = \frac{M_o}{I_z},\ \bar{M}_{\delta_z} = \frac{M_{\delta_z}}{I_z} \tag{3}$$

Next, the actuator dynamics is integrated with the vertical flight dynamics. This paper considers a second order actuator dynamics is given by the following transfer function:

$$G_{act}(s)=\frac{\delta_z(s)}{\delta_{zc}(s)}=\frac{K_{act}}{T_{act}^2 s^2+2T_{act}\xi_{act}s+1} \tag{4}$$

where δ_z is the control surface deflection, and K_{δ_z}, T_{δ_z}, ξ are respectively mean the constant gain, the time constant, the damping rate.

2.2 Integrated Guidance and Control

Differentiating (1b) respect to time, together with (1c) yields:

$$\ddot{q}=-\frac{\dot{R}}{R}\left(2\dot{q}+\frac{g}{V}\cos\gamma-\bar{L}_\alpha\alpha-\bar{L}_o\right) \tag{5}$$

The $\boldsymbol{x}=[x_1,x_2,x_3,x_4,x_5]^T=[\dot{q},\alpha,\omega_z,\delta_z,\dot{\delta}_z]^T$ is chosen as a state vector , and the control input is depicted as $u=\delta_{zc}$. Then the state space model of IGC is built as follows:

$$\begin{aligned}\dot{x}_1&=x_2+f_1+d_1\\ \dot{x}_2&=x_3+f_2+d_2\\ \dot{x}_3&=bx_4+f_3+d_3\\ \dot{x}_4&=x_5\\ \dot{x}_5&=f_{act}+\Delta f_{act}+b_{act}u+d_{act}\end{aligned} \tag{6}$$

where $f_i(t,x)$, $i=1,2,3$ are nonlinear functions with aerodynamics uncertainties, $f_{act}(t,x)$ is a linear function derived by the actuator transfer function, and $\Delta f_{act}(t)$ means its uncertain part. $d_i(t)$, $i=1,2,3$ and $d_{act}(t)$ are time-varying disturbances in different channels.

$$f_1(x)=-\frac{\dot{R}}{R}\left(2x_1+\frac{g}{V}\cos\gamma-\bar{L}_o-\left(\bar{L}_\alpha+\frac{\dot{R}}{R}\right)x_2\right) \tag{7a}$$

$$f_2(x)=\frac{g}{V}\cos\gamma-\bar{L}_\alpha x_2-\bar{L}_o \tag{7b}$$

$$f_3(x)=\bar{M}_\alpha x_2+\bar{M}_\omega x_3,\ b=\bar{M}_\delta \tag{7c}$$

$$f_{act}(x)=-\frac{1}{T_{act}^2}x_4-\frac{2\xi_{act}}{T_{act}}x_5,\ b_{act}=K_{act} \tag{7d}$$

To complete the IGC design, we have two assumptions about the nonlinear functions. Firstly, there exist constants $B>0$, $F>0$, $b_u>b_l>0$ such that $|f_i(t,0)|\le B$, $i=1,2,3$, $|\Delta f_{act}|\le F$, $b\in[b_l,b_u]$ hold for all $t\ge 0$. Secondly, assuming that $f_i(t,x)$, $i=1,2,3$ are uniform bounded and partial derivatives are semi-global uniform bounded. Then the following upper bound is yielded:

$$\|f_i(x,t)\|_{L_\infty}\le L_{\rho_r}\|x\|_{L_\infty}+B \tag{8}$$

where L_{ρ_r} is the bound of partial derivatives of the nonlinear functions $f_i(t,x)$, $i=1,2,3$. Besides, the time-varying disturbances are bounded.

Remark 1: The nonlinear function $f_i(t,x)$, $i=1,2,3$ can be linearly parameterized in two time-varying parameters using $\|(x_1,x_2,x_3)\|_\infty$ as a regress-or on any finite time interval.

$$f_i(t,x)=\theta_i(t)\|x(t)\|_\infty+\sigma_i(t),\ i=1,2,3 \tag{9}$$

where $\theta_i(t)$, $\sigma_i(t), i=1,2,3$ are unknown time-varying functions subject to:

$$\begin{aligned}&|\theta(t)|<\theta_\rho,\ |\dot{\theta}(t)|<d_\theta\\ &|\sigma(t)|<\sigma_b,\ |\dot{\sigma}(t)|<d_\sigma\end{aligned} \tag{10}$$

Then the IGC system can be approximated as follows:

$$\begin{aligned}\dot{x}_1&=\theta_1\|x\|_\infty+\sigma_1+x_2\\ \dot{x}_2&=\theta_2\|x\|_\infty+\sigma_2+x_3\\ \dot{x}_3&=\theta_3\|x\|_\infty+\sigma_3+bx_4\end{aligned} \tag{11}$$

In the terminal phase, the objective of IGC is to design a robust controller $u(t)$ in the presence of uncertainties and disturbances to ensure the system output $y=x_1=\dot{q}$ to converge to zero as fast as possible.

3 Controller Design

In this section, an adaptive command filtered controller is firstly developed based on the approximated model of (11), then a terminal sliding mode controller is designed for the actuator to track the virtual control input derived by the adaptive controller. The fundamental architecture of the adaptive controller is shown in Figure 2. The state predictor specifies the desired closed-loop dynamics, low-pass command filters $C(s)$ are employed to construct three virtual control inputs in the different frequency of uncertainties and disturbances, and the parameters in (11) is estimated by the adaptation.

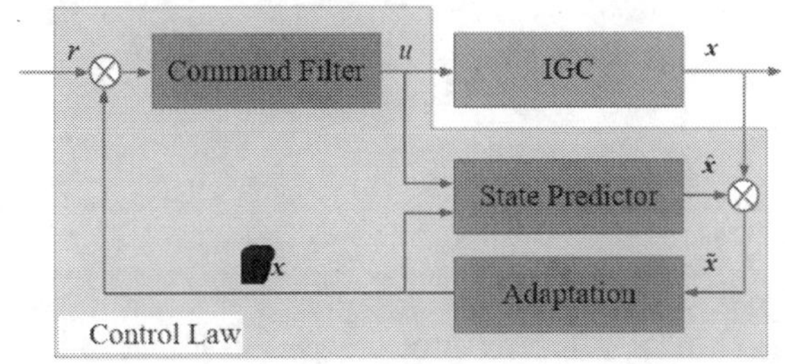

Figure 2. Architecture of the adaptive controller

3.1. Command Filters

In this case, the adaptive controller involves three command filters:

$$\boldsymbol{C}(s)=\begin{bmatrix} C_1(s) & & \\ & C_2(s) & \\ & & C_3(s) \end{bmatrix} \tag{12}$$

The filters have unity DC gains: $C_i(0)=1, i=1,\ 2,\ 3$, and $C_1(s)$, $C_2(s)$ have at least 2 relative degrees.

$$C_i(s)=\frac{k_i D_i(s)}{1+k_i D_i(s)},\ i=1,\ 2$$
$$C_3(s)=\frac{bk_3 D_3(s)}{1+bk_3 D_3(s)} \tag{13}$$

where $k_i>0,\ i=1,\ 2,\ 3$ and $D_i(s),\ i=1,\ 2,\ 3$ are strictly proper transfer functions.

In order to produce a magnitude, rate and bandwidth limited signal $v(t)$, a nonlinear command filter is designed as following:

$$\bar{v}(s)=\bar{G}(s)sat(e_{sat}(s)) \tag{14a}$$

$$\bar{G}(s)=\frac{\omega_{n,\max}^2}{s^2+2\xi_{\max}\omega_{n,\max}s+\omega_{n,\max}^2} \tag{14b}$$

where $e_{sat}=v_{sat}-\bar{v}$ and it is saturated with a maximum value $e_{\max}=\dfrac{2\xi_{\max}rate_{\max}}{\omega_{n,\max}}$. $rate_{\max}$ is the rate limit of actuator and $\xi_{\max}$, $\omega_{n,\max}$ are respectively the maximum values of damping rate and natural frequency in (4). The saturation function in (14a) is defined as following:

$$z_{sat}=sat(z)=\begin{cases} z_{\max}\operatorname{sgn}(z), & |z|\ge z_{\max} \\ z, & |z|<z_{\max} \end{cases} \tag{15}$$

3.2 State Predictor

Based on (11), the following state predictor is considered:

$$\dot{\hat{x}}_1=-a_1\tilde{x}_1+\hat{\theta}_1\|x\|_\infty+\hat{\sigma}_1+x_2$$
$$\dot{\hat{x}}_2=-a_2\tilde{x}_2+\hat{\theta}_2\|x\|_\infty+\hat{\sigma}_2+x_3$$
$$\dot{\hat{x}}_3=-a_3\tilde{x}_3+\hat{\theta}_3\|x\|_\infty+\hat{\sigma}_3+\hat{b}x_4 \tag{16}$$

The predictive states are $\hat{x}_1$, $\hat{x}_2$, $\hat{x}_3$, the prediction error can be depicted as $\tilde{x}_1=\hat{x}_1-x_1$, $\tilde{x}_2=\hat{x}_2-x_2$, $\tilde{x}_3=\hat{x}_3-x_3$. $a_1,\ a_2,\ a_3>0$ are positive constants specifying a desired closed-loop dynamics. Then we have two matrices:

$$\boldsymbol{A}_m=\begin{bmatrix} -a_1 & & \\ & -a_2 & \\ & & -a_3 \end{bmatrix},\quad \boldsymbol{A}_g=\begin{bmatrix} -a_1 & 1 & \\ & -a_2 & 1 \\ & & -a_3 \end{bmatrix} \tag{17}$$

Remark 2: The transfer function of the adaptive closed-loop system is expressed as follows:

$$G(s)=H(s)(\boldsymbol{I}-\boldsymbol{C}(s)),\quad H(s)=(s\boldsymbol{I}-\boldsymbol{A}_g)^{-1} \tag{18}$$

If the filters $\boldsymbol{C}(s)$ verify the following L1-norm upper bound, the adaptive closed-loop system is stable.

$$\|G(s)\|_{L_1}<\frac{\rho_r-P}{K_1\rho_r+K_2} \tag{19}$$

where K_1, K_2, P are constants related to ρ_r and B in (8).

3.3 Adaptation

The parameters $\hat{\theta}_1$, $\hat{\theta}_2$, $\hat{\theta}_3$, $\hat{\sigma}_1$, $\hat{\sigma}_2$, $\hat{\sigma}_3$, $\hat{b}$ in (16) are estimated by the following project law:

$$\dot{\hat{\theta}}_1=\Gamma\operatorname{Pr}oj\left(\hat{\theta}_1,-\tilde{x}^T\boldsymbol{P}[1,0,0]^T\|x\|_\infty\right)$$
$$\dot{\hat{\theta}}_2=\Gamma\operatorname{Pr}oj\left(\hat{\theta}_2,-\tilde{x}^T\boldsymbol{P}[0,1,0]^T\|x\|_\infty\right)$$
$$\dot{\hat{\theta}}_3=\Gamma\operatorname{Pr}oj\left(\hat{\theta}_3,-\tilde{x}^T\boldsymbol{P}[0,0,1]^T\|x\|_\infty\right)$$
$$\dot{\hat{\sigma}}_1=\Gamma\operatorname{Pr}oj\left(\hat{\sigma}_1,-\tilde{x}^T\boldsymbol{P}[1,0,0]^T\right)$$
$$\dot{\hat{\sigma}}_2=\Gamma\operatorname{Pr}oj\left(\hat{\sigma}_2,-\tilde{x}^T\boldsymbol{P}[0,1,0]^T\right)$$
$$\dot{\hat{\sigma}}_3=\Gamma\operatorname{Pr}oj\left(\hat{\sigma}_3,-\tilde{x}^T\boldsymbol{P}[0,0,1]^T\right)$$
$$\dot{\hat{b}}=\Gamma\operatorname{Pr}oj\left(\hat{b},-\tilde{x}^T\boldsymbol{P}[0,0,1]^T x_4\right) \tag{20}$$

where $\Gamma\in\mathrm{R}^+$ is the adaptation gain, the symmetric positive definite matrix $\boldsymbol{P}=\boldsymbol{P}^T>0$ solves the Lyapunov equation $\boldsymbol{A}_m^T\boldsymbol{P}+\boldsymbol{P}\boldsymbol{A}_m=-\boldsymbol{Q}$ for arbitrary $\boldsymbol{Q}=\boldsymbol{Q}^T>0$. The projection operator $\mathrm{Proj}(\ )$ ensures that $|\hat{\theta}_i(t)| \le \theta_b$, $\hat{\sigma}_i \le \Delta,\ i = 1,\ 2,\ 3$ and $\hat{b}(t) \in \Omega$.

3.3 Control Law

The adaptive controller is designed step-by-step for three subsystems. Firstly, the following subsystem with virtual control input $\bar{x}_2$ is considered:

$$\dot{x}_1 = \theta_1 \|x\|_\infty + \sigma_1 + \bar{x}_2 \tag{21}$$

The virtual control input is designed in the following form:

$$\bar{x}_2 = -a_1 x_1 - \hat{\eta}_{1C}(t) \tag{22}$$

where $\hat{\eta}_{1C}(t)$ is verified by the transfer function of $\hat{\eta}_{1C}(s) = C_1(s)\hat{\eta}_1(s)$, and $\hat{\eta}_1(t)$ is depicted as follows:

$$\hat{\eta}_1(t) = \hat{\theta}_1(t)\|x(t)\|_\infty + \hat{\sigma}_1 \tag{23}$$

Secondly, choosing the following subsystem:

$$\dot{x}_2 = \theta_2 \|x\|_\infty + \sigma_2 + \bar{x}_3 \tag{24}$$

then the virtual control input $\bar{x}_3$ has the following transfer function:

$$\bar{x}_3(s) = -k_2 D_2(s)\hat{\eta}_\alpha(s) \tag{25}$$

where $\hat{\eta}_\alpha(s)$ is the Laplace transform of $\hat{\eta}_\alpha(t)$.

$$\hat{\eta}_\alpha(t) = \hat{\eta}_2(t) + a_2(x_2(t) - \bar{x}_2(t)) - \dot{\bar{x}}_2(t) \tag{26a}$$

$$\hat{\eta}_2(t) = \bar{x}_3(t) + \hat{\theta}_2(t)\|x(t)\|_\infty + \hat{\sigma}_2(t) \tag{26b}$$

Thirdly,

$$\dot{x}_3 = \theta_3 \|x\|_\infty + \sigma_3 + b\bar{x}_4 \tag{27}$$

The control input of the adaptive controller is finally depicted as follows:

$$\bar{x}_4(s) = -k_3 D_3(s)\hat{\eta}_u(s) \tag{28}$$

where $\hat{\eta}_u(s)$ is the Laplace transform of $\hat{\eta}_u(t)$.

$$\hat{\eta}_u(t) = \hat{\eta}_3(t) + a_3(x_3(t) - \bar{x}_3(t)) - \dot{\bar{x}}_3(t) \tag{29a}$$

$$\hat{\eta}_3(t) = \hat{b}(t)u(t) + \hat{\theta}_3(t)\|x(t)\|_\infty + \hat{\sigma}_3(t) \tag{29b}$$

Lemma 1 The error dynamics can be depicted as follows:

$$\dot{\tilde{x}}(t) = A_m(t)\tilde{x}(t) + \begin{bmatrix} \tilde{\theta}_1 \|x\|_\infty + \tilde{\sigma}_1 \\ \tilde{\theta}_2 \|x\|_\infty + \tilde{\sigma}_2 \\ \tilde{\theta}_3 \|x\|_\infty + \tilde{\sigma}_3 + \tilde{b}u \end{bmatrix} \tag{30}$$

If $\|x_0\|_\infty < \rho_0$ and the adaptive gain satisfies $\Gamma > \Gamma_{\min}$:

$$\Gamma_{\min} = \frac{12\theta_\rho^2 + 12\sigma_b^2 + (b_u - b_l)^2 + 18\frac{\lambda_{\max}(\boldsymbol{P})}{\lambda_{\min}(\boldsymbol{Q})}(\theta_\rho d_\theta + \sigma_b d_\sigma)}{\lambda_{\min}(\boldsymbol{P})\gamma^2} \tag{31}$$

Then the states error over $t \in [t_0, \tau]$ holds the following bound:

$$\|\tilde{x}_\tau\|_{\mathcal{L}_\infty} < \gamma \tag{32}$$

Next, a terminal sliding mode controller is designed for the second order actuator to track $\bar{x}_4$. To get the second order derivatives of $\bar{x}_4$, a three order nonlinear differentiator is implemented:

$$\begin{aligned} \dot{x}_{thod1} &= x_{thod2} \\ \dot{x}_{thod2} &= x_{thod3} \\ \dot{x}_{thod3} &= \frac{4}{\varepsilon^3}\left(-2^{\frac{5}{3}}\left(|v|^{\frac{1}{3}}\operatorname{sgn}(v)\right) - |\varepsilon^2 x_{thod3}|^{\frac{5}{3}}\operatorname{sgn}(\varepsilon^2 x_{thod3})\right) \end{aligned} \tag{33}$$

where

$$v = x_{thod1} - x_{4u} + \left|\frac{1}{\varepsilon}x_{thod2}\right|^{\frac{9}{7}}\operatorname{sgn}(\varepsilon x_{thod2}),\ x_{thod1}(0) = \bar{x}_4(0).$$

The error between x_{thod1} and $\bar{x}_4$ can be enough small through choosing suitable ε. Then x_{thod2} and x_{thod3} are respectively the first and second order derivatives of $\bar{x}_4$.

Defining the tracking error $e_{act} = x_4 - \bar{x}_4$ and error vector $E = [e_{act}, \dot{e}_{act}]^T$, then the following nonlinear sliding mode surface is designed:

$$\sigma = CE - CP \tag{34}$$

where $C = [c_1,\ c_2]$ is a constant vector, and $P = [p,\ \dot{p}]^T$ is determined by the following nonlinear function:

$$p(t) = \begin{cases} e_{act} + \dot{e}_{act}t + \frac{1}{2}\ddot{e}_{act}t^2 - \left(\frac{10}{T^3}e_{act} + \frac{6}{T^2}\dot{e}_{act} + \frac{3}{2T}\ddot{e}_{act}\right)t^3 + \\ \left(\frac{15}{T^4}e_{act} + \frac{8}{T^3}\dot{e}_{act} + \frac{3}{2T^2}\ddot{e}_{act}\right)t^4 - \\ \left(\frac{6}{T^5}e_{act} + \frac{3}{T^4}\dot{e}_{act} + \frac{1}{2T^3}\ddot{e}_{act}\right)t^5, & 0 \le t \le T \\ 0, & t > T \end{cases} \tag{35}$$

It's obvious that T is the convergence time of this nonlinear sliding mode controller. Finally, the control input of integrated guidance and control is designed as follows:

$$u = -\frac{1}{b_{act}}\left(f_{act}(x) - x_{thod3} - \ddot{p} + \frac{c_1}{c_2}(\dot{e}_{act} - \dot{p})\right) - \frac{1}{b_{act}}\operatorname{sgn}(c_2\sigma)(F + D + K) \tag{36}$$

4 Numerical Simulations

To verify numerically the performance of the proposed method, series of 6 DOF numerical simulation are performed. The aerodynamics coefficients are chosen based on CAV-L (see Appendix A).

4.1 Simulation Parameters

The initial flight states are set as follows:

$$R_0 = 20km,\ V = 1200m/s,\ g = 9.8m/s^2 \\ \gamma_0 = 0\deg,\ \alpha_0 = 0\deg,\ q_0 = -60\deg \tag{37}$$

The actuator dynamics is given by the following parameters:

$$K_{act} = 0.9531,\ T_{act} = 0.0144,\ \xi_{act} = 0.4764 \tag{38}$$

For all scenarios, the adaptive command filtered controller is designed without reconfiguration. The command filters are set as follows:

$$C_1(s) = \frac{3}{s^2 + 1.1s + 3} \tag{39a}$$

$$C_2(s) = \frac{100}{s^2 + 15s + 100} \tag{39b}$$

$$C_3(s) = \frac{10}{s + 10} \tag{39c}$$

The state predictor is confirmed by:

$$a_1 = a_2 = a_3 = 2 \tag{40}$$

$$\boldsymbol{Q} = diag(15,\ 10,\ 1) \tag{41}$$

The projection bound is set as $\Omega = [0.1,\ 3]$, and the gain of projector is $\Gamma = 10^4$. The rate limit is 300deg/s, $\xi_{\max} = 0.7$, $\omega_{n,\max} = 20\,s^{-1}$ and the maximum value of control surface deflection is 30deg. The nonlinear differentiator in (33) is set with $\varepsilon = 0.015$, the terminal sliding mode controller is designed with $T = 0.5s$ and $C = [15,\ 50]$.

4.2 Simulation Scenarios

Three scenarios of different flight dynamics are set to test the proposed controller:

Scenario 1:

$$\bar{L}_\alpha = 1.1,\ \bar{L}_o = -1.5, \\ \bar{M}_o = 0.1x_2 - 0.1x_3,\ \bar{M}_{\delta_z} = 1 \tag{42}$$

Scenario 2:

$$\bar{L}_\alpha = 1.1 - 0.1x_2, \\ \bar{L}_o = -1.5 + 0.1x_2, \\ \bar{M}_o = 0.1x_2 - 0.1x_3,\ \bar{M}_{\delta_z} = 1 \tag{43}$$

Scenario 3:

$$\bar{L}_\alpha = 1.1 - 0.1x_2 - 0.05x_4, \\ \bar{L}_o = -1.5 + 0.1x_2, \\ \bar{M}_o = 0.1x_2 - 0.1x_3,\ \bar{M}_{\delta_z} = 1 \tag{44}$$

Here the aerodynamics moment coefficient is related to both AOA and pitch angle rate. In Scenario 2, the unified lift is determined by nonlinear function related to AOA. In Scenario 3, the coupling of aerodynamics lift and structural dynamics is considered that control surface deflection contributes to aerodynamics coefficient.

The unmatched time-varying disturbances in different channels are set as follows:

$$d_1(t) = 0.005\sin(2t) \\ d_2(t) = 0.005\cos(2t) \\ d_3(t) = 0.005\sin(t)\cos(2t) \\ d_{act}(t) = \sin(t) \tag{45}$$

4.3 Results and Discussion

The numerical simulation results are presented through two comparisons of S1 and S2 in Figure 3-9, S2 and S3 in Figure 10-16. The influences of both nonlinearity of aerodynamics coefficients and states coupling are shown in the simulations. Figure 3 shows the flight profile comparison between S1 and S2, which is comparatively large because of high speed. The curves of S2 are steep. It can be seen that the LOS rate converges to zero fast from Figure 4. Figure 6 indicates that the nonlinearity of lift coefficient have a great effect on AOA. The increase of lift in S2 makes the performance better than S1.

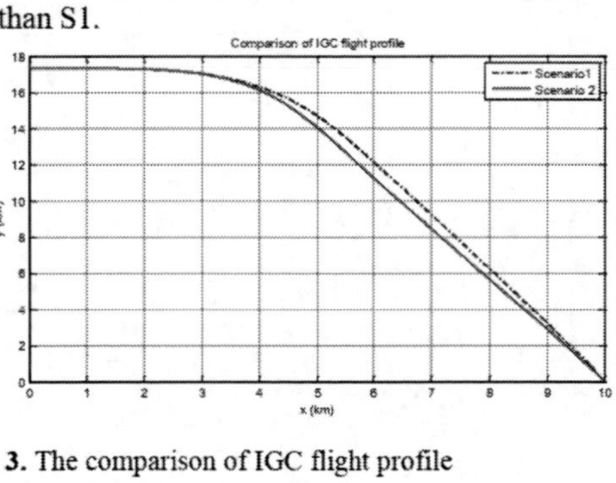

Figure 3. The comparison of IGC flight profile

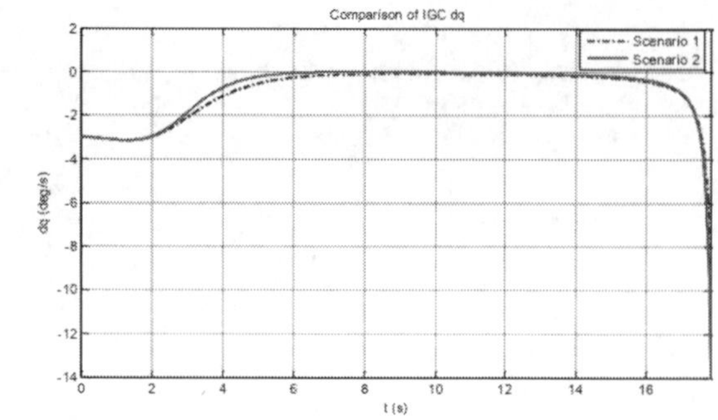

Figure 4. The comparison of $\dot{q}$ between S1 and S2

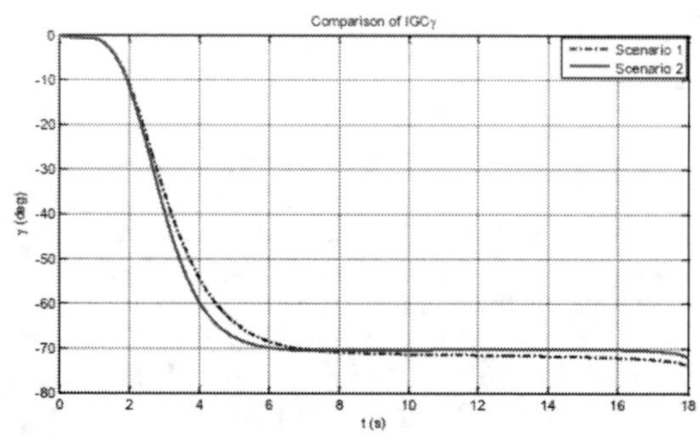

Figure 5. The comparison of γ between S1 and S2

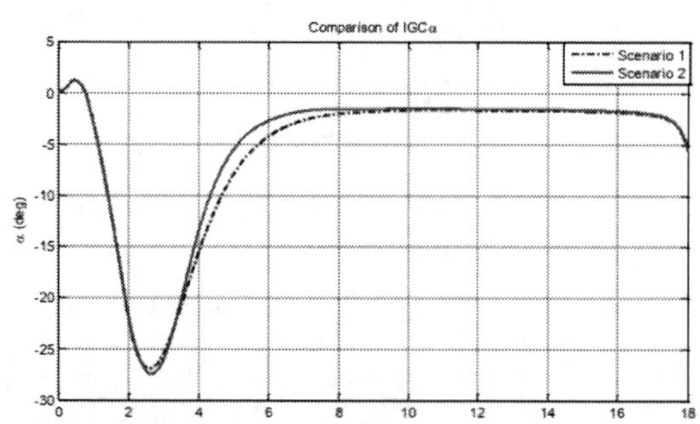

Figure 6. The comparison of α between S1 and S2

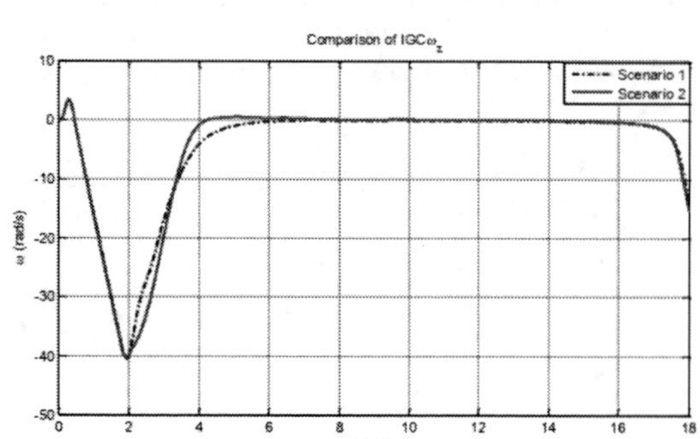

Figure 7. The comparison of ω_z between S1 and S2

Figure 8 and Figure 9 shows the curves of control surface deflection and its rate. The limiting conditions don't result in great chasing, and the proposed controller makes all the states convergence well.

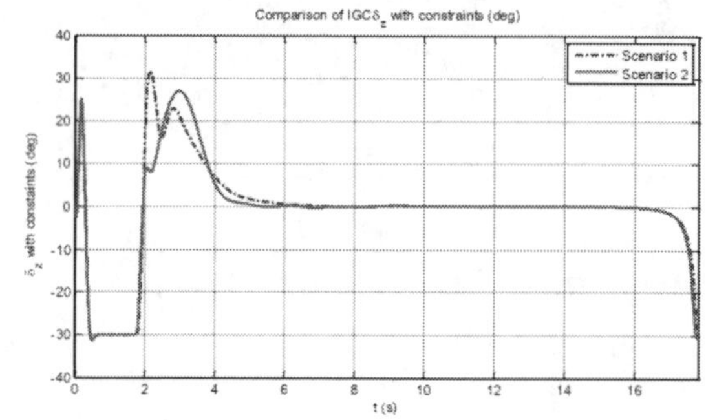

Figure 8. The comparison of δ_z between S1 and S2

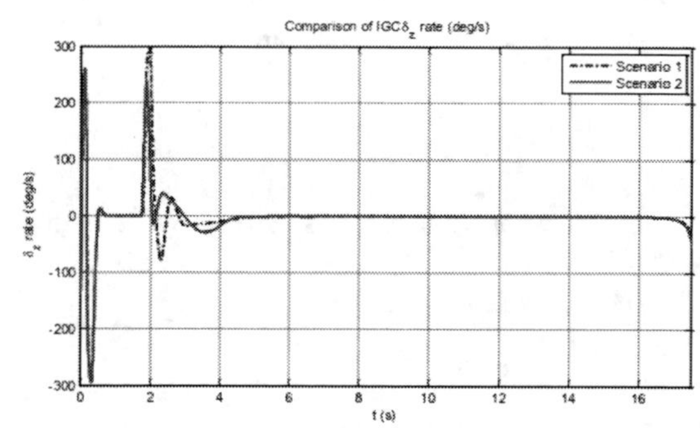

Figure 9. The comparison of IGC δ_z rate

Figure 10-16 show the comparison simulation results between S2 and S3. The difference in flight profile is small from Figure 10.

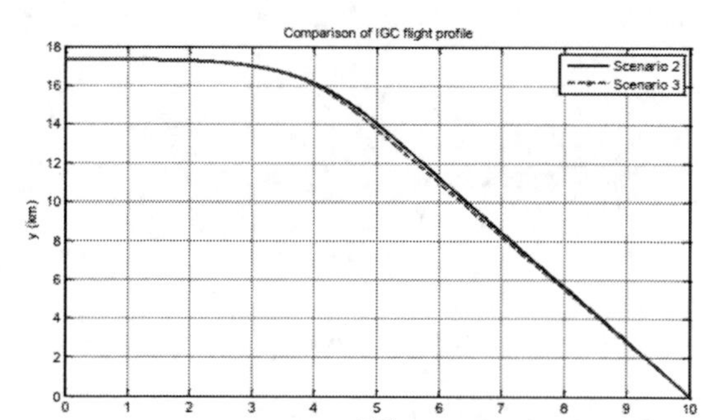

Figure 10. The comparison of IGC flight profile

However the main differences happen in pitch angle rate and control surface deflection. The state coupling between AOA and control surface deflection makes the curves chatter in a small range. All the states finally converges well under limited δ_z.

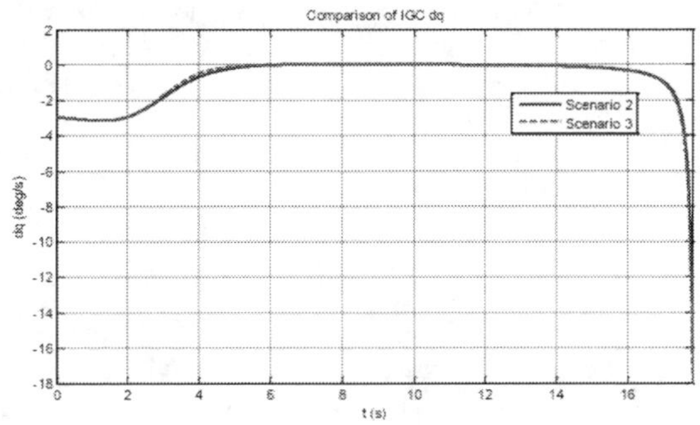

Figure 11. The comparison of $\dot{q}$ between S2 and S3

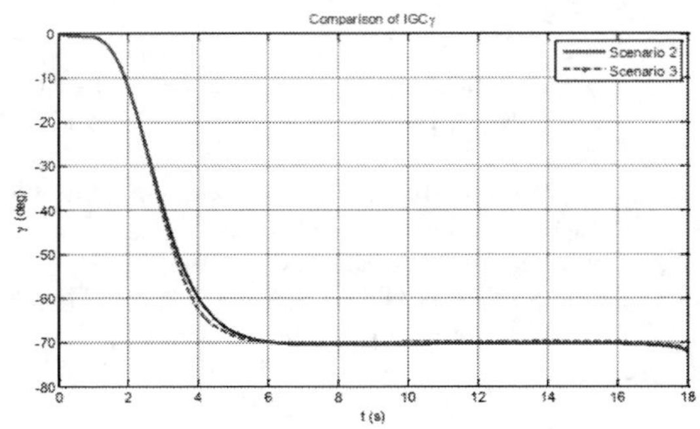

Figure 12. The comparison of γ between S2 and S3

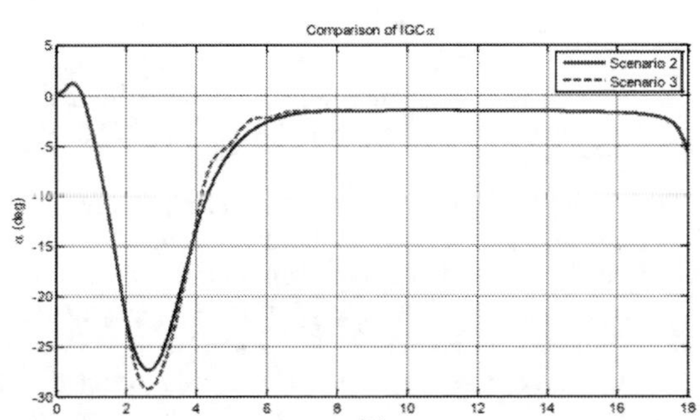

Figure 13. The comparison of α between S2 and S3

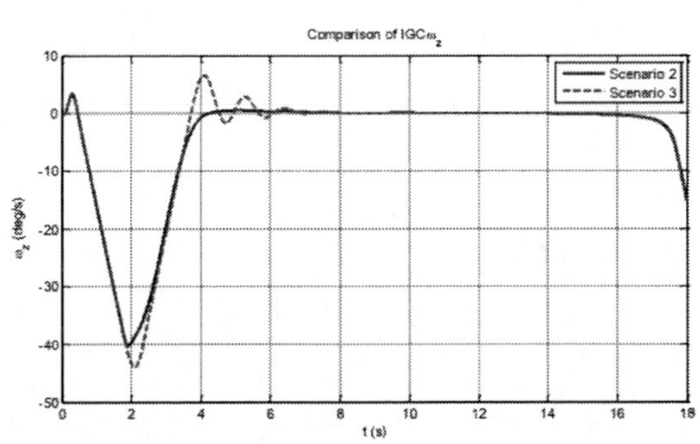

Figure 14. The comparison of ω_z between S2 and S3

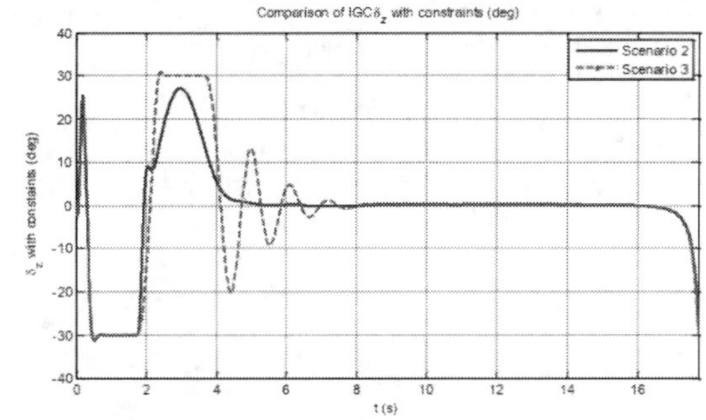

Figure 15. The comparison of δ_z between S2 and S3

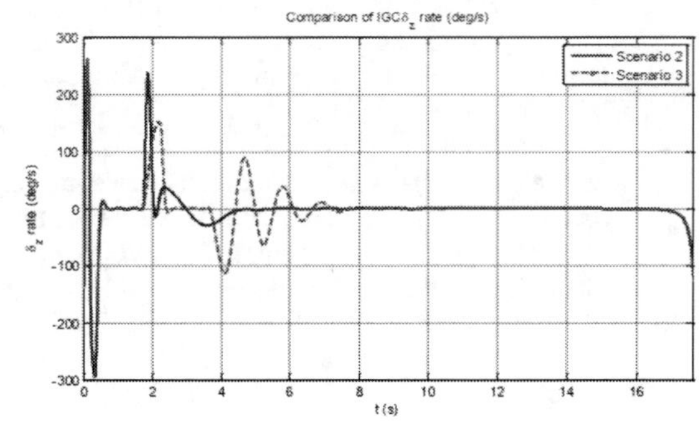

Figure 16. The comparison of IGC δ_z rate

5 Conclusions

This paper outlines the design of integrated guidance and control scheme for hypersonic vehicle in terminal phase. The adaptive command filtered controller is able to cancel out the great aerodynamics uncertainties and disturbances in flight dynamics, nonlinear tracking differentiator is employed to avoid directly differentiating the control surface deflection command, which alleviates the peaking phenomenon. The terminal sliding mode guarantees fast tracking ability for actuator dynamics. Comparison simulations are carried out to effectively confirm the superiority and feasibility of the proposed IGC design. The consideration of actuator dynamics is more practical in applications, additionally magnitude, rate and bandwidth constraints are synthesized in the proposed method.

References

1. D. Sziroczak, H. Smith., "A review of design issues specific to hypersonic flight vehicles," Progress in Aerospace Sciences, 2016, pp.1-28.
2. Serrani A., "Nonlinear flight control systems design for hypersonic vehicles: results and open problems," 8th IFAC Symposium on Nonlinear Control Systems, University of Bologna, Italy, 2010, pp.713-720.
3. Levy, M., Shima, T., Gutman, S., "Single versus two-loop full-state multi-input missile guidance," Journal of Guidance, Control, and Dynamics, 2015, 38, (5), pp. 843-853.

4. Levy, M., Shima, T., Gutman, S., "Linear quadratic integrated versus separated autopilot-guidance design," Journal of Guidance, Control, and Dynamics, 2013, 36, (6), pp. 1722-1730.
5. Xianghua Wang, Jinzhi Wang, "Partial integrated missile guidance and control with finite time convergence," Journal of Guidance, Control, and Dynamics, 2013, 36, (5), pp. 1399-149.
6. Xianghua Wang, Jinzhi Wang, "Partial integrated guidance and control for missiles with three-dimensional impact angle constraints," Journal of Guidance, Control, and Dynamics, 2014, 37, (2), pp. 644-657.
7. Han Yan, Yinzi He, "Adaptive integrated guidance and control based on backstepping for the landing of reusable launch vehicles," IFAC (International Federation of Automatic Control), 2015, pp.496-501.
8. Mingzhe Hou, Guangren Duan, "Adaptive dynamic surface control for integrated missile guidance and autopilot," International Journal of Automation and Computing, 2011, 8, (1), pp. 122-127.
9. Wasseem Aslam Butt, Lin Yan, Amezquita S., Kendrick, "Adaptive dynamic surface control of a hypersonic flight vehicle with magnitude, rate and bandwidth constraints," 18th IFAC World Congress, Milano, Italy, 2011, pp.5341-5346.
10. B. J. Bialy, J. Klotz, J. Willard Curtis, W. E. Dixon, "An adaptive backstepping controller for a hypersonic air-breathing missile," AIAA Guidance, Navigation, and Control Conference, AIAA paper 2012-4468, 2012.
11. Gongjun Li, Bin Meng, "Actuators coupled design based adaptive backstepping control of air-breathing hypersonic vehicle," IFAC (International Federation of Automatic Control), 2015, pp.508-513.
12. Yongping Pan, Huiming Wang, Xiang Li, etc. "Adaptive command-filtered back-stepping control of robot arms with compliant actuators", IEEE Transactions on Control Systems Technology (2017), DOI:10.1109/TCST.2017.2695600.
13. Yongping Pan, Haoyong Yu. (2015). "Dynamic surface control via singular perturbation analysis." Automatica, 57(2015):19-33.
14. Yongping Pan, Haoyong Yu. (2016). "Composite learning from adaptive dynamic surface control", IEEE Transactions on Automatic Control, 61(9):2603-2609.
15. Jay Farrell, Marios Polycarpou, Manu Sharma. (2006). "Adaptive back-stepping with magnitude, rate, and bandwith constraints: Aircraft longitude control." Air Force Research Laboratory, Wright Patterson AFB, OH, 45433.
16. Bin Xu, Shixing Wang, Daoxiang Gao, etc. (2014). "Command filter based robust nonlinear control of hypersonic aircraft with magnitude constraints on States and Actuators." Journal of Intelligent Robot System, 73:233-247.
17. Dongkyoung Chwa. (2011). "Global tracking control of under-actuated ships with input and velocity constraints using dynamic surface control method", IEEE Transaction on Control Systems Technology, 19(6):1357-1370.
18. Xianghua Wang, Jinzhi Wang. (2013). "Partial integrated missile guidance and control with finite time convergence." Journal of Guidance, Control, and Dynamics, 36(5): 1399-1409.
19. Xianghua Wang, Jinzhi Wang. (2014). "Partial integrated guidance and control with input constraints and impact angle." AIAA Guidance, Navigation, and Control Conference, National Harbor, Maryland.
20. Changyun Wen, Jing Zhou, Zhitao Liu, Hongye. (2011). "Robust adaptive control of uncertain nonlinear systems in the presence of input saturation and external disturbance." IEEE Transactions on Automatic Control, 56(7):1672-1678.
21. Xargay, E., Hovakimyan, N., Dobrokhodov, V., "L1 adaptive control in flight," Advances in Intelligent and Autonomous Aerospace Systems, 2012, pp.129-172.
22. Erdos, D., Shima, T., Kharisov, E., Hovkimyan, N., "L1 adaptive control integrated missile autopilot and guidance," AIAA Guidance, Navigation, and Control Conference, AIAA paper 2012-4465, 2012.
23. Peter, F., Holzapfel, F., Xargay, E., and Hovakimyan, N., "L1 adaptive augmentation of a missile autopilot," AIAA Guidance, Navigation, and Control Conference, AIAA paper 2012-4832, 2012.
24. Prime Z., Doolan, C., Cazzolato, B., "Longitudinal L1 adaptive control of a hypersonic re-entry experiment," 15th Australian International Aerospace Congress (AIAC15), Australian International Aerospace Congress, Melbourne, VIC, 2013, pp.717-726.
25. Banerjee, S., "L1 adaptive control augmentation for the longitudinal dynamics of a hypersonic glider," Journal of Guidance, Control, and Dynamics, 2016, 39, (2), pp. 275–291.
26. Banerjee, S., Creagh, M., Boyce, R., Wang Z., Baur B., Holzapfel F., "L1 adaptive control augmentation configuration for a hypersonic glider in the presence of uncertainties," AIAA Guidance, Navigation, and Control Conference, AIAA paper 2014-0453, 2014.
27. Banerjee, S., Creagh, M., Boyce, R., Wang Z., Baur B., Holzapfel F., "L1 augmentation configuration for a lateral/directional maneuver of a hypersonic glider in the presence of uncertainties," 19th AIAA International Space Planes and Hypersonic Systems and Technologies Conference, AIAA Paper 2014-2533, 2014.

Research on Human-Error Factors of Civil Aircraft Pilots Based On Grey Relational Analysis

Yundong Guo , Youchao Sun and Si Chen

College of Civil Aviation, Nanjing University of Aeronautics and Astronautics, Nanjing, China

Abstract. In consideration of the situation that civil aviation accidents involve many human-error factors and show the features of typical grey systems, an index system of civil aviation accident human-error factors is built using human factor analysis and classification system model. With the data of accidents happened worldwide between 2008 and 2011, the correlation between human-error factors can be analyzed quantitatively using the method of grey relational analysis. Research results show that the order of main factors affecting pilot human-error factors is preconditions for unsafe acts, unsafe supervision, organization and unsafe acts. The factor related most closely with second-level indexes and pilot human-error factors is the physical/mental limitations of pilots, followed by supervisory violations. The relevancy between the first-level indexes and the corresponding second-level indexes and the relevancy between second-level indexes can also be analyzed quantitatively.

1 Introduction

With the rapid technological development, the number of civil aviation accidents caused by the faults of the electromechanical equipment has declined sharply. To the contrary, aviation accidents caused by inappropriate pilot operation is increasing year by year. Human-error factors have accounted for 60-80% of the whole accident causes, becoming one of the main causes of civil aviation accidents[1]. By now, the research on human-error factors mostly concentrates on qualitative analysis, such as cause & effect analysis, event tree analysis, simulated investigation analysis, fishbone diagram analysis[2-5], and little improvement in the research on quantitative analysis has been realized. Quantitative research on cockpit human-error factors needs a great number of samples for mathematical statistics analysis, with high requirements for data integrity and data accuracy, and has not shown the relationship between human error and safety factors clearly. To reduce the aviation accident rate in China, it is of great importance to use the big data of aviation accidents to analyze the occurrence rules of human-error factors for higher pilot efficiency.

Based on the method of grey relational analysis, this paper researches the human-error factors of pilots in civil aircraft cockpits. There are many human-error factors that affect the task execution, the interactive environment and the interactive process are very complex and the relationship between human-error factors and other factors is still in an unclear grey system. Using grey relational analysis for the human-error factor research can remedy the limitations of traditional statistics analysis and provide theoretical basis for prohibiting and reducing human errors and mitigating flight risks.

2 Method

2.1 grey relational analysis

Grey correlation means uncertain correlation between objects, factors or between factors and main acts. The main idea is to judge how closely those objects are related based on geometry similarity between referential data sequences and comparative data sequences, which shows the relevance between curves. Finding out the main factor that affects the target values by calculating the relational degree is an easy and practical system analysis method, which can make up the shortages of traditional mathematical statistics method, such as demand of large samples, large amount of calculation and inconsistence between the results of quantitative analysis and qualitative analysis. The calculation process of grey relational analysis is as follows:

1)Determine the referential sequence, denoted as:

$$X_0(k), \quad k = 1, 2, \cdots, m \tag{1}$$

2)Determine the comparative sequence, denoted as:

$$X_i(k), \quad i = 1, 2, \cdots, n. \tag{2}$$

3)Calculate the correlation coefficient between the referential sequence and the comparative sequence, denoted as:

$$\xi_i(k)=\frac{\min_i\min_k|X_0(k)-X_i(k)|+\rho\max_i\max_k|X_0(k)-X_i(k)|}{|X_0(k)-X_i(k)|+\rho\max_i\max_k|X_0(k)-X_i(k)|} \tag{3}$$

$\rho \in (0,1)$ means the identification coefficient, and normally we take the value of 0.5.
4)Calculate the relational degree, denoted as:

$$\gamma_i=\frac{1}{n}\sum_{k=1}^{n}\xi_i(k) \tag{4}$$

γ_i means the relational degree between the sequence X_i and the sequence X_0, and the higher value means the closer relationship.
5)Rank the relevancies according to their values, and find out the factors that affect the main act.

2.2 Construction of human error index system

Contingency exists in the occurrence of civil aviation accidents, and both dominant factors and recessive factors can result in human error, so it is hard to quantify the relationship between human-error factors. Based on the Human Factors Analysis and Classification System (HFACS)[6], we can classify the factors that affect the pilot-error incidents as 4 first-level indexes (organizational influences, unsafe supervision, preconditions for unsafe acts and unsafe acts) and 18 second-level indexes. According to the cognitive process of aircraft operation, we can build the human error index system (see Fig 1) by analyzing the degree to which these second-level factors interact with each other.

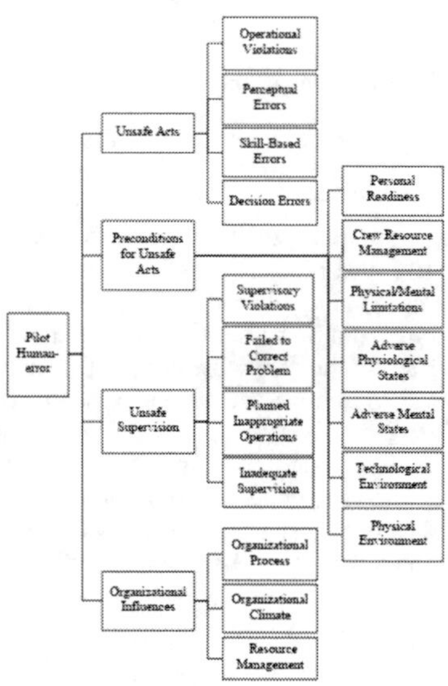

Figure 1. human error index system

3 A Case study

Use the materials of 88 civil aviation accidents happened worldwide in the recent four years: 23 accidents in 2008, 46 accidents in 2009, 17 accidents in 2010, and 2 accidents in 2011[7]. Based on HFACS model, analyze the human factors in those 88 accidents and find out that there are 347 human-error events, then summarize the analysis results for every year (see Table 1).

Table 1. Analysis Results of Human-error Events in Civil Aircraft Accident at Home and Abroad From 2008 to 2011

Human Error Index	2008	2009	2010	2011
Organizational Influences X_1	12	25	8	4
Resource Management X_{11}	5	13	4	2
Organizational Climate X_{12}	2	2	0	0
Organizational Process X_{13}	5	10	4	2
Unsafe Supervision X_2	23	36	16	5
Inadequate Supervision X_{21}	7	12	5	2
Planned Inappropriate Operations X_{22}	5	5	1	0
Failed to Correct Problem X_{23}	6	8	5	1
Supervisory Violations X_{24}	5	11	5	2
Preconditions for Unsafe Acts X_3	28	52	20	5
Physical Environment X_{31}	3	11	6	0
Technological Environment X_{32}	15	20	4	1
Adverse Mental States X_{33}	1	5	2	1
Adverse Physiological States X_{34}	1	3	1	1
Physical/Mental Limitations X_{35}	1	2	1	0
Crew Resource Management X_{36}	7	8	6	1
Personal Readiness X_{37}	0	3	0	1
Unsafe Acts X_4	21	52	37	4
Decision Errors X_{41}	6	11	9	1
Skill-Based Errors X_{42}	6	16	9	1
Perceptual Errors X_{43}	5	10	9	1
Operational Violations X_{44}	4	15	10	1
Number of Human-error Events X_0	84	165	80	18

3.1 Grey relational analysis of human errors and first-level indexes

Set the amount of human error events (X0) in civil aviation accidents as the referential sequence, and set organizational influences (X1), unsafe supervision factors (X2), precondition for unsafe acts (X3) and unsafe acts (X4) as comparative sequences. Use the initial value method to non-dimensionalize these sequences, and the results are as follows :

$$X_0' = X_0/X_0(2) = (0.5091, 1.0000, 0.4848, 0.1139)$$
$$X_1' = X_1/X_1(2) = (0.4800, 1.0000, 0.3200, 0.1600)$$
$$X_2' = X_2/X_2(2) = (0.6389, 1.0000, 0.4444, 0.1389)$$
$$X_3' = X_3/X_3(2) = (0.5385, 1.0000, 0.3846, 0.0962)$$
$$X_4' = X_4/X_4(2) = (0.4038, 1.0000, 0.7115, 0.0769)$$

Use the data in Table 1 and the relational degree formula to calculate the absolute difference $\Delta_i(k)$, the first-order minimum difference $\min_k \Delta_i(k)$, the first-order maximum difference $\max_k \Delta_i(k)$, the second-order minimum $\min_i \min_k \Delta_i(k)$, and the second-order maximum $\max_i \max_k \Delta_i(k)$, and $\Delta_i(k) = |X_0(k) - X_i(k)|$, $\Delta_i(k) = |X_0(k) - X_i(k)|, X_0(k) = X_0', X_i(k) = X_i'(k), i = 1,2,3,4$ T he results are shown in Table 2.

Table 2. Calculation Results of the Absolute Difference and the First-order Minimum/Maximum Difference

Δ_i \ k	1	2	3	4	$\min_k \Delta_i(k)$	$\max_k \Delta_i(k)$
Δ_1	0.0291	0.0000	0.1648	0.0461	0.0000	0.1648
Δ_2	0.1298	0.0000	0.0404	0.0250	0.0000	0.1298
Δ_3	0.0294	0.0000	0.1002	0.0177	0.0000	0.1002
Δ_4	0.1053	0.0000	0.2267	0.0370	0.0000	0.2267

$$\min_i \min_k |X_0(k) - X_i(k)| = 0$$

$$\max_i \max_k |X_0(k) - X_i(k)| = 0.2267$$

The correlation coefficients and the results of gray relational grade calculation are as follows:

$$\xi_1 = (0.7957, 1.0000, 0.4075, 0.7109)$$

$$\xi_2 = (0.4662, 1.0000, 0.7372, 0.8193)$$

$$\xi_3 = (0.7940, 1.0000, 0.5308, 0.8649)$$

$$\xi_4 = (0.5184, 1.0000, 0.3333, 0.7539)$$

$$\gamma_1 = 0.7285, \gamma_2 = 0.7556, \gamma_3 = 0.7974, \gamma_4 = 0.6514$$

Apparently, the precondition for unsafe acts is the main factor that affects human error in civil aviation accidents, rather than unsafe supervision or unsafe organization. Notably, inadequate organizational management typically involves operational processes such as inadequate or nonexistent procedures, directives, standards, and requirements, or in the case of commuter/on-demand operations, inadequate surveillance of operations.

3.2 Grey relational analysis of first-level indexes and second-level indexes

Set 'organizational influences' X1 as the referential sequence, and set 3 second-level indexes of the first-level index X1 as the comparative sequence. Use the same method as above and we can get the relational degrees, respectively, between 3 second-level indexes (inappropriate resource management, bad organizational environments, mistakes in operational process) and the organizational influences:

$\gamma_{11} = 0.9156, \gamma_{12} = 0.6001,$

$\gamma_{13} = 0.8900$, and the ranking is $\gamma_{11} > \gamma_{13} > \gamma_{12}$.

The data shows that inappropriate resource management is the factor that has the most important effects on organizational influences. The reason why bad organizational environment has less impact may be that the accident reports are relatively simple. Some involved departments only include immediate causes and do not explore the root causes reflected by accidents. Organizational culture plays a key role in any flight accident.

The relational degrees between second-level indexes (inadequate supervision, planned inappropriate operations, failure to correct known problems and supervisory violations) and the first-level index 'unsafe supervision' are $\gamma_{21} = 0.8750, \gamma_{22} = 0.5810, \gamma_{23} = 0.6710, \gamma_{24} = 0.8120$.

and the ranking is $\gamma_{21} > \gamma_{24} > \gamma_{23} > \gamma_{22}$.

The data confirm the influences of unsafe supervision related with operational process and show that failure of known type may have something to do with involved operational type. For example, accidents of civil aviation are usually associated with the way procedures or instructions are delivered, but such accidents have closer relationship with the lack of organizational supervision. Actually, these databases can provide directions for solving the question how unsafe supervision threaten aviation safety.

Relational degrees between second-level indexes (physical environments, technology environments, adverse mental states, adverse physiological states, physical/mental limitations, aircrew resource management, and personal readiness) and the first-level index 'preconditions for unsafe acts' are $\gamma_{31} = 0.5633$,

$\gamma_{32}=0.6099,\gamma_{33}=0.6633,\gamma_{34}=0.5919,\gamma_{35}=0.6957,$

$\gamma_{36}=0.5716,\gamma_{37}=0.4312.$ The ranking is $\gamma_{35}>\gamma_{33}>$

$\gamma_{32}>\gamma_{34}>\gamma_{36}>\gamma_{31}>\gamma_{37}.$

The calculation results show that pilot mental/physical limitations result in most human errors in civil aviation accidents, and the main reason is that mental/physical limitations disable pilots to carry out flight control tasks successfully when failure or emergencies happen.

Relational degrees between second-level indexes (decision errors, skill-based errors, perceptual errors, violations) and the first-level index 'unsafe acts' are $\gamma_{41}=0.8187,\gamma_{42}=0.8650,\gamma_{43}=0.7411,\gamma_{44}=0.8614.$ The ranking is $\gamma_{42}>\gamma_{44}>\gamma_{41}>\gamma_{43}$.

When considering pilot personal conditions, the factor 'skill-based errors' has the highest relational degree with unsafe acts, such as unintentionally touching control device that is irrelevant to the current task or operating the process in wrong order. The followed is violations which include not conducting instructions from air traffic control department, unauthorized approaches and so on. Decision errors and perceptual errors are also important factors that can lead to accidents, for example, making risky decisions when time is limited.

3.3 Grey relational analysis of human errors and second-level indexes

Set X0 as the referential sequence, and set all the second-level indexes as comparative sequences. Use the initial value method to calculate relational degrees and we can get the ranking:

$$\begin{aligned}&\gamma(X_0,X_{35})=0.8293>\gamma(X_0,X_{24})=0.7684>\gamma(X_0,X_{43})\\&=0.7664>\gamma(X_0,X_{13})=0.7631>\gamma(X_0,X_{21})\\&=0.7230>\gamma(X_0,X_{41})=0.7181>\gamma(X_0,X_{42})\\&=0.6901>\gamma(X_0,X_{23})=0.6610>\gamma(X_0,X_{11})\\&=0.6459>\gamma(X_0,X_{31})=0.6213>\gamma(X_0,X_{36})\\&=0.6032>\gamma(X_0,X_{33})=0.5988>\gamma(X_0,X_{32})\\&=0.5655>\gamma(X_0,X_{34})=0.5389>\gamma(X_0,X_{22})\\&=0.4955>\gamma(X_0,X_{12})=0.4717>\gamma(X_0,X_{44})\\&=0.4501>\gamma(X_0,X_{37})=0.4268\end{aligned}$$

All the second-level factors are at the same level and what can be seen from the data are the influences that different types of factors exert on human errors. The most influential one is 'mental/physical limitations', one of the factors subjected to 'preconditions for unsafe acts', and the followed is 'supervisory violations' belonging to the factor 'unsafe supervision'.

3.4 Grey relational analysis of second-level indexes

Complex interactions exist among factors of the same level, and we can analyze the correlations between different factors by setting one factor as the referential sequence and setting other factors of the same level as comparative sequences. Set 'decision errors' as the referential sequence and set 'skill-based errors', 'perceptual errors', and 'violations' as comparative sequences, and the relational degrees are

$$\gamma(X_{41},X_{43})=0.8050>\gamma(X_{41},X_{42})=0.6265>\gamma(X_{41},X_{44})=0.4831$$

The results indicate that perceptual errors have the most influences on decision errors, followed by skill-based errors and violations. Therefore, we can take control of the main factors to achieve the auxiliary control of cognitive errors on the impact of unsafe behavior, according to the degree of influence of cognition and skill on decision making.

4 Conclusions

Based on civil aviation accidents happened between 2008 and 2011, grey relational analysis can work out the main human error factor and the relational degrees between factors, offering the quantitative relationship between different human-error factors. The result of the analytic hierarchy process shows that, the major factor of human error is 'preconditions for unsafe acts', followed by unsafe supervision. According to the importance of factors, we can take targeted prevention or control measures to reduce human errors. Based on HFACS, grey relational analysis of human errors in civil aviation accidents can also be applied to accidents in nuclear plants, automobiles and collieries. Relational analysis can provide the order of importance of all the factors that contribute to the accident and will provide theoretical foundation of prevention measures.

Acknowledgement

This work was supported by National Natural Science Foundation of China (U1333119); Postgraduate Research & Practice Innovation Program of Jiangsu Province (KYCX17-0273).

References

1. Dayong D, Jinhai Y, Baofeng L, et al. Acta Aeronaut. Astronaut Sin., **37(1)**:310-316 (2016)
2. Pengcheng L, Licao D, Li Z, et al. China Safety Science Journal, **20(4)**:75-82 (2010)
3. Kolny Olesiak J, Weller H. Acs Appl. Mat. Interfaces, **5(23)**:12221-12237 (2013)
4. Woodcock K, Drury C G, Smiley A, et al. Appl. Ergon., 36(**1**):1-12 (2005)
5. Ciocoiu C N, Ilie G. Knowledge Management Research & Practice, **2(1)**:1-20 (2010)
6. Shappell S, Detwiler C, Holcomb K, et al. Human Factors, **49(2)**:227 (2007)
7. MinQin W, Bozhi G. *Civil aircraft and incident statistics analysis manual* (China Aviation Industry Press, 2015)

A method of coding for aerospace product quality DNA

Yao Jiayu[1]

[1] *Tongji University, School of Mechanical Engineering, No.4800 Caoan Highway, Shanghai, China.*

Abstract: Aiming at the problem that the manufacturing process of our aerospace products is relatively discrete and the lack of appropriate quality monitoring and feedback mechanism, a method of coding for aerospace product quality DNA was proposed. Based on the structure of biological DNA and the theory of quality assessment, equipment diagnosis and quality traceability, the biological DNA structure was transformed into the structure of aerospace product quality DNA, and the concept of aerospace product quality DNA was defined, including the genetic and variation characteristics of aerospace product quality DNA. The coding rules of aerospace product quality DNA were designed, and the designed encoding rules are applied to the case of welding of wall panels in the manufacturing process of aerospace products. The results show that the coding method can monitor and feedback the core information related to quality in the manufacturing process of aerospace products.

1 Introduction

Aerospace manufacturing industry is an important part advanced manufacturing industry in China, but also an important application area of national scientific and technological innovation and traction forces. It is a symbol of comprehensive national strength, reflecting a country's ability to use space and the development of space technology. In recent years, although our aerospace manufacturing industry has made great progress, there is still a big gap compared with the advanced level of the world[1,2].

Aerospace products are mostly aluminum alloy welded structure, with large size, light, thin wall, complex and other typical features, including machining, welding and riveting three manufacturing processes. The quality standard of aerospace products is very high and requires zero defects. However, due to the fact that the manufacturing process of our aerospace products is relatively discrete and the information level is low, the lack of appropriate quality monitoring and feedback mechanism[3] leads to the difficulty of tracing product quality information and low production efficiency. These problems seriously affect the development of our aerospace manufacturing industry[4].

In view of the above problems, this paper proposes a method of coding for aerospace product quality DNA based on biological DNA, which is used to monitor and feedback the core information related to quality in the process of manufacturing aerospace products, including product quality assessment information, equipment diagnosis information, product quality traceability information, in order to trace back quality information, predict product performance, improve the intelligent diagnosis of manufacturing process, achieve the goal of zero defect quality of aerospace products.

Product quality DNA is evolved from the concept of product DNA. In 2007, the University of Michigan proposed the concept of product DNA. They believe that the product lifecycle management has a lot of manufacturing and process data. In order to improve the process, predict the performance[5,6], they defined product DNA. Compared with product lifecycle management, product DNA will be more accurate to guide the optimization of product processing[7,8] and quality defect diagnosis[9,10].

In order to gain a better understanding of the connotation of the product DNA, it is necessary to proceed from the similarity between biological and mechanical products. At present, many scholars at home and abroad demonstrate that there are similar properties between biological and mechanical products from a variety of perspectives[11,12].From the origin, the creature is created by nature through physical and chemical reactions, and the mechanical product is created by the manufacturer through a series of processing methods, so the biological and mechanical products are similar in origin, but the difference is that the creature is created by nature, and the mechanical product is created by mankind. In terms of composition, a highly evolved creature usually contains many independent systems, each consisting of different cells; Mechanical products usually contain many components, each component consists of many parts, so mechanical products and biological are similar in composition.

In this paper, based on the structure of biological DNA and the theory of quality assessment, quality diagnosis and quality traceability, the biological DNA structure was transformed into the structure of aerospace product quality DNA, and the concept of aerospace product quality DNA was defined, including the genetic and variation characteristics of aerospace product quality DNA. The coding rules of aerospace product quality DNA were designed, and the designed encoding rules are applied to the case of welding of wall panels in the manufacturing process of aerospace products.

2 Composition of aerospace product quality DNA

2.1 Composition of biological DNA

DNA, also known as deoxyribonucleic acid, is a kind of genetic information with biological macromolecules. It is composed of four deoxyribonucleotides linked by phosphodiester bonds, with genetic and variation characteristics. The basic unit of DNA is deoxynucleotides, consisting of phosphoric acid, deoxyribose, nitrogen-containing bases. There are four nitrogen-containing bases, namely: A-adenine, G-guanine, C-cytosine, T-thymidine.

DNA molecules are double helix structures, from two reverse parallel deoxynucleotides long chain hovering. DNA molecules in the phosphoric acid and deoxyribose are alternately connected, they are arranged on the outside, the base arranged in the inside. The bases on the two long chains are linked by hydrogen bonds, which follow the principle of complementary pairing to form base pairs, A with T, C with G. The genetic information of DNA is expressed by the order of the bases, as shown in Figure 1. Under the influence of the environment, DNA information in the genetic process will be mutated, sometimes accompanied by the emergence of defects.

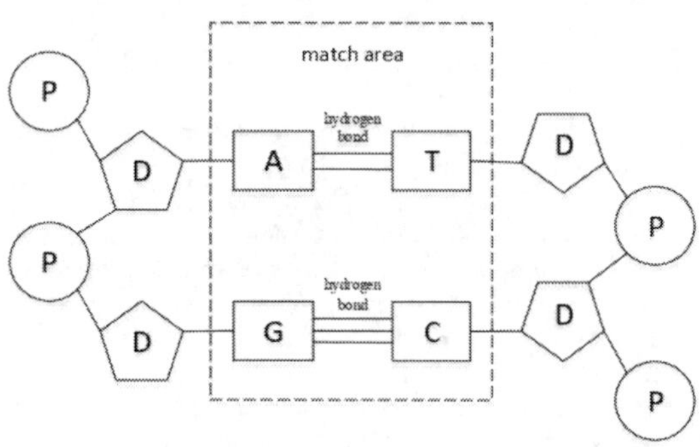

Figure 1. Composition of biological DNA

2.2 Evolution of aerospace product quality DNA

Aerospace product quality DNA evolution process is divided into two steps. The first step is to transform the phosphoric acid and deoxyribose in the biological DNA into the identification code and signature code. "Identification code + signature code " constitutes the code of aerospace product quality DNA. The identification code is composed of part number, and the signature code is composed of evaluation code, diagnostic code and traceability code.

The second step is to change the pairing principle of base pairs in biological DNA into the principle of comparison between standard manufacturing information and actual manufacturing information, and then change the hydrogen bond into the result of comparison, as shown in Figure 2. The value of the signature code is determined by the result of the match.

The value of the evaluation code is determined by the result of the comparison between the product process standard and the processing result. The value of the diagnostic code is determined by the result of the comparison between the equipment process standard and the real-time processing information. The value of the traceability code is determined by the result of the comparison between the defect case library and the product manufacturing information, as shown in Figure 3.

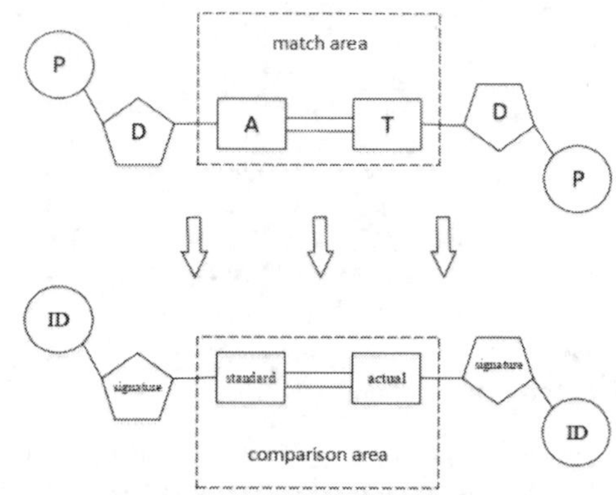

Figure 2. Evolution of aerospace product quality DNA

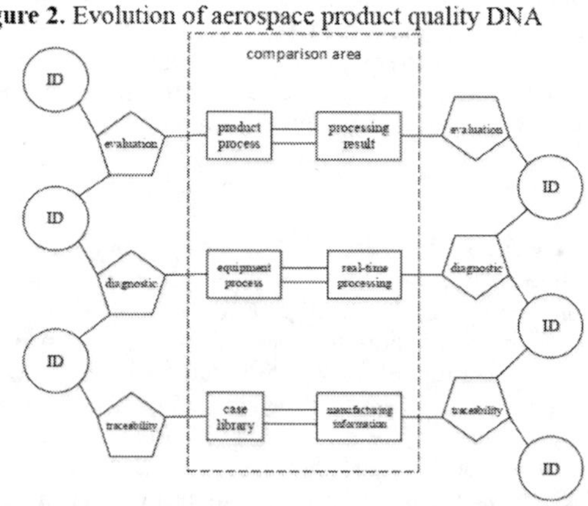

Figure 3. Composition of aerospace product quality DNA

Therefore, aerospace product quality DNA has a similar structure with biological DNA, including quality assessment model, equipment diagnostic model, quality traceability model and related code, and genetic and mutation characteristics. The quality of aerospace products is inherited on the basis of the previous processing quality, continue to carry out the current processing, and then the current processing quality genetic to the subsequent processing. Under the influence of the environment, aerospace product quality DNA will be mutated in the genetic process, sometimes accompanied by the emergence of defects.

3 Coding rules of aerospace product quality DNA

3.1 Category of code

There are many types of code. According to their attributes, they can be divided into two categories: identification code and signature code. The identification code is used to identify the identity of the object, and the signature code is used to identify the characteristics of the object.

Code according to whether it has meaning can be divided into meaningless code and meaning code. The meaningless code includes sequence code and disorder code, and the meaning code includes series sequential code, numerical alphabetical order code, hierarchical code, feature combination code and composite code. The hierarchical code is a code that follows the order of the object's dependencies and hierarchies. It is from left to right, that the level from high to low level. The leftmost code is the highest level code, and the rightmost code is the lowest bit level code. Each level of code can be used in sequence code or series of sequential code. Feature combination code is the object according to its characteristics into a number of "face" and then were encoded separately. There is no hierarchical relationship between the "face" and "face" code, and there is no affiliation.

3.2 Design of the coding rules

Aerospace product quality DNA coding consists of identification code + signature code. The identification code is part number, and the signature code consists of evaluation code, diagnostic code and traceability code, as shown in Figure 4.

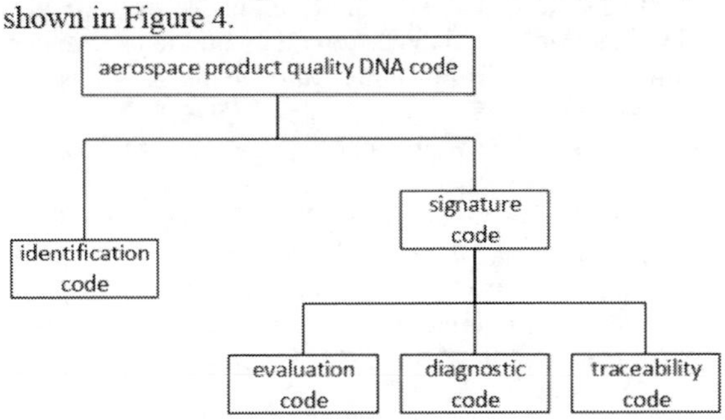

Figure 4. Composition of aerospace product quality DNA code

The evaluation code consists of the results of evaluating machining, welding and riveting. The diagnostic code consists of the diagnostic results of the equipment. The traceability code consists of the results of tracing machining, welding and riveting.

The signature code is encoded in alphabetical and digital hybrid code. The length of the code is 19 bits, as shown in Figure 5.

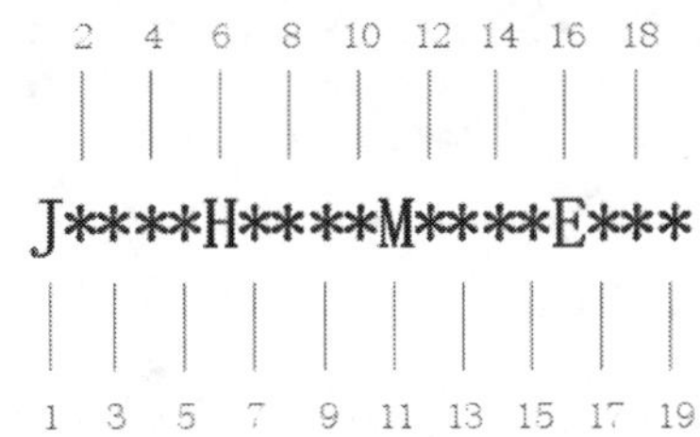

Figure 5. Length of the signature code

Evaluation code and traceability code are composed of "J****", "H****", "M****" composition. They indicate the results of the evaluation and traceability of the product quality after machining, welding and riveting. 1 to 5, 6 to 10, 11 to 15 bits are encoded in hierarchical code + feature combination code. The second to fifth bits are subordinate to the first bit, the 7th to 10th bits are subordinate to the sixth bit, and the 12th to 15th bits are subordinate to the 11th bit.The first, sixth, and eleventh bits are the classification codes, which are represented by capital letters. They are J (machining), H (welding) and M (riveting). 2, 7, and 12 bits are the evaluation status codes, which are the result of the evaluation of the 1st, 6th and 11th places. 3 to 5, 8 to 10, 13 to 15 bits are traceability status codes, which are the result of the first, sixth, and 11th tracebacks. There is a hierarchical relationship between the evaluation status code and the traceability status code, and there is no hierarchical relationship and no affiliation.

The evaluation status code (bits 2, 7, 12) is encoded in series sequential code, indicated by the numbers 0 to 3. "3" indicates that the evaluation result is "good", "2" indicates that the evaluation result is "normal", "1" indicates that the evaluation result is "poor" and "0" means "no evaluation".

The traceability status code (bits 3 to 5, 8 to 10, 13 to 15) is encoded in hierarchical code. Bits 4, 9, 14 are subordinate to bits 3, 8, 13, and bits 5, 10, 15 are subordinate to bits 4, 9, 14. The 3rd, 8th and 13th bits indicate the number of defects, denoted by the numbers 0 to 9. "0" means no defects, "1 ~ 8" indicates the number of defects, "9" indicates that the number of defects is 9 and above. The 4th, 9th, and 14th bits represent one of the most serious defects of the previous one, denoted by lowercase letters a to z. "A ~ x" represents a common 24 defects, "y" indicates a defect other than the previous 24 defects, and "z" indicates no defects (if bits 3, 8 and 13 are "0"). The 5th, 10th, and 15th bits indicate the cause of the previous defect, and are represented by the numbers 0 to 9. "0" indicates no reason (if the 3rd, 8th and 13th bits are "0"), "1 ~ 8" indicates the eight common causes of defects at the 4th, 9th and 14th positions. "9" indicates other reasons than the previous eight reasons.

The diagnostic code (bits 16 to 19) is encoded in hierarchical code + signature code. Bits 17 to 19 are subordinate to bit 16. The 16th bit is a class code, which is represented by the capital letter E (equipment). The 17th to 19th bits are diagnostic status codes, which are represented by the numbers 0 to 9. The 17th bit indicates the number of equipments whose diagnostic result are "good". The 18th bit indicates the number of equipments whose diagnostic result are "normal". The 19th bit indicates the number of equipments whose diagnostic result are "poor". "0 ~ 8" indicates the number of equipments. "9" indicates that the number of equipments is 9 and above.

4 Case study

The coding method is applied to the case of welding of wall panels in the manufacturing process of aerospace products. Table 1. shows the manufacturing information of the wall panels.

Table 1. Welding process of the wall panels

Process Elements	Standard	Actual
chamfering	40°	40°
gap(mm)	0.5~1	0.54
misalignment(mm)	0~0.5	0.28
weld width(mm)	14~18	16.32
weld reinforcement(mm)	0.5~2	1.33
leakage(mm)	0.5~1	0.93
groove width(mm)	3~4	3.56
groove depth(mm)	0.5~1	0.62

The quality of the welding of the wall panels was evaluated by fuzzy analytic hierarchy process. The results are shown in Table 2.

Table 2. Evaluation results

<table>
<tr><th>Process Elements</th><th colspan="3">Evaluation Results</th></tr>
<tr><td>chamfering</td><td>good</td><td rowspan="3">good</td><td rowspan="8">good</td></tr>
<tr><td>gap(mm)</td><td>normal</td></tr>
<tr><td>misalignment(mm)</td><td>good</td></tr>
<tr><td>weld width(mm)</td><td>good</td><td rowspan="2">good</td></tr>
<tr><td>weld reinforcement(mm)</td><td>good</td></tr>
<tr><td>leakage(mm)</td><td>normal</td><td rowspan="3">normal</td></tr>
<tr><td>groove width(mm)</td><td>good</td></tr>
<tr><td>groove depth(mm)</td><td>normal</td></tr>
</table>

Therefore, the evaluation code for the wall panels is "H3***". Table 3. shows the detection information of welding of the wall panels.

Table 3. Detection information of welding

Detection Process	Detection Results
radiographic	no defects
penetrant	no defects
ultrasonic	no defects

So the traceability code is "H*0z0". Finally, the evaluation and traceability code for the wall panels is "H30z0".

5 Conclusions

Aiming at the problem that the manufacturing process of our aerospace products is relatively discrete and the lack of appropriate quality monitoring and feedback mechanism, a method of coding for aerospace product quality DNA based on biological DNA was proposed.The coding method is applied to the case of welding of wall panels in the manufacturing process of aerospace products.The results show that the coding method can monitor and feedback the core information related to quality in the manufacturing process of aerospace products.

The research work of this paper is concentrated on the method of coding for aerospace product quality DNA, which is the first step to develop the entire framework of the concept. In order to complete the whole product quality DNA framework for different purposes of applications, a number of future research may be taken up following the studies in this paper. For instance, product quality DNA based performance analysis and prediction, and identify appropriate prediction tools to achieve product performance prognosis.

References

1. Xia Weiqiang, Integration of two forces to promote Aerospace Intelligent Manufacturing[N]. China Aerospace News, 2015-08-05A03.
2. Meng Guang, Analysis of the status of aerospace industry by the intelligent manufacturing development[J]. Shanghai Economy, 2015. 09: 30-31.
3. Bogue, R., New high speed imaging CMM system aids car production. Assembly Automation, 2005. 25(4): 273-284.
4. Li, L.-W., S. Dong, and K. Cheng, 3D surface topography formation in ultra-precision turning. Journal of Harbin Institute of Technology (New Series), 2004. 11(1): 82-86.
5. Zhou, S., et al., Diagnosability study of multistage manufacturing processes based on linear mixed-effects models. Technometrics, 2003. 45(4): 312-325.
6. Wang, H., Q. Huang, and R. Katz, Multi-operational machining processes modeling for sequential root cause identification and measurement reduction. Journal of Manufacturing Science and Engineering, Transactions of the ASME, 2005. 127(3): 512-521.
7. Zhang, X., C. Krewet, and B. Kuhlenkotter, Automatic classification of defects on the product surface in grinding and polishing. International Journal of Machine Tools and Manufacture, 2006. 46(1): 59-69.
8. Hu, X., Z.-Z. Wang, and X.-M. Ren, Classification of surface EMG signal with fractal dimension. Journal of Zhejiang University: Science, 2005. 6B(9): 844-848.
9. Muralikrishnan, B. and J. Raja, Functional filtering and performance correlation of Plateau honed surface profiles. Journal of Manufacturing Science and Engineering, Transactions of the ASME, 2005. 127(1): 193-197.
10. Muralikrishnan, B. and J. Raja, Inference engine for process diagnostics and functional correlation in surface metrology. Wear, 2004. 257(12 SPECISS): 1257-1263.
11. Chen Mei,Wang Zongyan, Research and Realization of Variation Design of Mechanical Products Based on Genetics[J]. Mechanical Design, 2011. 28(6): 70-73.
12. Cao Xuyang, Ten Ruming, Integration of Mechanical Product Hybrid Design and Construction of Product Gene Bank[J]. China Construction Machinery Journal, 2012. 10(3): 1-8.

AUTHOR INDEX

Akimov, I. O. 121
Alikhani, Alireza 110
Anand, Suraj 27
Bahrami, Salman 33
Bai, Hui 115
Barara, Daamanjyot 27
Chand, Dharmahinder S. 27
Chang, Yoon S. 126
Chen, Chung-De 1
Chen, Ming 80
Chen, Shulin 47
Chen, Si 157
Chen, Zhaoyue 47
Cheng, Chun-Der 17
Cheng, Keming 66
Choe, Yeong H. 126
Da-Poian, Victoria 121
Ganesh, Gautam 27
Gao, Chao 61
Goh, J. W. 101
Gu, Yunsong 66
Guo, R. 101
Guo, Yundong 157
Hamada, Shohji 71
He, Guoyi 57
Hui, Zou 86
Iljukhin, Stepan N. 121
Jegarkandi, Mohsen F. 110
Kashitani, Masashi 37
Kokuytseva, Tatiana V. 121
Koryanov, Vsevolod V. 121
Li, Chao 41
Li, Cong 115
Li, Feng 61
Li, Maoquan 75
Li, Qing 142
Li, Shihao 75, 132
Li, Wan-Li 105
Lin, Jinkwan 90
Liu, Bing-Fei 86
Liu, De-Shan 86
Liu, Haoyu 132
Liu, Huixiang 57
Liu, Huo-Xing 41
Liu, Jiangshan 80
Liu, Li 47
Liu, Thomas J.-C. 17
Liu, Xiang-Yong 105
Liu, Yingzheng 66
Liu, Yinli 12
Lu, Cunyue 115
Luo, Yun 57
Ma, Chengyu 57
McAndrew, Ian 96
Mohamado, Noa 121
Moradi, Rouzbeh 110
Mun, Sun J. 126
Nishiyama, Masanori 37
Peng, Di 66
Peng, Ke 148
Phyoe, S. M. 101
Pi, Chenhuan 90
Pichumani, Sivachidambaram 7
Rahimian, Mohammad H. 33
Ramamoorthi, Venkatraman 7
Ren, Xudong 61
Saadon, S. 21
San, Khin T. 126
Shi, Xiaofeng 137
Srinivasan, R. 7
Stone, Cheng 90
Sun, Youchao 157
Sun, Zhijun 66
Taguchi, Masato 37
Toporkov, Alexey G. 121
Tseng, Ji-Fu 17
Vishnevskaya, Elena 96
Wang, Donghui 148
Wang, Dongmei 12
Wang, Hongyu 115
Wang, Liang 148
Wang, Qi 57
Wen, Jingqian 142
Witcher, Kenneth 96
Xie, Mingjuan 52
Xu, Zhi-Qiang 105
Yao, Jiayu 161
Yao, Minghui 12
Yin, Xinglu 52
Yokoi, Yoshifumi 71
Zhang, Baomin 115
Zhang, Chunxiao 137
Zhang, Chunyan 52
Zhang, Wei 12
Zhang, Weihua 148
Zhang, Yunfei 75, 132
Zhao, An-Min 86
Zhao, Qian 142
Zhao, Xun 115
Zhao, Zijie 61
Zhong, Z. W. 101
Zhou, Zhi-Hong 41

EDP Sciences – Web of Conferences
17, Avenue du Hoggar
Parc d’Activité de Courtabouf
BP 112
F-91944 Les Ulis Cedex A
France

ISBN 978-1-5108-5873-2

CURRAN ASSOCIATES INC.
proceedings.com

9 781510 858732